# Automation Control Systems & Process Control for Industry 4.0

# Automation Control Systems & Process Control for Industry 4.0

Editor

**Sergey Y. Yurish**

Basel • Beijing • Wuhan • Barcelona • Belgrade • Novi Sad • Cluj • Manchester

*Editor*
Sergey Y. Yurish
IFSA Publishing
Castelldefels
Spain

*Editorial Office*
MDPI AG
Grosspeteranlage 5
4052 Basel, Switzerland

This is a reprint of articles from the Special Issue published online in the open access journal *Processes* (ISSN 2227-9717) (available at: https://www.mdpi.com/journal/processes/special_issues/automation_control_process_control).

For citation purposes, cite each article independently as indicated on the article page online and as indicated below:

Lastname, A.A.; Lastname, B.B. Article Title. *Journal Name* **Year**, *Volume Number*, Page Range.

**ISBN 978-3-7258-1869-3 (Hbk)**
**ISBN 978-3-7258-1870-9 (PDF)**
**doi.org/10.3390/books978-3-7258-1870-9**

# Contents

# About the Editor

**Sergey Y. Yurish**

Dr. Sergey Y. Yurish has served as the president of the International Frequency Sensor Association (IFSA)—one of the major professional associations serving the sensor industry and academia—for more than 25 years. Dr. Yurish is a founder of three companies. He is Editor-in-Chief of the international peer-reviewed journal Sensors & Transducers and an editor for several open access multivolume special issue series. Dr. Yurish obtained his PhD degree in 1996 from the National University Lviv Polytechnic (UA). He has published more than 180 articles and papers in international peer-reviewed journals and conference proceedings. Dr. Yurish holds nine patents and is the author and co-author of 12 books. He has delivered more than 90 speeches, tutorials and keynote presentations at industries, peer institutions, and professional conferences in 30 countries. Dr. Yurish was a Marie Curie Chairs Excellence Investigator at the Technical University of Catalonia (UPC, Barcelona, Spain) from 2006 to 2009, where he led and developed one of the most successful projects in the UPC on Smart Sensors Systems Design (SMARTSES), totaling EUR 425,000. His professional accomplishments also include a Senior Research Fellowship at the Open University of Catalonia (UOC, Barcelona, Spain) where he spent a year in 2009–2010. Dr. Yurish has over 30 years of research and academic experience, during which he has developed numerous projects on an international level with various programmers, including NATO, FP6 and FP7.

# Preface

According to a recent market study, the Global Industry 4.0 Market was valued at USD 102.94 billion in the year 2022 and is projected to reach a value of USD 433.84 billion by the year 2030. The Global Market is expected to grow, exhibiting a Compound Annual Growth Rate (CAGR) of 19.70 % over the forecast period.

Industry 4.0 is defined as the integration of intelligent digital technologies into manufacturing and industrial processes. It encompasses a set of technologies that include industrial IoT networks, artificial intelligence, big data, robotics, automation and control.

The primary driving factor of the Industry 4.0 market is the increasing adaptation of automated equipment and tools, rising investment for additive manufacturing, and the emergence of digital technologies such as IoT, AI, ML, Cloud Services, 5G, Blockchain, Extended Reality, Digital Twins, and 3D Printing. The key market restraints are economic constraints in emerging countries and business market obstacles.

This reprint, entitled "Automation Control Systems & Process Control for Industry 4.0", contains extended papers from the 2nd IFSA Winter Conference on Automation, Robotics & Communications for Industry 4.0 (ARCI' 2022), held on the 2nd–4th February 2022 in Andorra la Vella, Andorra. This conference addressed the following topics: Automation Control Systems, Process Control, Smart Manufacturing and Technologies.

The reprint contains 10 chapters written by 40 authors from 7 countries: China, Germany, Italy, Mexico, Peru, Poland and Venezuela. This book will inform readers about cutting-edge developments, in the field and provide a road map for further research and development. All chapters follow a similar structure: firstly, there is an introduction to the specific topic under study, and secondly, there is a description of the field, including applications. Each chapter ends with a list of references, including books, journals, conference proceedings and websites.

The reprint is intended for researchers and scientists from academia and industry, as well as for graduate and postgraduate students.

**Sergey Y. Yurish**
*Editor*

 *processes*

*Article*

# Technical Considerations for the Conformation of Specific Competences in Mechatronic Engineers in the Context of Industry 4.0 and 5.0

**Eusebio Jiménez López [1,*], Francisco Cuenca Jiménez [2], Gabriel Luna Sandoval [3], Francisco Javier Ochoa Estrella [4,*], Marco Antonio Maciel Monteón [3], Flavio Muñoz [4] and Pablo Alberto Limón Leyva [5,*]**

[1]  Research and Development Department, Universidad Tecnológica del Sur de Sonora-Universidad La Salle Noroeste, Ciudad Obregón 85190, Mexico
[2]  Engineering Design Department, Universidad Nacional Autónoma de México, Ciudad de México 04510, Mexico; fracuenc@gmail.com
[3]  Department of Industrial Engineering and Manufacturing, Universidad Estatal de Sonora, San Luis Río Colorado 83500, Mexico; gabriel.luna@ues.mx (G.L.S.); marco.maciel@ues.mx (M.A.M.M.)
[4]  Postgraduate and Research Subdirection, Tecnológico Nacional de México/ITS de Cajeme, Ciudad Obregón 85024, Mexico; fmunoz@itesca.edu.mx
[5]  Electromechanics Department, Instituto Tecnológico de Sonora, Ciudad Obregón 85000, Mexico
*  Correspondence: ejimenezl@msn.com (E.J.L.); fochoa@itesca.edu.mx (F.J.O.E.); pablo.limon@potros.itson.edu.mx (P.A.L.L.)

**Citation:** Jiménez López, E.; Cuenca Jiménez, F.; Luna Sandoval, G.; Ochoa Estrella, F.J.; Maciel Monteón, M.A.; Muñoz, F.; Limón Leyva, P.A. Technical Considerations for the Conformation of Specific Competences in Mechatronic Engineers in the Context of Industry 4.0 and 5.0. *Processes* **2022**, *10*, 1445. https://doi.org/10.3390/pr10081445

Academic Editor: Sergey Y. Yurish

Received: 13 May 2022
Accepted: 4 July 2022
Published: 24 July 2022

**Publisher's Note:** MDPI stays neutral with regard to jurisdictional claims in published maps and institutional affiliations.



**Abstract:** The incursion of disruptive technologies, such as the Internet of Things, information technologies, cloud computing, digitalization and artificial intelligence, into current production processes has led to a new global industrial revolution called Industry 4.0 or Manufacturing 4.0. This new revolution proposes digitization from one end of the value chain to the other by integrating physical assets into systems and networks linked to a series of technologies to create value. Industry 4.0 has far-reaching implications for production systems and engineering education, especially in the training of mechatronic engineers. In order to face the new challenges of the transition from manufacturing 3.0 to Industry 4.0 and 5.0, it is necessary to implement innovative educational models that allow the systematic training of engineers. The competency-based education model has ideal characteristics to help mechatronic engineers, especially in the development of specific competencies. This article proposes 15 technical considerations related to generic industrial needs and disruptive technologies that serve to determine those specific competencies required by mechatronic engineers to meet the challenges of Industry 4.0 and 5.0.

**Keywords:** Industry 4.0; competency-based education; cyber-physical systems; specific competencies; engineering education

## 1. Introduction

Today's industrial production, characterized by globalization and uncertainty, is being affected by the rapid development and application of various technologies, including Information and Communication Technologies (ICT) [1], and is also pressured by market demands for increasingly specialized and differentiated products. The incursion of frontier technologies in production lines makes the changes in industries and in society in general accelerated, and has wide repercussions in industry value chains and in contemporary cities. It can be said that the world is undergoing a transition between two major industrial events. To meet the challenges facing companies, a new industrial paradigm known as "Industry 4.0" (I4.0) has emerged. This new industrial production proposal involves generating new organizations and proposals for control of high value-added systems [2].

Consequently, the industrial world is in a phase of accelerated transition between the industrial revolution characterized by electronics, computing and automation (Manufacturing 3.0 (I3.0)) to another industrial revolution characterized by digitalization, cloud computing, the internet of things and cyber-physical systems (CPS), and in another less rapid transition phase between I4.0 characterized by the displacement of humans from production systems and Industry 5.0 (I5.0) which seeks closer collaboration between operators and machines.

The effects of I4.0 are felt in companies, universities, cities and modern society in general. I4.0, so conceived in 2011 in Germany [3], is similar to the era of mechanization characterized by steam power; it is also similar to the era where production lines and electricity were the engine of the economy, as well as to the era where computing and automation improved and optimized production systems. This means that I4.0 is a new industrial revolution and its effects are global and disruptive, mainly in companies and production processes.

Changes in industries are driven by the incursion of new technologies, such as the Internet of Things (IoT), Collaborative Robotics, Artificial Intelligence (AI), Augmented Reality (AR), CPS and Digital Twins (DT), among others [4].

I4.0 is a philosophy of the large-scale integration of methods, tools, systems, knowledge and technologies whose purpose is to enhance production chains by improving and optimizing processes. In fact, an important objective of the fourth industrial revolution is to optimize the third computerized industrial revolution (I3.0).

I4.0 proposes the digitization of the entire value chain through the integration of physical infrastructure into systems and networks associated with frontier (disruptive) technologies to create added value [5]. In this way, a production plant that is ready for I4.0 may be conceived as a system of systems, where elements of those partners that are part of the value chain (e.g., suppliers, manufacturers, and factory employees) must operate together to achieve the stated goals [6].

I4.0 is used to mutually interconnect the following three factors [7]:

1.  The process of integration and digitization of simple technical-economic relationships into complex networks.
2.  The process of digitalization of product and service offerings.
3.  In venturing into new market models.

Today many activities performed by mankind are interconnected in various ways with the help of communication systems. It is highly likely that I4.0 will improve the lives of human beings in many aspects and in various areas of opportunity. I4.0 is initiating various dynamic changes in the way companies envision and the ways products are manufactured, which will involve changes at all levels of manufacturing and supply chains, as well as changes in manufacturing line workers, engineers, and CPS developers, as well as changes in customers [8]. In the same way, education, and especially engineering education, will have to adjust to the new industrial paradigm seeking to provide graduates capable of facing the challenges required by the companies of today and the near future.

On the other hand, digitization is a process of high value and interest within I4.0. However, the core technologies of this new industrial revolution are CPSs [5]. These systems are of utmost importance in I4.0, since they act as a means to relate the physical world, integrated by elements such as mobile devices, sensors, mechanical systems and actuators, with the Internet seeking to simulate or reflect the events of reality in a digital or computational environment called cyberspace, with the aim of processing inspection and time management [9].

Currently, CPSs have diverse applications in critical situations where it is required to have reliable, protected and safe functions, and where the synchronization is subject to specific and strict requirements, such as in homes and universities or in automated systems where human operators work, or in modern production systems where there are collaborations between humans and robotic systems [10]. CPSs represent the core technology of I4.0, so the characterization and study of CPSs are of great importance for industry and engineering education.

Digital Twin (DT) technology, similarly to CPS, has an extremely important role in I4.0. Since their conception, it was announced that DTs would give a revolutionary technological boost to companies and industries, since with them various operations and processes can be simplified [11]. A DT is a virtual copy or replica of a dynamic system composed of physical elements that fulfill a specific function. The physical system and its replica are directionally interconnected, so that there is a feedback between them composed of information from the physical elements (e.g., sensors), and from processed or evaluated information from the DT to the physical part.

DTs are the most important technologies with the greatest applications today in I4.0, providing industrial processes with efficiency and optimization, among other important benefits. The study and analysis of DTs are important and necessary for those companies or industries seeking to improve their processes under the I4.0 philosophy. It is important to note that the technologies of the DTs and CPS should be taught in universities to engineers as they represent the central basis of the fourth industrial revolution so that the design, construction and operation of the same will be crucial for the current and future industry.

From an operational point of view, I4.0 seeks to optimize I.30. In this sense, IA is a discipline that offers various algorithms to achieve the optimization of production systems. In fact, according to [12] the employment of Industrial AI towards process optimization in manufacturing is gaining rapid traction, enabling smarter, more efficient data-driven decision-making by leveraging both historical and real-time data. Thus, the application of Industrial AI for process optimization can contribute to make manufacturing processes more profitable, while also being more sustainable and efficient. Artificial Intelligence (AI) and its various algorithms are computational tools that greatly assist the development of intelligent equipment. In fact, it can be stated that AI is a cognitive science that is used for research, in image and natural language processing, in robotic systems and for automatic learning, among other applications related to I4.0 to conceive, design, manufacture and control systems considered intelligent. In addition, AI, in conjunction with Big Data, IoT, cloud computing, DTs and CPS, among other technologies, will make companies and industries of today and the future operate in an optimal, efficient, flexible, lean and green way. Currently, industrial AI (Artificial Intelligence with applications in industries) is in an initial or incipient phase, so it is necessary to study and propose techniques, structures, frameworks and methodologies for its correct use and implementation in the industrial sector [13].

AI is a field of knowledge that is not new, but that will now be a necessary input for those companies that want to operate under the context of the I4.0 and that, due to its great importance, should be a subject of formal study in engineering education. The adaptation of companies and universities to I4.0 must be in a certain sense accelerated because many companies already handle various disruptive technologies and therefore put pressure on other companies that make up the value chain to upgrade. Some countries, such as Germany and France are in an accelerated process of industrial reconversion, but other countries such as Mexico or Brazil do not present a substantive dynamic of change. The urgency for change is based on two facts: (1) Many companies have not measured the implications of I4.0 and, therefore, have not started upgrades and (2) It has begun to be discussed that there is already a fifth industrial paradigm known as Industry 5.0 (I5.0), which seeks to integrate human beings back into industrial processes, improve the environment, obtain better social benefits and implement resilient systems, among other innovations.

I5.0 focuses on the following basic elements [14]: (1) On human agency, (2) On sustainability, and (3) On the ability of a system to maintain important functions and processes in the face of stress by resisting and then recovering or adapting to change (resilient system). I5.0 will bring about relationships between systems of different classes and technological configurations associated with I4.0 that are linked together for mutual benefit and between skilled operators (symbiotic relationship between technology and humans), to create workplaces and work environments where the human being is at the center of the work and is able to generate high value-added, top quality and customized products. The I4.0 is characterized by the implementation of state-of-the-art technologies and with this, better and high performances are achieved; on the other hand, the I5.0 seeks to establish highly cooperative relationships of the synergistic type between production systems improved with new technologies and social systems, with the aim of seeking a more personalized and massive production of parts, products, solutions and services [15]. I5.0 should be considered in the education of today's engineers, since, similar to I4.0, I5.0 represents technological changes and challenges in companies and society in general.

Industry 4.0 is changing the world and teaching processes, so higher education needs to be innovated. In fact, higher education institutes are contributing to these changes through research and education [16]. I4.0 causes engineering education to undergo significant changes, requiring teachers to modify traditional engineering education. Teachers have to train their students under the guidelines of the new industrial paradigm so that they can cope with I4.0 and have the ability to continue research on the subject under lifelong learning conditions. In other words, it is required to establish such tangible engineering skills, both in processing and thinking that can be applied to emerging technologies [17].

The I4.0 vision seeks to train engineers who are capable of performing more complex activities, since robots with intelligence will replace operators and engineers in various industrial activities. Therefore, the education of professionals must be oriented towards obtaining information and improving their skills in tasks and activities that cannot be performed by robots. I4.0 causes a displacement of humans from production systems due to high automation and AI. This distortion of I4.0 in the activities of professionals causes education to seek mainly skills and abilities, and specialized development in engineers, for example, in the mastery and applications of AI and in the handling of large amounts of data (Big Data). In this sense some of the skills required today to face I4.0 are advanced analytics, digital security and IoT. It should be noted that in order to align with the challenges of I4.0, education is moving towards the digitization of its processes [18].

I4.0 seeks to ensure that intelligent production processes generate customized results through more efficient, sustainable and environmentally friendly production strategies. However, in order to achieve the proposed objectives (producing efficiently and under sustainable criteria), entrepreneurs must face certain challenges, such as the lack of personnel with the necessary knowledge and skills to enable them to develop, implement and manage high-tech systems. Therefore, the implementation of I4.0 requires the labor market to change and to promote the training of professionals with the necessary knowledge and skills so that companies can achieve their objectives within this new industrial paradigm [19].

In this context, engineering education must seek approaches, methods or educational models that allow it to train the engineers of the present and future to meet the challenges imposed by the I4.0 and the fifth industrial revolution, without forgetting that in many countries the transition from manufacturing 3.0 (I3.0) to I4.0 is taking place. Competency-Based Education (CBE) may be an ideal educational approach for training engineers in the context of I4.0, as it fosters comprehensive training, flexibility and self-management, promotes active learning, develops technical and social skills, and encourages engineers to solve problems in complex situations [20], among other relevant features of this approach. Competency is considered substantially important and is given high consideration because of its ability to provide competitive advantage to organizations. One of the challenges that professionals and operators face in the workplace is to a constantly changing environment, whether brought about by management or by industrial change in technology or ideology. These changes have a significant impact on the skill and competency needs of workers [21].

On the other hand, the specific competencies associated with the job are essential to find those professionals required by companies, so a meticulous identification and description of them helps employers not only to obtain employees in line with their requirements, but also provides information to determine the measures to be adopted in the company. Specific competencies are the guidelines for universities to develop their curricula. I4.0 and cutting-edge technologies have caused progressive companies to seek professionals trained in information technologies and digital systems [22].

CBE is both goal-oriented and outcome-oriented. Competency-based education focuses on standards or norms that involve all the processes involved in the models and modalities that adopt it, namely: standardization, assessment, certification, recognition, training and validation. In this arrangement, competency standards can motivate the learning of individuals and groups, but also limit and even discourage it [23].

The I4.0 promotes changes in engineering education, but with more emphasis on Mechatronics Engineering, so it is necessary that education in mechatronics and, in general, mechatronics as an area of knowledge, must be reinvented. There is no single path that indicates how mechatronics should be reinvented, but it is possible to know important aspects of its evolution. In this sense, today's mechatronic systems have very advanced capabilities that are mainly based on the evolution of mechatronic enabling technologies and of the mechatronic design methodology itself. It is possible to observe the improvement in intelligence of mechatronic systems and, above all, the increase in their complexity. The changes that are occurring in the technologies and in the mechatronic design methodology itself drive the new processes, products and systems that are generated under the vision of mechatronics to possess new properties and capabilities, which drive the generation of new systems that support productive systems [24]. These devices or products evolved from simple monitoring to self-optimization of their performance.

Mechatronics Education is essential in I4.0, so it is necessary to apply or propose appropriate educational models or approaches for the training of engineers that allow their incorporation into the working world whose dynamics is being affected by the integration of various disruptive technologies. In this sense, this article proposes a general conceptualization of the technologies and processes of I3.0, I4.0 and I5.0, mainly related to CPS, DT and technological conditioning (Retrofitting), with the objective of proposing a series of technical considerations through which it is possible to build specific competences for mechatronic engineers. The importance of CBE and active methodologies in the training of the engineer in the vision of I4.0 and I5.0 is discussed.

Although all the competencies (basic, generic, specific) are important in the training of the mechatronic engineer, this article only focuses on proposing some technical considerations that help to shape the specific competencies. It does not propose an in-depth study or the methodological construction of these competencies, because in reality these are shaped taking into account the local, regional or national needs of companies and industries.

## 2. Materials and Methods

This section presents the methodology used in this research work. The type of research was descriptive-qualitative, because the generalities of I4.0 and I5.0, and their relationships with engineering education, especially with Mechatronics Engineering and CBE, were analyzed. This part of the paper consists of two general blocks: (1) The first block is related to the industrial aspects of I4.0 and I5.0 and (2) The second block is associated with the aspects of engineering education. The first block describes some aspects related to industrial revolutions, I4.0 and I5.0 and the technologies that support them, as well as the definitions of CPS, DT, AI and simulation. Block 2 describes some important aspects of engineering education and Mechatronics Engineering education in the vision of I4.0, CBE and active methodologies. With the analysis of both blocks, 15 guides for the conformation of specific competences in Mechatronics Engineering are proposed.

### 2.1. Industries 4.0 and 5.0

This section presents various concepts and definitions that shape I4.0 and I5.0, as well as the transition process between I3.0 and I4.0. Some disruptive technologies, CPS and DT, are described, and the concept of technology overhaul is discussed. The understanding and analysis of these technologies will allow to know those technical characteristics important for the shaping of specific competences in mechatronic engineering.

### 2.1.1. General Aspects of Industrial Revolutions

Throughout history, each industrial revolution has changed the way of life of the human race; in fact, the industrial paradigms that are currently being presented provoke a broad reflection on what human beings can do in terms of technology and on the capacity and enormous importance of people's creativity [25]. An industrial revolution or industrial paradigm is generally conceived by human beings as a single occurrence or an isolated event, but in reality paradigms in a given time bring about changes and innovations that give impetus to companies and improve world economies. History has shown that the changes brought about by industrial paradigms are progressive, i.e., they are modernizations of previous ones [26].

Each industrial revolution has one or more core technologies and processes that distinguish them. The rise of the textile industry was the main feature of the first industrial revolution (IR1.0); the introduction of electricity, production lines for mass production and internal combustion systems were the core technologies of the second industrial revolution (IR2.0) and the introduction of automation and computers were the focus of the third industrial revolution (IR3.0) [27]. The last two technological paradigms appeared just 12 years ago and their core technologies are CPS and DTs, for the case of I4.0 (fourth industrial revolution) and human-machine collaboration systems and personalization, for the case of IR5.0 (fifth industrial revolution). Figure 1 shows each known industrial revolution and its core technologies [28].

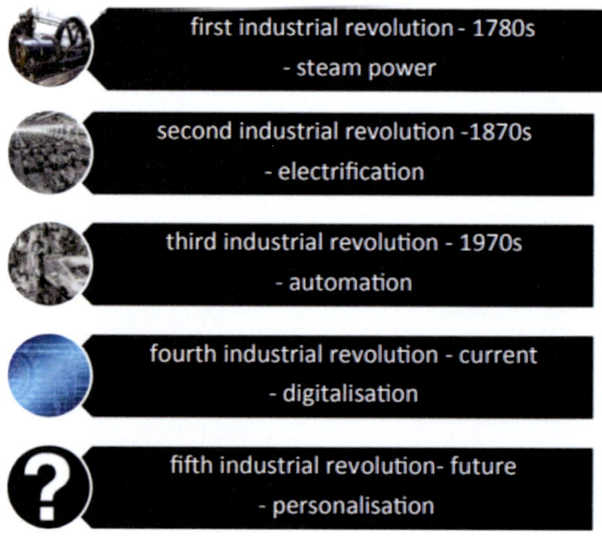


**Figure 1.** Industrial revolutions and their core technologies [28].

Steam engine, electrification, automation, digitization and customization describe one of several characterizations of industrial revolutions (see Figure 1). Another such characterization up to I4.0 is the following [29]:

IR1.0: Steam systems marked the beginning of IR1.0 in human history.

- IR2.0: Assembly lines for mass production of cars involved the systematization of manufacturing processes.
- IR3.0: The introduction of the computer increased production volumes and reduced the importance of human labor.

- I4.0: The CPS are identified as the central part of the I4.0 technologies, since developments in computing and in the conformation of high-powered information systems allow productive systems to operate in an interconnected way.

According to Ratanlal [30] IR5.0 should perfectly be the elaboration of the modern manufacturing framework to enable man and artifact to perform tasks hand in hand, combining the specific and cognitive knowledge of workers and the precise and specialized knowledge of robots to introduce an ultramodern way of life in care.

### 2.1.2. Industry 4.0 and Industry 5.0 Definitions

In order to know the field of action of the mechatronic engineer, it is necessary and important to know the definitions and general aspects of I4.0 and I5.0.

According to Dilmé [31], in some studies conducted by universities and companies, they found that 90% of the data in the world today has been created only in the last two years, 30% of companies started to monetize their data assets in 2017, and according to S&P 500, the average lifespan of an industry has shrunk by 50 years in the past century, from 67 years in 1920 to 15 years today; 86% of CEOs consider digital their first priority and 76% of millennials believe innovations are their most valuable trait. These studies show the changes that are occurring at an accelerated rate in industries and that have to do mainly with the changing demands of consumers and that are related to the applications of novel (disruptive) technologies in production processes.

I4.0 will bring about major changes in companies, from technical to organizational issues [32]. It is possible to exemplify some visions of the changes that I4.0 will bring about:

- New level of social-technical interaction: It is possible to plan high value-added processes between organizations using autonomous and self-organized production resources.
- Intelligent products: The operating parameters of the production lines and of the generated products are known data and the information of both is exchanged. The production of the products can be optimized if it is possible to form them into technological groups.
- Individualized production: Flexibility in production systems allows the specifications or characteristics required by customers and the products themselves to be taken into account during the design of the product life cycle.
- Autonomous control: Operators will be able to take control and reconfigure intelligent technology assets taking into consideration the highly sensitive objectives of today's environment.
- Product design controls product-related data: Product information is key and crucial for product life cycle management and development.

On the other hand, the World Economic Forum applied a survey to 371 companies of high global relevance, to find out which were the main technological resources that are driving innovations in current jobs, in competencies and that will be determinant in future jobs [33]. The results obtained from this survey were as follows:

1.  Systems that increase computational capacity and the use of large amounts of data (Big Data).
2.  The Internet that allows connection between mobile devices and cloud computing.
3.  New energy supplies and technologies.
4.  The IoT that makes possible the connection through sensing elements, communications and better processing of information and energy in industrial equipment and domestic systems.
5.  Open collaboration, collaborative economy and peer-to-peer (P2P) networking.
6.  Collaborative and advanced robotics, as well as independent (autonomous) transportation.
7.  AI and machine learning.
8.  Modern advanced manufacturing and additive manufacturing.
9.  Novel and advanced materials, biotechnology and genomic technology.

The technological resources mentioned above can be applied to various processes in companies and factories with the aim of improving efficiency and optimizing production systems by injecting added value. I4.0 was conceived with the purpose of being able to systematize and understand the changes in industries in recent years arising from innovations and improvements made by companies, the incursion of novel technologies and the implications of digitalization in industries. I4.0 describes a set of methods that are used to move from manufacturing conducted primarily by machines to digital manufacturing [34]. I4.0 can be understood as the integration of novel and disruptive technologies whose main objective is process optimization. I4.0 is a trending concept, promising remarkable results for industries while profoundly changing organizations in many terms. The changes start in the way business models are established throughout the entire production process up to the end point when the customer receives the product [35]. One of the goals of every company is to decrease the costs of parts or processes and to have processes that are more flexible and capable of manufacturing small but complex batches of products to meet the demands of highly specialized and differentiated markets. In this sense, I4.0 focuses on improving competitiveness by reducing costs and increasing flexibility in decentralized production systems to offer customized products, which is an advantage to satisfy customer markets [36].

The I4.0 promotes the use of a wide variety of technologies to encourage companies to implement changes in their production processes, including AI, Big Data, additive manufacturing, among others [37], which make it possible to form CPS since it is possible to create significant information flows in real time. In addition, the I4.0 makes it possible for industries framed in the vision of the I3.0 to be highly interconnected and to seek the computerization of their processes. I4.0 seeks to achieve several objectives, including ICT driven production customization, autonomous flexibility of production lines, systematic monitoring and communication of components, products and machines, and promoting methods where human-machine operations need to be implemented, as well as promoting the application of optimization methods in production lines using the IoT and offering industrial models in which there is high interaction in high-value production chains [38].

The I4.0 paradigm seeks to link physical and digital systems in new interaction models. There is a consensus among the visions of scholars and industries related to the projections of production for the future based on I4.0. These visions agree that there will be [39]: (1) Smart industry, (2) Products or items with certain degrees of intelligence, (3) New models for doing business, and (4) Buyers and customers.

In 2011, the concept of I4.0 was first proposed in Germany and was considered as a new initiative or strategic projection with the objective of being the first country to push its manufacturing industry to modern development [40]. A formal definition of I4.0 is as follows [41]:

*I4.0 is the technical integration of CPS in manufacturing and logistics and the use of IoT and services, for applications in industrial processes. This integration will have implications for value creation, business models, downstream services and work organization.*

I4.0 is defined according to Rodriguez and Bribiesca [42] *as the digital transformation driven by connected technologies to build a cyber-physical organization.*

The common term between the two definitions described above is the cyber-physical entity or CPS, so it can be said that CPS is the core of I4.0. I4.0 can be considered as an industrial paradigm that promotes high automation for process optimization. Somehow this industrial paradigm displaces the human being from the center of production. However, it is well known that AI applied to industry makes it possible for humans and devices or machines to interrelate with each other. Many machines, such as robots, can now learn many operations performed by operators. It is from these collaborative practices between cybernetic machines and humans that the term Industry 5.0 (I5.0) is conceived.

I5.0 promotes the use of robotic systems to collaborate with operators, particularly in mitigating risks in companies. These systems may be able to understand and in some way sense the operators, as well as understand the tasks they perform. The aim is for robotic systems to be able to observe, sense and learn the tasks that workers perform so that they can help them perform them. I5.0 promotes the integration of AI into the lives of human beings with the main purpose of improving their capabilities [43]. In most applications, I5.0 has shown an important connection between intelligent systems and humans through precise manufacturing automation with critical thinking skills [44]. In addition, I5.0 will put workers on another level by moving from manual work to cognitive work, which will involve adding more value to work activities in modern industries. I5.0 will be based on decision making, human creativity, innovations and critical thinking, which will generate more personalized products, articles and services with higher added value, while robotic systems will perform repetitive, high-risk and labor-intensive tasks [45].

I5.0 promotes integration and greater collaboration between operators and intelligent machines through the application of highly specialized and precision automation techniques accompanied by the power of human critical thinking. The idea of I5.0 is to empower organizations to reach high levels of competitiveness with the help of novel and efficient tools, and that companies can use industrial recycling seeking to have rapid change capabilities without economic or capital investment [46].

Another contribution of I5.0 is the mass customization of products, where users have the facility to prefer products made to their liking or to their specific needs. I5.0 will lead to a significant increase in the efficiency of production and will make possible a great versatility between humans and machines, which will allow for task responsibility and constant supervision [47].

There are several proposed definitions of I5.0, two of which are described below:

**Definition 1.** *I5.0 is considered as an initial evolution of human-operated industry based on the principles of the 6Rs (Recognize, Reconsider, Realize, Reduce, Reuse and Recycle) of industrial upcycling. The 6Rs represent a technique of waste prevention and logistic efficiency design to value standard of living, innovative creations and to produce customized items developed with improved quality [48].*

**Definition 2.** *The I5.0 brings together the most elementary and strongest concepts of the so-called SCP and the power of the intelligent mind of human beings for the conformation of productive systems that operate synergistically [49]. Because I4.0 weakens the direct work of operators, government officials promote a design of products and systems that is innovative, ethical and where man must be the center of the proposals.*

I5.0 not only refers to the integral cooperation between cybernetic machines and human beings, but also involves aspects of sustainability and social considerations. The I5.0 paradigm promotes the recognition that companies have the power to achieve more far-reaching social goals beyond the benefits of work and economic growth, and that they can be resilient and prosperous providers, that make it possible for production systems to respect the limits of the planet and that put workers at the center of production [50].

2.1.3. Base Technologies of I4.0. and I5.0

Knowledge of the technologies that support both I4.0 and I5.0 is fundamental for the design of specific competencies in Mechatronics Engineering. I4.0 is supported by nine technological pillars, which are shown in Figure 2 [51].

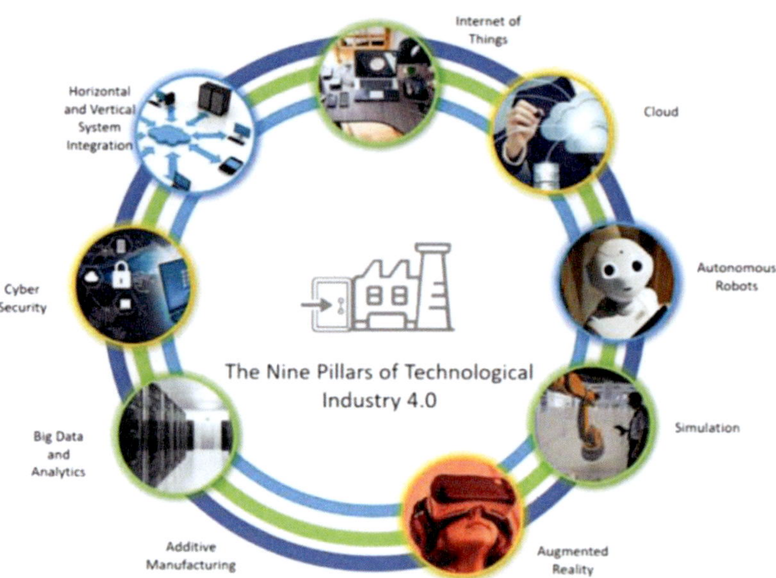


**Figure 2.** Technologies supporting I4.0 [51].

Each of the technologies shown in the figure above are described briefly below [51]:

- Autonomous robots: These systems have been improving over time and when incorporated into production lines they make them more flexible and more autonomous, allowing close collaboration between humans and other robotic systems.
- Simulation: This technology allows the creation of DTs that are used to optimize processes, and is also applied for virtual product design, for material selection and to mimic the behavior of production lines where machines, products and humans coexist. It is possible to virtually build complete production systems with the help of simulation.
- Horizontal and vertical systems integration: This technology seeks to computerize supply chains by forming networks that are capable of integrating information and data between external and internal systems (cross-functional type).
- Industrial IoT: This technology allows the integration of mobile devices with IT systems with the objective of sharing data and information in real time.
- Cybersecurity: This technology is necessary in I4.0 in order to protect the sensitive information of industries and so that productive systems can be guaranteed the exchange of reliable information.
- Cloud: Many of the services of different IT sites that rely on the storage and exchange of information and data will be managed in the cloud.
- Additive Manufacturing (AM): This technology is used for the design and manufacture of parts and products that can be customized to customers' needs. Because AM produces lightweight designs and small batches of products, product warehousing is reduced as well as logistics costs.
- Augmented Reality (AR): AR can be used for virtual training of companies' human resources, for equipment maintenance and to perform better work techniques and procedures on production lines.

I4.0 can be implemented by companies seeking to be competitive. However, to achieve this they must know what I4.0 is, the pillars on which it rests and their interrelationships. Figure 3 shows a proposed conceptual map based on three major sets of connections (inherent connections, cybersecurity connections, and interpillar connections) [52].

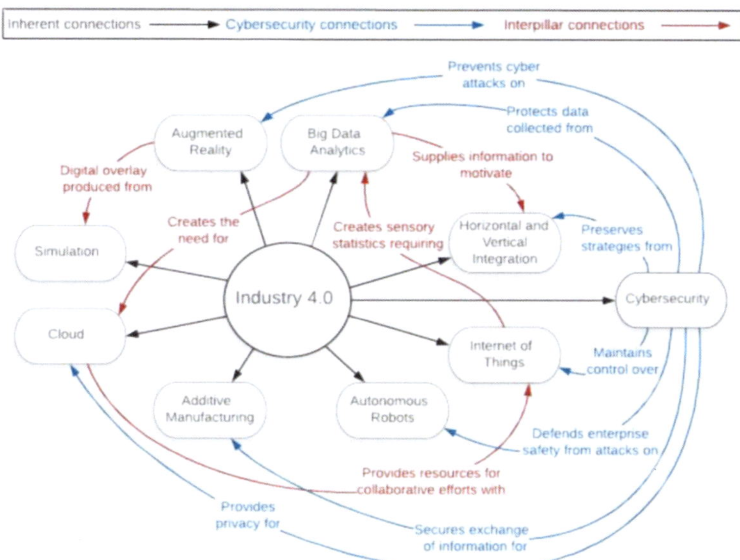



**Figure 3.** The relationships between the three pillars of connections of I4.0 technologies [52].

The map shown in Figure 3 does not represent fixed relationships between technologies and more sets of connections can be considered, as the rapid pace of technology innovation may pose different connections. On the other hand, Figure 4 shows a proposal of the technologies that integrate I5.0 [53].

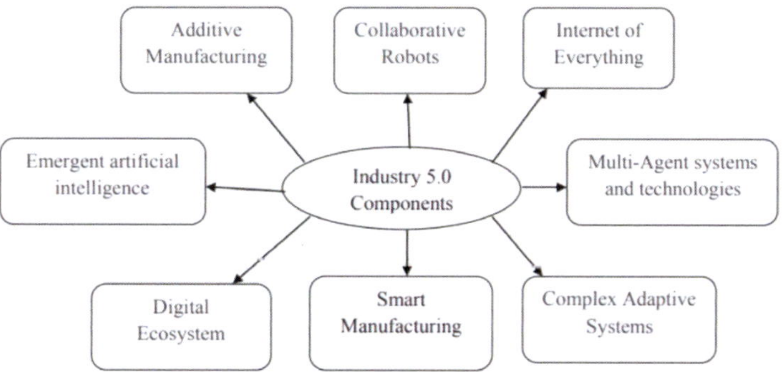



**Figure 4.** Technologies supporting I5.0 [53].

The I5.0 associates aspects of environmental care and social aspects, as well as the interaction between men and machines, in fact, from this interaction is generated the concept of Operator 4.0, which seeks the conformation of new manufacturing systems called Human-Cyber-Physical Production Systems (H-CPPS) whose objective is to generate the necessary conditions for workers to improve and expand their capabilities [54]. The development of trust-based relationships between operators and devices or machines is a goal of the Operator 4.0 idea. Such relationships would enable smart industries to maximize their strengths in their context by including machinery endowed with intelligence. At the same time, these human-machine interactions would enable empowering and augmenting the skills and capabilities of workers to achieve more far-reaching goals and opportunities that are generated by the implementation of I4.0 [55].

Figure 5 presents the technological resources associated with the Operator 4.0 idea [44].

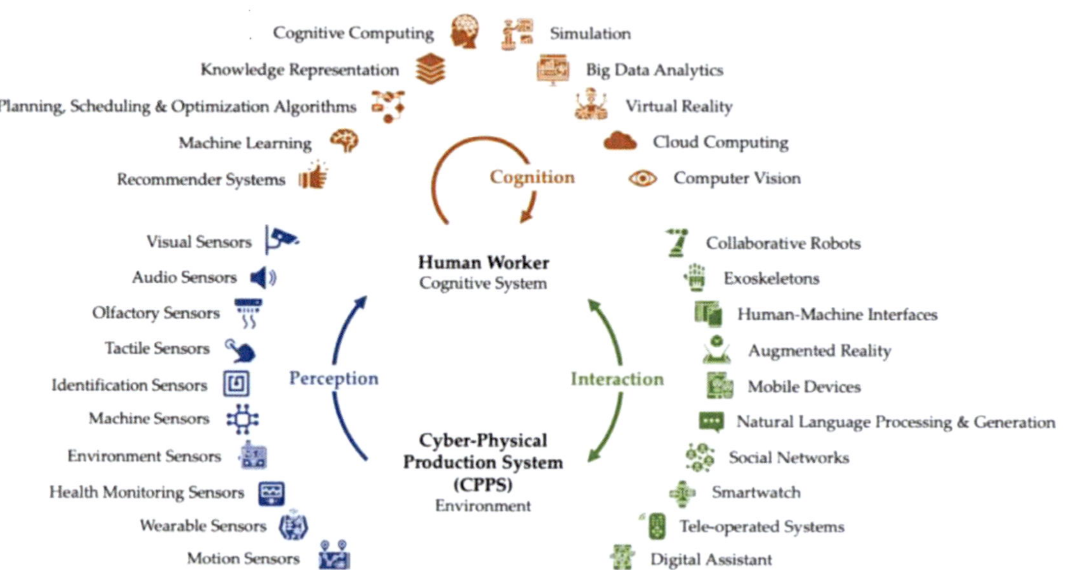



**Figure 5.** Human-machine interactions (symbiosis) of I5.0: Technologies increase people's capabilities and skills due to a cognitive-type process [44].

### 2.1.4. Technological Transitions: Technical Aspects

The knowledge of the information of industrial processes related to the technological transitions between one revolution and another is fundamental to be able to design specific competencies for Mechatronic Engineering. This information allows determining which knowledge should be taken into account and which should not, since the industrial world is actually undergoing an industrial transition that implies gradual and/or disruptive changes in production systems.

Transitions between industrial revolutions are often complex and pose significant economic challenges, especially for companies, as they must adapt to change in order to be competitive. For several companies in the world, especially those whose production systems are framed in the I3.0 and that do not have sufficient capital, the technological transition (from I3.0 to I4.0) has become a serious problem, as they face the challenge of deciding whether a technological upgrade (an almost total change in production systems) that involves a considerable economic investment or a technology reconversion that involves a substantial improvement to the production systems they already have and, therefore, less economic investment, is relevant [20]. The Industry 3.5 (I3.5) concept is associated with the representation of the technological transition from I3.0 to I4.0, and can be considered a hybrid strategy, not only for a technological transition in production systems, but also for managing any disruptive impact, such as total resource management for sustainability [56]. The concept of I3.5 becomes an overall strategic framework that concatenates high-tech applications, IoT, big data analytics, resource allocation, improvement and optimization, seeking to develop a basis for smart production [57].

One of the main objectives of I4.0 is to promote the development of productive systems endowed with intelligence where devices or machines have special characteristics, such as adaptability, a high degree of flexibility, learning by themselves (self-learning) and being able to be self-adaptable [57]. However, I3.5 is only interested in ensuring that the benefits and improvements made to production systems are short term, and for this purpose it must use various tools, including those used for analysis and those applied to process optimization. These tools are applied to the development of company operating plans and supply chain management, plant and quality controls, so that highly automated operating systems conceived under the I3.5 approach can perform various tasks, including decision making. An expected result with the incursion and improvement of IoT, DT, CPS, data processing and physical assets (infrastructure), is the effective and efficient communication between the different and diverse devices, machines and production lines conceived under the I3.5 approach, so that a conception towards the intelligent industry of the I4.0 can be derived from this communication. Table 1 shows a comparison between I3.0, I3.5 and I4.0 [58].

**Table 1.** Comparison between I3.0, I3.5 and I4.0 [58].

| Features | Industry 3.0 | Industry 3.5 | Industry 4.0 |
| --- | --- | --- | --- |
| Core Concept | Highly automated system | Decision making ability with the improvement of existing environments | Smart factory with CPS and IoT |
| Production Strategy | Mass Production | Flexible Manufacturing (diverse products with small lot size) | Mass Customization |
| Quality Control | Statistical Process Control | Advances Process Control | Self-aware; Self-prediction |
| Resources Management | Materials Management; Human Resource Management, etc. | Total resource Management | Self-configure; Self-optimize |
| Development Priorities | Investment of hardware | Integration of ability of data analysis and experience of management | Construction of CPS and IoT |

Another challenge faced by I3.5 is the decision making of companies to decide how to introduce disruptive technologies to their production processes, especially those companies that have difficulties investing in new equipment. Technological reconversion or also called technological upgrading may be an option. However, it is not an easy task to upgrade a conventional system or machine to align with the demands and requirements of I4.0, because components and systems, such as actuators, sensing systems and computer systems generally have communication protocols that prevent, among other things, multidirectional interconnection both internally and externally. In other words, although an I3.0 CNC machine is a CPS, its transformation into a CPS developed under the I4.0 approach is complicated due to its technological limitations. In this sense, the retrofitting of machinery and systems can be a suitable method for technological reconversion.

Retrofitting can be classified into: (1) Traditional retrofitting and (2) Intelligent retrofitting. The first one refers to the replacement of parts or subsystems to achieve the optimization of some process variables such as, for example, the reduction in maintenance time, the speed of machines and processes, as well as the accuracy of various tasks, among others; while the second one focuses on the low-cost adaptation of subsystems, machines or equipment that already exist in the companies [59]. In fact, Intelligent Retrofitting studies and analyzes production lines and machinery whose design was not conceived to operate in I4.0, so its priority objective is to achieve the transfer of the most important aspects of the I4.0 paradigm to the machinery and production systems of companies in a shorter time and at a low cost. This type of reconditioning can be supported with the help of the Lean philosophy [60]. I4.0 requires both reconditioning of machines or production systems to achieve its purposes.

Technological reconditioning does not have a single methodology, so there are several proposals for machines or CPS, as well as for processes. Figure 6 shows a proposal for the reconditioning of a process for its operation under the I4.0 philosophy.

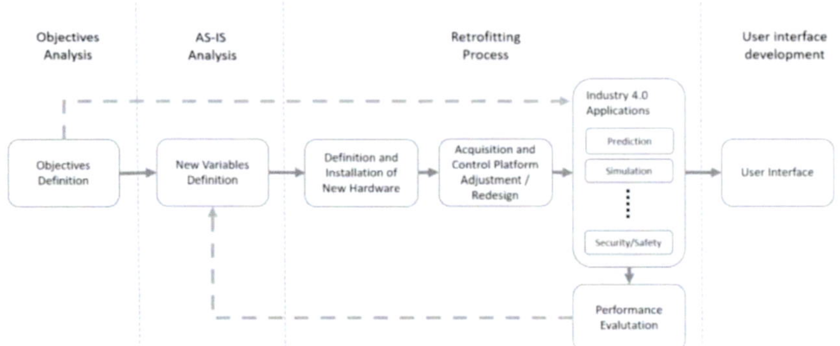

**Figure 6.** Approach adopted for the retrofitting process [61].

### 2.1.5. The Importance of CPS

History shows that in general there is one or several technologies on which each industrial revolution is based. It is crucial for the formation of specific competences in Mechatronics Engineering to know the generalities of the CPS since they represent the technological basis of the I4.0.

Cyber-physical systems (CPS) are very complex systems that are integrated with collaboration of communication, computation, and control together termed as 3C technology [62]. Cyber–physical systems (CPS) are composed of physical and computer-based (i.e., cyber) parts, which are highly interconnected [63].

I4.0 has CPS as its technological backbone. In fact, there are a large number of researchers who claim that CPS is the critical element [64] and the core [65] of I4.0. The idea of CPS was initially described and coined by the National Science Foundation (NSF) in the USA in 2006 [66]. CPSs are described as any entity composed of physical and cyber elements that interact autonomously with each other, with or without human supervision [67]. There is a range of cyber-physical entities that can be currently observed in industrial activities, e.g., robots, coordinate measuring machines, manufacturing cells, CIM (Computer Integrated Manufacturing) systems, machining centers, data acquisition systems, and SCADA (Supervisory Control and Data Acquisition) systems, among others. Figure 7 shows a CPS represented by the iCIM 3000 (FESTO DIDACTIC, Denkendorf, Germany) didactic manufacturing cell conceived under the I3.0 philosophy and located in the facilities of Universidad la Salle Noroeste, Mexico.

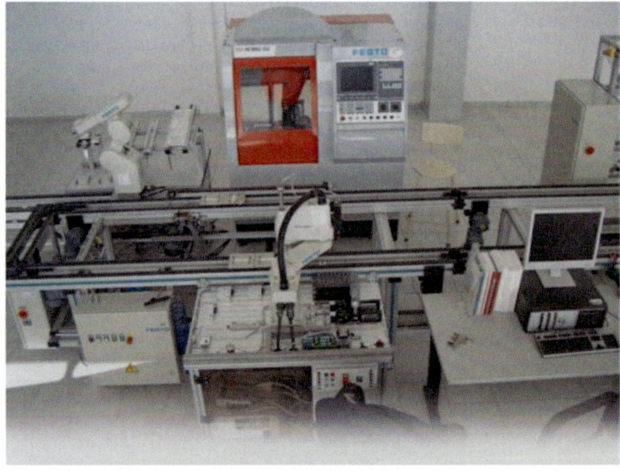

**Figure 7.** Traditional CPS: iCIM3000 didactic cubicle.

CPSs are characterized by high computational integration, physical assets that perform a large number of tasks and operational networks. It can be stated that the most important particularity of the CPS is the high integration between different hardware components and software systems. In such integration, actions and operations of computation, effective control and efficient communications coexist, which are taken into account at the same time and together with physical assets for the design of production systems [68]. CPS represent a generation of new digital systems that are based on complex interdependencies and relationships, as well as high integration between physical assets and the digital world [69]. The CPS is an intelligent computer system that uses controlled mechanisms and different algorithms to connect software and hardware parts so that it can function and display a variety of forms and approaches.

The CPS is an intelligent computer system that uses controlled mechanisms and different algorithms to connect software and hardware parts so that it can function and display a variety of forms and approaches [69]. The general structure of a CPS is made up of two major groups of elements or subsystems. The first of these corresponds to the physical assets and the conditions imposed by the environment surrounding them, while the second subsystem relates to the computer control system whose function is to analyze and control the state variables of the processes by sending information via sensors to the physical assets so that they adjust if the parameters of the environment change. There is no person who is able to observe with his sensory system the environment around him and who can develop a mental map model of the tasks that are being presented and at the same time is able to provide an intelligent buffer between the system of physical assets and the cyber-physical world [70].

A CPS is integrated by two layers: (1) The cyber layer is composed of many intelligent monitoring nodes (including people, servers, information sites or various mobile devices) and their communication links and (2) The physical layer that integrates various interrelated physical entities. Under the interaction both layers, the system realizes information interaction and decision making by 3C (Computation, Communication and Control) technology [71]. The design of CPSs is a challenge nowadays and engineers must take into account variables such as: security, scalability, interoperability and robustness, among others, to develop reliable and trustworthy CPSs. CPSs are interrelated between computational processes and physical processes, and can have the potential and ability to perceive the environment and have autonomous control, as well as make important decisions. Therefore, the deep integration of discrete computation of discrete computation and continuous physical process is one of its typical characteristics of CPSs [72], i.e., they integrate a series of dynamic models and with hybrid characteristics.

The study of CPS requires knowledge of various models, such as computational, physical and network models. The former can be considered as analytical discrete transition models that number states (e.g., data flow models, event models and state machines, among others), while the latter can be represented by differential equations (continuous-time models) and their solution methods [73]. CPSs are associated with various highly integrated systems such as equipment, various devices, offices and buildings, transportation systems and production lines, among others, however, CPSs also integrate various logistics, process management and coordination processes and operations.

CPSs collect, manage and analyze data through the support of various signaling elements, while actuators are used to react to production or organizational changes and communicate with the other components. CPSs can be implemented to manage different issues such as production, logistics, quality, planning and scheduling activities within the factory [74]. There are many different types of CPSs in everyday life: industrial control system CPSs, smart grid CPSs, medical CPSs, smart vehicle/automobile CPSs, domestic CPSs, aerospace CPSs, and defense CPSs, among others [75].

The development and operation of a CPS requires consolidating and integrating various systems and actors, such as mathematical and computational models, analytical methods and techniques, as well as a great diversity of tools. In order to increase the probability of operation of a CPS (reliability), it is necessary to consider two aspects: (1) To have the ability to consider the consequences and implications of the evolution of the elements that make up the system and (2) To have the ability to look for and point out those changes that, although considered as optimal, may jeopardize the reliability and overall reliability. In this sense, for the conception, design and operation of a CPS it is necessary to take into account several descriptive type models (generated during the process in which the CPS sub-systems are designed) and inductive type models (obtained from the databases that are generated when the system is in operation) or combinations of such models. By considering a combination of the aforementioned models, there can be a huge potential for the conception, design and operation of a DT that learns and optimizes processes and that can be integrated to a CPS so that it can make various decisions either when it is operating online or offline [76].

CPSs are capable of performing a variety of tasks [77]:

- Capture data from sensors and store it on local servers or in cloud architectures;
- Drive physical processes by means of actuators;
- Connect and operate with other CPS;
- Interacting with machines and humans;
- Provide real-time response to stimuli generated by both the surrounding environment and the CPS itself.

Some key features of CPSs include [78]:

- They are considered as a system of systems: This means that CPSs are made up of various subsystems that operate and interact with each other in many ways and in complex ways.
- CPS requires new and novel relationships between computing, control and communication systems to be considered; that is, a robust and integrated design needs to be considered in order to develop tasks and operations that work on their own and with a high level of automation.
- CPS requires that there is a strong relationship or articulation between physical assets and cyber systems that must be considered depending on the specific application.

It is possible to describe the main elements that make up the CPS [79–81]:

1.  Supporting technologies: IoT is a necessary technology as it provides and enables machine-to-human and machine-to-machine communications, ubiquitous computing, embedded systems, fuzzy systems and cloud computing, among others.
2.  Physical assets: actuators, sensors, numerical control machines, control systems, robotic systems, mechanical production systems, server technology, intelligent devices or machines, data handling and processing systems, interfaces and data transmission systems, among others.
3.  Elements of the information environment: PLCs (Programmable Logic Controller), software for cloud implementation, SCADA systems and other systems such as: component and data lifecycle management and administration, and planning systems, among others.

Figure 8 shows a structure of a CPS in terms of the organization and behavior of the subsystems that comprise it.

The construction of its architecture is the first step in the research and development of a CPS. There are several architectures to represent a CPS, such as the 3C and the 5C type. A 5C architecture is shown in Figure 9 [83].

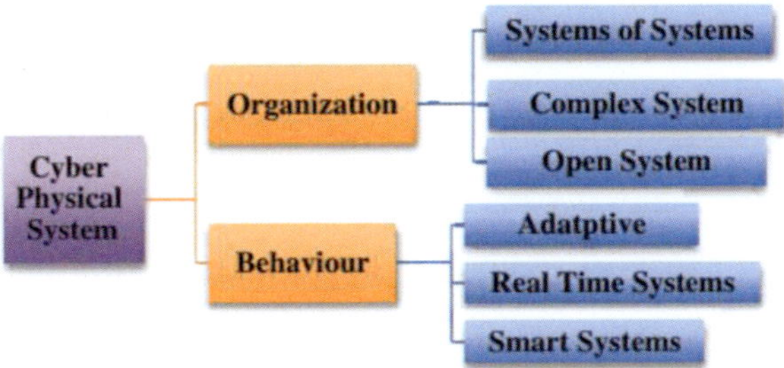



**Figure 8.** Configuration of a CPS [82].

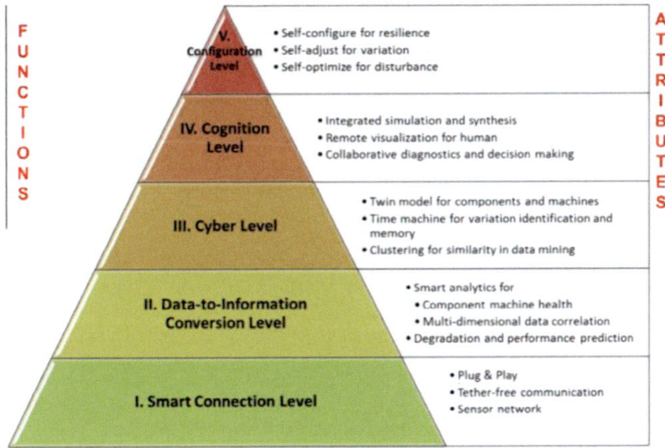



**Figure 9.** 5C architecture for the implementation of a CPS [83].

The 5-tier structure of a CPS proposed in Figure 9 (the 5C architecture) allows a step-by-step design, operation and implementation of a CPS. The five levels are explained below [83,84]:

- Level I: At this level, physical asset data is generated, collected and sent to the next level where the data will be further converted. The data must be obtained as accurately and reliably as possible.
- Level II: Collection of data sent from Level 1 and selection and conversion of important and significant information for subsequent application at Level III in tasks such as prevention, analysis and management.
- Level III: At this level the information for the design of the system configuration is centralized. It concentrates the information of each physical asset connected to the network.
- Level IV: Once the information from other physical assets has been collected through the network layer, the system is supervised taking into consideration historical information and various predictive models to determine any machinery failure.
- Level V: At this level the company's managers make decisions by reviewing the supervision, monitoring and control systems, taking into account feedback and interactions from the digital space to the physical assets.

One class of CPS are Cyber-Physical Production Systems (CPPS); the architecture of one of these systems is shown in Figure 10, which, among other functions, is used to diagnose and anticipate the quality of molten metals, as well as to have a control of manufacturing operations which are developed in novel technologies such as computational simulation, large database management, AI and IoT, among others [85]. The extended explanation of the CPS shown in Figure 10 can be found in [85].

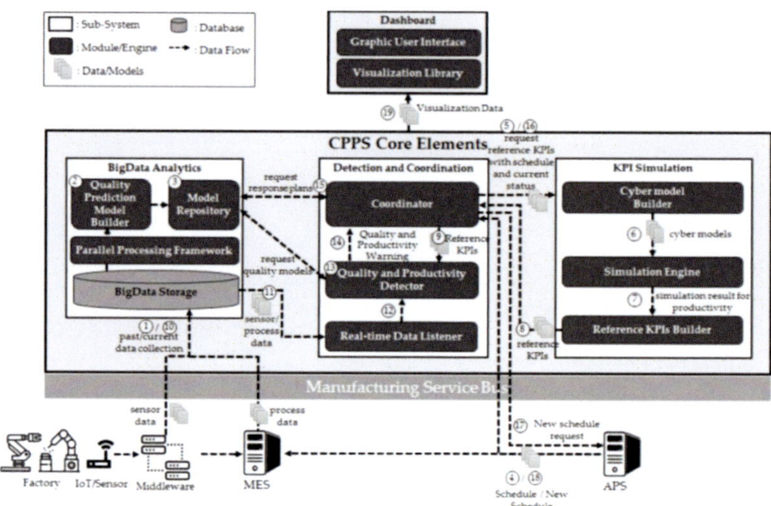

**Figure 10.** CPPS architecture [85].

I4.0 is not only applied in traditional manufacturing systems, but its approach is more general, for example, there is the concept of pharma industry 4.0 or Pharma 4.0 which means the digitization of the pharmaceutical industry [86]. In this sense the technology of CPSs and DTs are applied in the pharmaceutical industry. Figure 11 shows a CPS used in the production of the pharmaceutical industry [87].

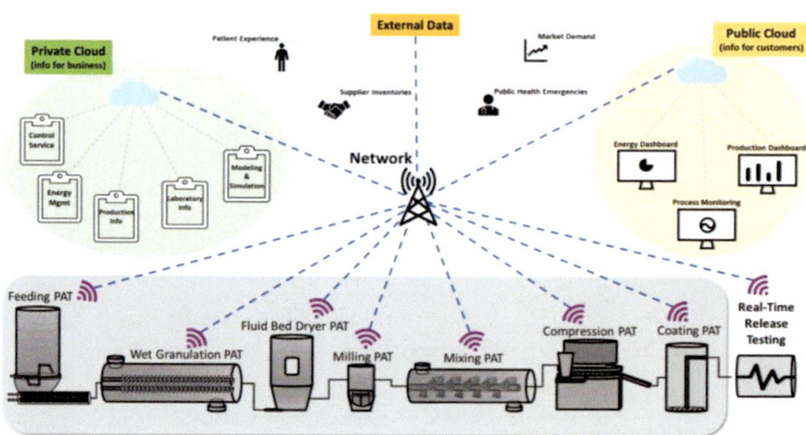

**Figure 11.** SCP for pharmaceutical manufacturing in I4.0. [87].

The CPS shown in the figure above is made up of three major components:

1.  A public cloud that performs an integration of application services (used by customers).
2.  A private cloud that is responsible for information management, which is used to perform tasks of the main or upper layer that integrates SCADA systems, manufacturing, simulation, information and data from laboratories, process control, modeling and analysis, among other technologies. The first and second layers are related to the digitization of system states, which allows various on-the-fly predictions and process optimization studies to be carried out.

3. The manufacturing plant is integrated by various physical assets, process analytical technology (PAT) and RTRt (real-time release testing). The control and management of the production line is governed by the PAT while production quality, which is derived by taking and utilizing information from the manufacturing process, is performed by the RTRt system. Many of the operations such as blending, tablet coating, wet granulation, among others, are connected to the company's network systems and cloud systems through cyberspace [87].

There are various applications of the Pharma 4.0 concept, for example, Ouranidis et al. [88] conducted an inquiry on mRNA therapeutics for the treatment of SARS-CoV-2 and studied the laboratory production of therapeutic mRNA. The proposed process flow design releases a continuous and highly automated production that satisfies GMP (Good Manufacturing Practice) standards, in line with the standardization principles of the pharmaceutical industry 4.0. In another work [89] digital design tools focused on pharmaceutical 4.0 systems, i.e., convergent mass and energy balance simulations, Monte-Carlo machine learning iterations and spatial layout analysis were used, to design the related and integrated bioprocesses in scalable devices, compatible with the continuous operation of mRNA drugs. Similarly, in the work [90] an elastic tensor analysis was performed to quantify the stability of the API (Active Pharmaceutical Ingredient) during the process. In the same way the authors structured a thermodynamic model, represented by the anisotropic minimization of the Gibbs energy, of the stabilizer coated nanoparticles, to predict the system solubility of the material quantified by the application of PC-SAFT (equation of state that is based on statistical associating fluid theory) modeling. The comprehensive fusion of elastic tensor and PC-SAFT analysis in the systems-based Pharma 4.0 algorithm provided an integrated, validated, multi-level method capable of predicting critical material quality attributes and corresponding key process parameters.

Finally, CPS have a relationship with the so-called "embedded systems". These systems are integrated by complex electronics that admit information and send actions to physical assets and software elements using lists of instructions [91]. CPSs integrate digital systems, such as the following: SCADA system, Machine-to-Machine (M2M) systems and industrial automatic control systems [92,93]. CPSs focus on research and development in the area of embedded systems and the study of sensorics [94], in fact, in [95] defined a CPS as "the integration of embedded systems with global networks such as the Internet". A typical architecture of an embedded system is shown in Figure 12.

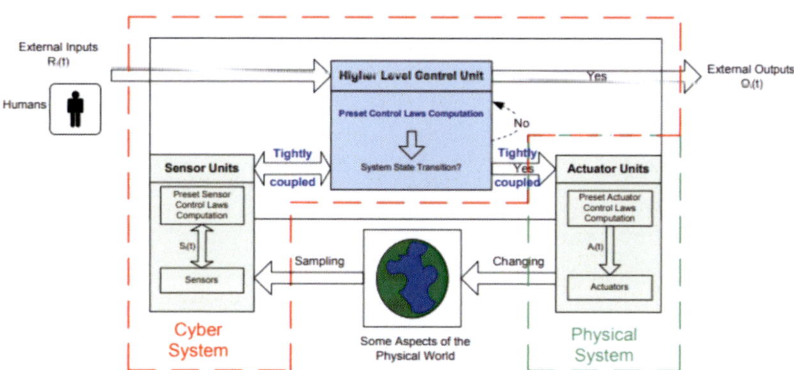


**Figure 12.** Typical architecture of an embedded system [96].

As can be seen in the figure above, sensors and actuators are tightly coupled and related to a control system that has a higher level. At each operating level of the system, variables are being monitored and synchronization is being measured so that the control loops and cycles operate correctly in functional and temporal terms.

### 2.1.6. Digital Twins

Another technology of utmost importance in the operation of I4.0 and I5.0 are the DTs. Knowledge of this technology is crucial for the education of the mechatronic engineer, since its design and operation are based on various fields of knowledge, mathematical models of various physical phenomena and computational models that include some AI tools.

DTs are equally important as CPSs in the I4.0 vision. A digital twin (DT) is a structure of interconnected digital replicas of physical entities [97]. A DT is *a formalized digital model that represents a system, entity, asset or process that collects attributes and properties of assets to establish a bidirectional communication, to form a storage process of the attributes and to perform certain information processing within a specific environment that is influenced by cyberspace* [98]. It can be said that DTs are functional connections between physical assets and the digital world and they use signaling systems (sensors) to collect data and information instantly from the physical asset. The information obtained is applied to generate a computational or digital duplicate of part or the entire asset, which facilitates the understanding of its operation. In this way, it is possible to carry out various analyses of its behavior, to have control over it and to implement different methods to optimize its functions, such as performance [99].

Semeraro et al. [100] conducted an extensive literature search related to DTs and concluded that: a DT can be considered as a set that collects several adaptive models that reproduce the behavior of a physical asset, process or system, in a digital or virtual system which is able to obtain data instantaneously in order to update or reconfigure itself throughout its life cycle. The DT virtually models the physical asset and is used for various applications such as, for example, detecting and fixing faults in the operation, obtaining instantaneous information to optimize processes and to analyze and evaluate events that occur unexpectedly during the operation of the physical asset. Between the DT and its physical replica there must be communication, i.e., [101].

- Between the physical asset and its digital replica.
- Between the digital replica and other different DTs that are located in the environment.
- Between the DT and the domain experts, who interact and operate digital replication, through a set of usable and accessible interfaces.

Originating from industrial design, the DT concept leverages object-specific data to simulate replicas in the virtual world for predictive analysis of security risks and testing of different optimization solutions [102]. DT has three constituent parts [103]:

1. An asset or a set of physical assets (a machine, some process or production system).
2. A digital model with which simulation processes can be performed in terms of the data present and which is integrated by: (a) Various types of algorithms that represent the model to be simulated, machine learning algorithms and data and information extraction procedures to extract special models from the collected data and implement them in the corresponding software and (b) Connectivity components and systems, such as IoT.
3. A collection of historical data or data being obtained instantaneously from the operation of physical assets. The data are the key elements in DT and are used to know whether the objectives of digital replication can be achieved.

The Figure 13 shows a graphical representation associated with a DT.

Tao et al. [105,106] proposed a complete framework of a DT composed of five parts: digital space, virtual space, connection, data, and service. The connection between the five is shown in Figure 14. The DT contains information of the static type (geometric dimensions, bill of materials, processes, and order data) and information that evolves over time (dynamic).

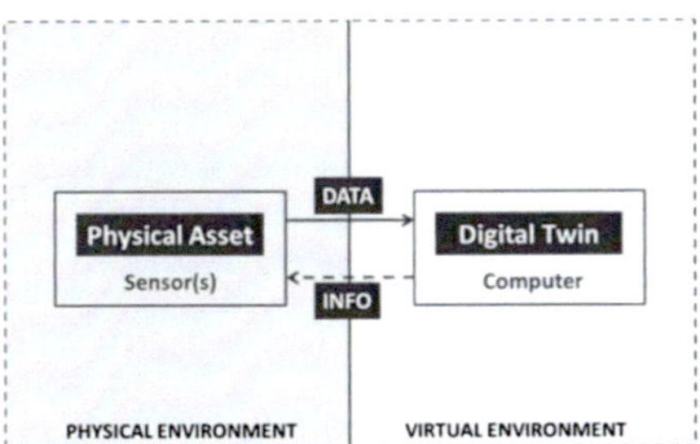

**Figure 13.** Conceptual representation of a DT and its association with a physical asset [104].

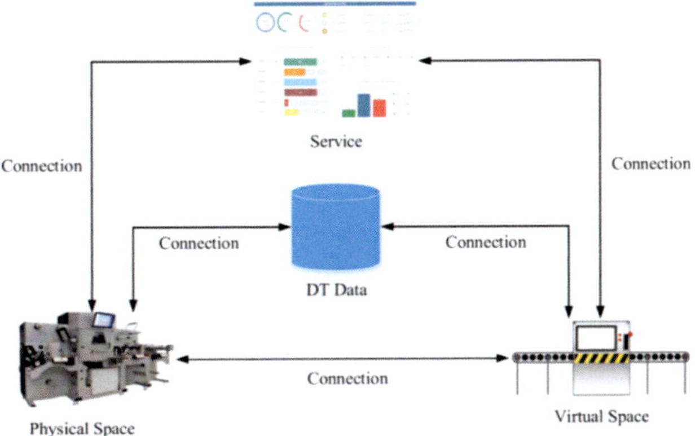


**Figure 14.** Five elements of a DT [105,106].

On the other hand, according to Malakuti [107], industries for many years have used digital representations to model the information of an asset or system (e.g., a device, a production cell, a plant) throughout its life cycle, but often do not study them from the view of DTs. Today, DTs are beyond information models and relate mainly to improvements in digital technologies, architecture development and standardization, interactions, new use cases and business models that enable DTs. Some important elements that integrate a DT are shown in Figure 15.

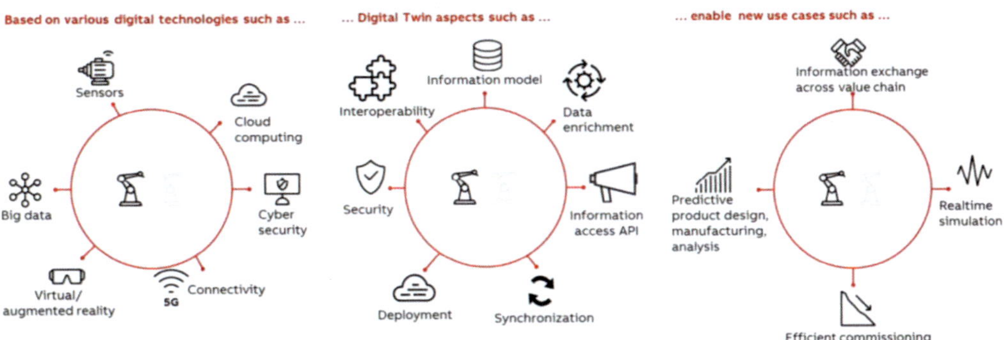


**Figure 15.** Overview of some concepts associated with a DT [107].

There are other concepts similar to DT, so it is necessary to know the differences between them:

1. Digital Model: this model only allows data to be exchanged manually, and no online status update or synchronization between the two objects is possible. This is the typical concept associated with the design phase.
2. Model called Digital Shadow: In this model, the data flow is automated and only occurs in one direction, namely from the physical to the virtual entity, so there is no feedback to the real system from its virtual replica. This model is adopted at most in the service and maintenance phase, and is used to track and predict the behavior of a product in its use phase.
3. The DT: This model is characterized by a complete, automated, bi-directional data flow between physical assets and cyberspace. This vision is best suited for manufacturing applications, such as product quality prediction, production planning or human-robot collaboration.

The differentiation described above between the models is a function of the degree of integration between the physical asset and its digital replica, and the nature and frequency of data and information exchange between the two entities [106]. A DT is endowed with the following capabilities that can be used as decision support [108]:

- Descriptive capabilities: These are based on data describing the current and past states of the production lines, considering the monitoring of the system's good condition and data describing the production that is generated and updated instantaneously.
- Predictive capabilities: These capabilities are developed from models that have the ability to deduce future states and conditions, and the productivity of the system based on hard data according to different circumstances. Some examples of this type of models are: (1) Those based on physics, and (2) Models that allow autonomous learning and are capable of producing predictions in short times so that it is possible to make decisions.
- Prescriptive capabilities: These capabilities serve to support various decision making processes in the physical context through the use of improved or optimized action plans transformed into production managers. The process to generate prescriptive capabilities uses various predictive and/or optimization models taking into account simulation or can also use single models. The results generated by the models must be obtained quickly so that they can be used to make decisions instantaneously or in operational terms and with the ability to react quickly to sudden and unexpected events.

Some of the requirements that apply to the design of DTs are summarized in the following points [109]:

1. Reusability: DT solutions need to be more portable and reusable to better leverage the "develop once, use many" approach.
2. Interoperability: The proposed solutions of a DT design must have the capability to interoperate with other DTs and other classes of DTs, as well as have the ability to interoperate with non-DT systems.
3. Interchangeability: The proliferation of DTs requires them to be designed in a modular way to facilitate their evaluation, easy replacement or updating.
4. Verification and validation capability: Because the DTs will be in common use and integrated into critical production systems, the capability of the DTs must be ratified and verified before they are incorporated into real applications.
5. Maintainability: An underrated feature of DTs is their maintainability, however, they are required to be able to operate and be maintained throughout their lifetime and usefulness.
6. Capability and accuracy: Capability and precision (accuracy) are two capabilities that DTs must have, as these capabilities are a consequence of the evolution of intelligent manufacturing.
7. Extensibility: It is required that the DTs can be extensible with the help of ICTs and that they are part of an ecosystem.

8. Support of a technology partnership: The design of a DT requires special and strict technology requirements, and to develop them there must be a better and greater collaboration between those actors that make applications of DTs in different areas and those existing technology partnerships.

DT is exploited by many industrial sectors such as [110]:

- Automotive industry: The concurrence of physical and virtual products has the potential to address many of the challenges that currently exist in the automotive value chain. DT in the automotive industry can enable the convergence of existing gaps between physical and virtual versions of product prototypes, the shop floor and the actual vehicle on the road.
- Aeronautics: The large number of sensors on commercial aircraft transmit asset data to improve system maintenance and operational status.
- Medical industry: Connected medical systems and tools provide assurances of product integrity and are able to measure patient outcomes.
- Manufacturing: Physical assets in digital factories can increase manufacturing uptime and throughput while potentially reducing repair and maintenance rates.
- Oil and gas: Remote platforms send system health data that limit routine inspections and maintenance.
- Rail: Vision of deployed locomotives and asset health optimizes scheduling and reduces maintenance time.
- Utilities: Digital models of power grid systems improve demand response and energy efficiency functions.

On the other hand, CPSs and DTs are similar, but not the same, in their description of cyber-physical integration, since both are composed of a physical and a digital part. Table 2 presents the differences between these technologies [111,112]. DTs, such as CPSs, achieve synergistic (cyber-physical) integration between physical assets and cyberspace, due to the fact that they support dynamic functionality of the bidirectional type between real systems and digital representations [113]. DTs are important and necessary because they enhance the capabilities of CPSs. DTs and CPSs seek to be coherent through the interrelationship between physical assets and digital entities. The central characteristic of the DT is to achieve one-to-one interactions or relationships, while the CPS works on one-to-many or many-to-many relationships. The CPS has interactions more with the digital part of the system, while the DTs have relationships and interact more with the physical assets. It is worth mentioning that communication systems, control and computation, enabled by physical assets (signaling elements and actuators), are the primary elements of the CPS. These elements make interactions and relationships between physical and digital assets easier due to information and data exchange processes, while DTs are dependent on the design of models that are generated from the data and the application of the data to perform predictive tasks and to have control over the actions or behavior of physical assets [112].

**Table 2.** General differences of CPS and DT ([111,112]).

| Criteria | CPS | DT |
| --- | --- | --- |
| Origin | Proposed by Helen Gill at NSF in 2006 | Presented by Michael Grieves in 2003 |
| Interaction type | Cyber and Physical interaction | Cyber and physical interaction |
| Interaction level | One-to-many components interaction | One-to-one component interaction |
| Core elements | Computation, communication, and control | Computation and communication |
| Control means | Models and actuators | Models |

There is a symbiotic role between IoT and CPS towards DTs. A DT is an entity that is built in an emergent manner, but has conceptual differences in relation to CPS and IoT. Similarly, to CPSs, DTs rely on communication to generate a highly coherent and synchronized digital image/representation of physical objects or processes. However, DT, in addition, uses softwarized embedded models in this accurate image to simulate, analyze, predict, and optimize its operation in real time through feedback [114].

### 2.1.7. IoT, Simulation, AI and Interoperability

In this section we describe other disruptive technologies that are involved in I4.0 and I5.0, with the objective of being able to understand their important considerations that allow us to establish that essential knowledge for the formation of the mechatronic engineer.

One of the important technologies in the implementation of I4.0 is IoT, which is about connectivity with objects rather than people. In fact, IoT is a new paradigm that is rapidly gaining ground in the modern wireless telecommunications scenario. The basic idea of this concept is the ubiquitous presence around us of a variety of things or objects such as sensors, actuators, cell phones, etc. that, through unique addressing schemes, are able to interact with each other and cooperate with their neighbors to achieve common goals [115].

IoT is a technology that is focused on the connectivity of many computing devices and appliances. Many applications can be realized if the devices can be combined or integrated with AI, autonomous learning models and various data mining techniques. It can be said that IoT makes the interaction between humans and computing systems easier. The IoT is a concept that allows us to have the idea that devices and devices can have the power to perform various inspections and that they can collect or lift information flows from the environment, and then introduce it into cyberspace and have the opportunity to perform a myriad of applications with that information [116]. IoT with industrial applications can transform data into novel information, depth knowledge, and intelligent systems. IoT applications in industries enable industries to improve their efficiency and increase reliability in processes and operations, because IoT technology is directed towards various industrial communications such as M2M, large database management, and autonomous learning. According to its use and customers IoT is divided into [117]:

1. Consumer IoT: This technology considers devices and appliances that are connected to the network, such as portable phones, games, smart cars and home appliances, among others.
2. IoT for commerce: This technology takes into account devices and appliances such as GPS, medical devices and inventory control systems, among others, that are connected to the network.
3. IoT for industry: This technology operates physical assets connected to the network, such as: robots, production lines, machines and wastewater management systems, among others.

IoT can be classified in terms of its functionality (see Figure 16) and different technologies can be related to each subclass [118].

**Figure 16.** IoT elements [118].

IoT is a technology that facilitates communication and cooperation between CPSs. It is worth mentioning that various physical assets, such as signaling elements, actuators and portable devices, among many others, are everywhere and are applied for countless tasks, such as in transportation tasks, in medical centers, in traditional households, in companies for product manufacturing, in agricultural production and in systems in critical infrastructures, such as in refineries or nuclear power generation plants [119]. Figure 17 shows a diagram showing the connection between IoT, CPSs and DT [114].

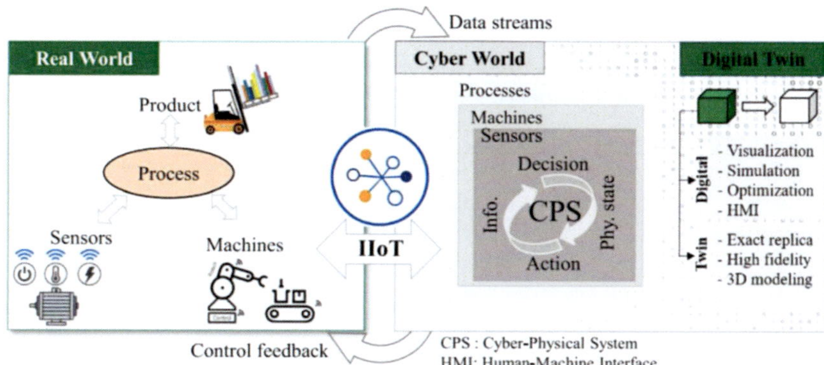



**Figure 17.** Functional relationships between industrial objects and processes from the real environment to cyberspace based on industrial IoT, data flows and DT [114].

On the other hand, one of the technologies on which the CPS is based is computational simulation, which is conceived as the imitation of the behavior of a system operating in the real world over time, and which takes as a reference the generation and observation of artificial histories, from which it is possible to draw inferences about the operation of the real environment [120]. Simulation technology modeling can be interpreted as a methodology that is based on models that can be physical, mathematical or other, which are related to the real or fictitious behavior of a system and which are used for various applications, among them to make predictions or to improve the operation of systems [121]. Simulation is the central basis of DTs and the support of CPSs. In fact, simulation technologies are taken into account for engineering processes and in decision making. Traditionally, simulation focuses on design phases and engineering processes. However, simulation will be more commonly used in production lines to make decisions and drive processes. For simulation to be used in industry as a useful and productive tool it must be transformed into a multipurpose or multidisciplinary simulation [122].

However, today there is no integrated simulation base or platform that has the power or capability to mimic the behavior of an entire CPS. This fact is of utmost importance due to the fact that there are several advanced tools and methods to build models and to perform the necessary testing of physical assets or data and communication networks. In order to enhance the simulation, several approaches can be used, such as co-simulation, which consists of taking two conventional simulators and combining them to perform the tasks, so that one of them is associated with the physical assets and the other one is related to the communication networks. Co-simulation is a good alternative since it can consider, on the one hand, technologies that already exist to perform simulations and, on the other hand, it can take advantage of the large number and variety of tools and methods that are ready to be used in deep and comprehensive studies [123]. Computational simulation is considered as one of the main tools used to study and evaluate the performance and efficiency of a system, and to assist humans and digital systems in decision making. Posada et al. [124] consider simulation as a base component for the successful implementation of I4.0, due to the fact that the following three dimensions can be evaluated: (1) The integration dimension: as an end-to-end digital engineering integration tool, (2) The product and production dimension: as a decision-making tool, and (3) The human factor level, as it can improve work organization and design. Simulation with mixed approaches is considered a primary technological tool of I4.0, in fact, DT is a derivative of simulation and is pointed out as one of the most hopeful modern technologies [125].

Simulation in general has diverse applications, for example, many production processes can be simulated before being taken to real operation; logistics is another important task in industries that can use simulation models for decision making. Computational simulation uses methodologies and tools of a technological nature with the purpose of performing various tasks, such as, for example, carrying out different experimentations, predictions, validations and design, among others. In the case of I4.0, simulation will face important challenges due to the fact that systems are increasing in complexity and must also integrate other tools such as big data, cloud computing and IoT, among others.

Just as CPSs are related to embedded systems, DTs are closely associated with simulation. Table 3 shows the evolution over time of modeling for simulation.

**Table 3.** Evolution of the simulation modeling paradigm [121].

| Individual Application | Simulation Tools | Simulation-Based System Design | Digital Twin Concept |
|---|---|---|---|
| Simulation is limited to very specific topics by experts, e.g., mechanics | Simulation is a standard tool to answer specific design and engineering questions, e.g., fluid dynamics. | Simulations allow a systemic approach to multi-level and multi-disciplinary systems with enhanced range of applications, e.g., model based systems engineering. | Simulations is a core functionality of systems by means of seamless assistance along entire life cycle, e.g., supporting operation and service with direct linkage to operation data. |
| 1960+ | 1985+ | 2000+ | 2015 |

DTs are used for various operations, such as physical asset control, and have the ability to process and manage data and information generated by various devices and appliances. A DT has the ability to operate in instantaneous time and is capable of making predictions of the possible effects of the operation. If considered a "software copy", the DT can be conceived as a simulation of the part, system or process being copied. The main tasks to be performed by a DT operating in a CPS, is to perform simulations in cyberspace and make various predictions, in this way business managers or AI methods will be able to know evaluated information and make decisions as required [126].

It is possible to consider that the most significant transformation related to the way products are manufactured is digitalization. I4.0 has as a priority to optimize I3.0. This needs the development of intelligent equipment and systems that have access to more data, thus becoming more efficient and productive by making decisions in real time [127]. The concept of AI is one of the most fundamental components of I4.0. AI is divided into two types: narrow and powerful. Narrow AI applications have been created to perform a single task in a single application area. Strong AI has human-level intelligence and problem-solving capability. In AI studies, there are complementary elements such as IoT, deep learning, autonomic learning, and neural networks [128,129]. AI has diverse applications, as shown in Figure 18.

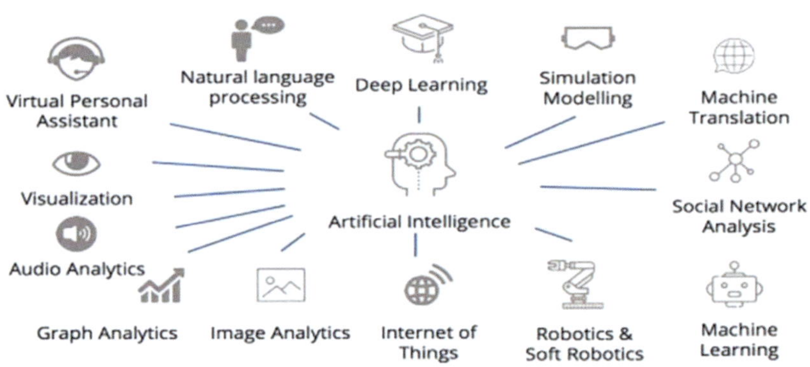


**Figure 18.** Applications of Artificial Intelligence [128,129].

A CPS is an intelligent system in which computer units and physical objects are highly integrated and interact in a networked environment [130]. There are various applications of AI in which other technologies are combined, such as IoT. Specifically, the automotive technology TESLA uses AI and various devices to know several variables that allow cars to drive safely, and are even able to make predictions about the physical movements of the environment. In future applications, the AI incorporated in CPSs will be used for various applications such as monitoring the health status of patients, operating robots and managing manufacturing systems, providing solutions to human problems and natural disasters, among other important tasks [131]. AI applied in DTs is considered as universally applicable theoretical and technical system with many uses, such as product design, equipment manufacturing, medical analysis, aerospace and other fields [132].

AI with applications in Industry has the following particularities [133]:

1. Infrastructures: with respect to hardware and software, there is a strong emphasis on real-time processing capabilities, ensuring reliability in industrial terms with high security and interconnectivity requirements;
2. Data: high volume and high speed data is required from various units, products, regimes, etc.
3. Algorithms: integration of knowledge of physical assets, digital and heuristic resources and high complexity derived from model management, implementation and governance is required.
4. Decision-making: in the industrial environment, tolerance to error must be very low, so a significant management of uncertainty is required. Similarly, to handle problems that require robust or large optimization, system efficiency must be taken into account.
5. Objectives: AI focuses its attention on shaping real value by taking into account various factors such as increased quality, decreased waste, multiplied operator capabilities and performances or accelerated times.

On the other hand, I4.0 can be described as a high integration of disruptive technologies whose main function is to optimize current manufacturing systems. To achieve the goal of integration, global interoperability is a property of utmost interest in I4.0. Interoperability has the capability or ability to make two or more software components or systems cooperate with each other despite their differences in interface, execution platform and language [134]. IEEE defines interoperability as the ability of two or more systems or components to exchange information and use the information that has been exchanged [135].

It is of paramount importance in the I4.0 vision that physical assets and cyberspace are linked or integrated so that effective and seamless collaboration can exist. However, connectivity and interoperability of information and communication technologies are challenging tasks for I4.0 implementation. Interoperability can be considered as an advantage in I4.0, but to make it efficient, standards or proprietary approaches must be homogenized and exchanged for open and standardized communication solutions. Then it can be deduced that the lack of standards is considered a major problem so it is necessary to direct research efforts in the definition of protocols, languages and standard type methodologies [136].

In smart manufacturing and production, interoperability takes two general forms. The first is associated with vertical type integration, e.g., interoperability between software for manufacturing and production, shop floor and design departments, processes and tasks performed by different teams, various shop floor systems, etc. [137]. The second is associated with horizontal type integration; for example, interoperability between different intelligent automation devices and appliances, cloud services, cloud platforms and enterprises [138].

The study of the concept of interoperability in I4.0 is very important, and several authors have researched on the subject. Yang [139] presents a concentrate of studies of the issues related to this concept. The studies agree that the architecture or structure of interoperability in I4.0 has four levels: (1) Operational (organizational), (2) Systematic (applicable), (3) Technical and (4) Semantic interoperability. The first of these relates to the general structures of concepts, standards, languages and relationships within CPS and I4.0. The second level identifies the guidelines and principles of methodologies, standards, domains and models. The third level articulates the tools and platforms for technical development, IT systems, ICT environment and related software. Finally, the fourth level ensures the exchange of information between different groups of people, malicious packages of applications and various levels of institutions [139]. Figure 19 shows a framework of I4.0 interoperability.

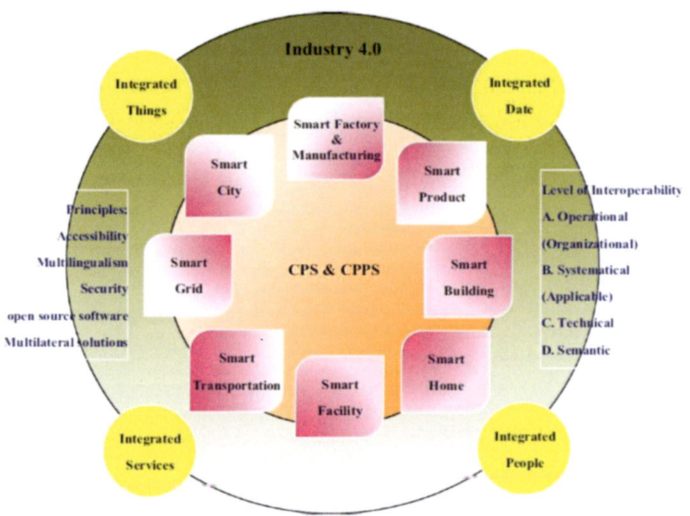


**Figure 19.** Interoperability framework for I4.0 [139].

## 2.2. Engineering Education in the Vision of the I4.0

In this section, some relationships between I4.0 and engineering education are presented. Then, mechatronic engineering and the basic concepts of Competency Based Education (CBE) are described, as well as active methodologies. Finally, 15 technical guides are presented, which will allow designing and building specific competencies for mechatronic engineers.

### 2.2.1. Engineering Education (IE), I4.0 and Mechatronics

It is important to know those relationships that exist between IE, I4.0 and mechatronics, as they are essential for the analysis and shaping of specific competencies of the mechatronic engineer.

The influence that I4.0 has on the industrial sector has been projected to the topic of engineering education. In recent years, there have been several and numerous works and researches on education topics in the vision of I4.0, ranging from qualification studies, analysis of topics that should be in the curriculum to adapt them to I4.0, in the search for an Education 4.0 similar to the philosophical framework of I4.0 to the conceptualization of practices in the laboratory [140]. While the impacts of I4.0 are still unquantifiable, the innovations it will bring will be too rapid and too profound, requiring higher education to respond to the challenges and opportunities posed by I4.0. Sakhapov and Absalyamova [141], state that I4.0 has already started due to industrial changes in IoT, integration of CPS in production processes and application of neural networks. For education, and especially for engineering education, this brings important implications such as individualization and digitization of education, empowerment of projects and multidisciplinarity of engineering education, as well as interaction of educational resources.

Due to the implementation of I4.0, current and future engineers will have to face a highly differentiated and specialized society operating with a globalized vision, and with highly automated production systems, with a digitized world connected to cyberspace, where diverse and novel business models and plans, intelligent artifacts, cutting-edge procedures and techniques, and optimized processes are dynamically generated [142]. As mentioned above, the CPS represents the technological center of the I4.0, which may represent several advantages, but also a cataclysm for many professional careers, engineering and technical specializations, because these systems may involve various changes not necessarily positive in education and universities since it should be noted that companies and industries are truly in a fundamental change in production and industrial manufacturing, and not as commonly thought that the I4.0 is only a technological improvement of services to industrial processes [141]. The innovations brought about by I4.0 involve the design of fully automated production systems, displacement of operators from the center of production, fundamental changes in the value chain and business in companies and industries, minimization of social implications, innovations in intellectual assets (patents and industrial secrets) and technical specialization of modern industrial processes [143].

It can be observed that due to the incursion of the I4.0, there are innovations of great importance in economic aspects, both in the processes and in the micro and macro economy of the countries. These changes or innovations are a consequence of the incursion of new knowledge generated by science and new disruptive technological resources that are producing significant social changes and are reducing the cycle and life time of professions, which motivates workers and in general the workforce to have a greater and better adaptability. Therefore, universities play a role of utmost importance as they must better prepare their engineers to meet the new requirements of I4.0 [144]. As previously mentioned, the I4.0 promotes the use of novel technologies (AI, IoT, CPS and DT, among others) in industrial processes; these technologies have disruption as their main characteristic and imply that students in study centers must be prepared, qualified, highly competent and master multiple technical skills such as: having leadership, possessing strategic thinking, mastering computer skills and capabilities to design cyber security systems, among others, to be able to operate in the vision of the I4.0 [145].

The I4.0 vision needs preparation and high training of engineering students so that they have the ability to solve problems and to face the challenges and challenges of I4.0. One of the key technologies, as mentioned above, is CPS, so engineers must be able to be trained under an inter- and multidisciplinary approach to master this technology so that, as a result, I4.0 can be implemented quickly and efficiently. [146].

Although all engineering education is influenced by the inertia of I4.0 implementation, Mechatronics Engineering stands out for being of an integrative nature (a process on which I. 4.0 itself is based). Mechatronics was conceived in IR3.0 and consequently, together with computing, informatics and robotics, brought a significant technological impact to the industrial world for a period of at least 50 years [20]. Today, the training of the mechatronics engineer faces the major challenges of transforming CIM systems (core of manufacturing 3.0) into modern CPSs that integrate several DTs and that are the basis of I4.0. While Mechatronic Engineering was first conceived as a synergistic integration of mechanical engineering with electronics and intelligent computer control in the design and manufacturing of industrial products and processes [147], over time there have been many changes in the functionality of the systems, due to the evolution of ICT, which has involved the development of much more complex and computationally intensive computing systems. Mechatronics is being strongly influenced by new technological developments related to CPS, DT, cloud computing and IoT. These technological developments have meant that there are better opportunities to make the changes and innovations, particularly in terms of supplying smart components and subsystems and configuring them. For mechatronics to be incorporated into I4.0 it must evolve in such a way that traditional mechatronic devices and systems must be transformed to provide the intelligent components and objects with which the new industrial revolution is being built [148]. This implies that Mechatronics Engineering Education must also be transformed to be able to integrate various disruptive technologies already present and ICT in modern industrial processes. Such transformation must be able to integrate new conceptual approaches, seeking as far as possible to preserve the basic theory of systems integration as a support for mechatronic engineering.

Undoubtedly, mechatronics from the 1960s to the present day has and continues to provide the tools for technological integration that enable the design and manufacture of complex production systems that require a convergence of technologies and knowledge. However, mechatronic systems have transitioned over time towards CPS and IoT. Figure 20 shows such a transition [149].

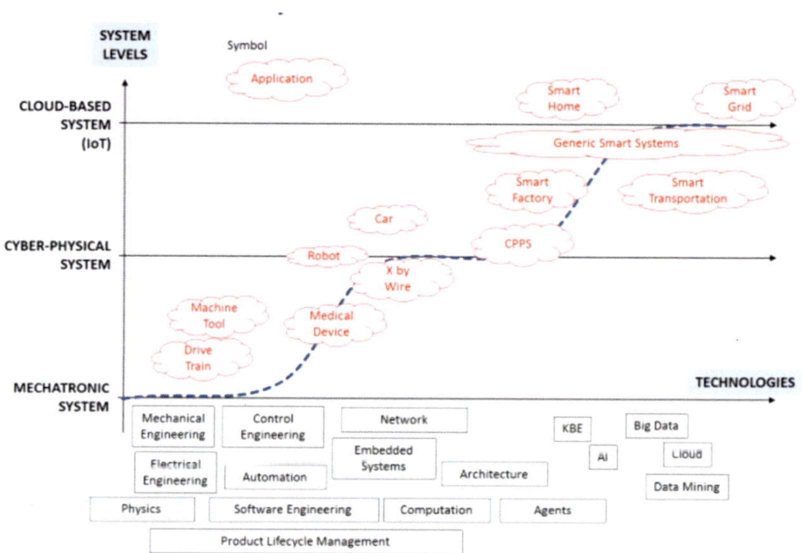


**Figure 20.** Evolution of mechatronic systems towards CPS and IoT [149].

The systems shown in Figure 19 are shaped or structured around a massive interconnection process of products, devices and mechatronic objects that have certain intelligence, and are associated with a large amount and variety of information in terms of parameters of the physical environment. The information collected has several uses: (1) To verify how the entities behave in time and space, (2) To inspect and control the relationships and interactions within the physical and computing environments, (3) To enable the analysis of the processed data and allow its visualization, to have a good support to enable the operation of the system and the interaction with the consumer or user [150]. Mechatronics can be considered as an evolution of electromechanical products and CPSs come from an evolution of cyber-systems. Thus CPSs have their origin in an information technology (IT) domain and in software development. This helps to explain why there is a strong software and communications dimension to CPS. The design methodology of CPS is strongly linked to systems engineering approaches [151].

The pressing industrial needs have led mechatronics to be considered as one of the best practical applications of engineering graduates. Since its origin, mechatronics was conceived as a field of knowledge that integrates diverse technological systems to achieve a greater optimized functionality of the systems it develops. It seeks a synergy between physical assets (mechanical, computational and electronic components) to optimize the operation of the system from its conception and throughout its life cycle, this advantage enables engineers to make complex decisions. Consequently, it can be stated that integration is the key concept in any mechatronic design and that the complexity of the design has shifted or transferred from mechanics to electronics and computation.

It is possible to consider mechatronics as part of evolutionary design which implies that there is a vertical and horizontal integration of various disciplines related to engineering and between design tasks and manufacturing activities. Some authors have proposed that mechatronics be considered as the new modern mechanics [152]. Mechatronics has gradually transformed due to the incursion of new technologies and cutting-edge knowledge. Today the new industrial revolution imposes new challenges to mechatronics engineers.

Faced with the constant challenges of I4.0, mechatronics and its teaching must be reinvented. In this sense, mechatronic engineers who develop products, processes and systems of any type and variety, must be able to know, master and apply a collection of principles to be successful in the implementation of I4.0. Some of these principles are (1) software, (2) hardware, (3) technology, (4) ubiquity, (5) connectivity, among others [153]. Mechatronics education should consider as essential and necessary the principles of mechatronics design and the application of novel technologies that cause disruptive changes. However, it is also necessary to take into account educational, economic and social aspects that directly impact I4.0. In this sense, education in Mechatronics Engineering, placed in the vision of I4.0, has several problems to be considered, for example:

- The speed of transition and adaptation of disruptive technologies is slower in universities than in companies, since the former have to go through cycles or generations to make changes in their curricula, while the latter need to adapt as quickly as possible.
- Educational models in educational institutions in general are different, which prevents the design of a single educational model for Mechatronics Engineering education.
- Another aspect to consider is that it is often planned, both entrepreneurially and in universities, in terms of the theoretical needs of the already established I4.0, when in reality a technological transition characterized by I3.5 is taking place.
- There are countries that are leaders in the implementation of I4.0, but the vast majority of the remaining countries are at some stage of technological transition and some even have manufacturing systems with technologies that are tending to obsolescence.
- The implementation of I4.0 among companies also varies in terms of technological strategy, as those with high economic capacity can replace their production systems quickly. Other companies with less economic capacity will opt for reconditioning methods to improve their production systems and adapt them to the value chains imposed by I4.0.
- Mechatronics education does not follow a universal model, as each country and each region defines the specific knowledge and skills required, according to: the regional industrial environment (type of companies and their needs), the regional educational environment (type of universities and teaching capabilities) and national educational policies.

It is possible to look at the world of education in the era of I4.0 from two different perspectives [154]: As an opportunity (the birth of a new business unit in the community that is able to penetrate the unlimited space by using information technology and vice versa) and as a threat (one of the threats is the result of automation of the many human jobs performed by machines, systems and robots that implies a displacement of operators and a loss of new businesses).

### 2.2.2. Educational Models in Mechatronics Engineering Education

This section describes some important aspects related to CBE and active methodologies. These educational models and approaches are being used today for the training of mechatronic engineers and in general for engineering education.

Engineering education has been characterized by having a teacher-centered pedagogical model. However, nowadays learner-centered and CBE models and approaches have been positioning themselves in universities around the world [155–157]. It is possible to describe CBE as an educational approach that is concerned with goals and outcomes. Adult-based learning theory is integrated into CBE and describes that it increases the likelihood that adult learners will be more interested, engaged, and work harder when it comes to meeting stated goals or very specific outcomes. This idea implies that the purposes of CBE in terms of teaching and learning methods are directed toward shaping skills that are already stated and described (usually by companies) and these are measured or graded by observing how learners perform those skills [158]. The competency-based approach can be an ideal way to train new engineers, especially mechatronic engineers for I4.0, as it fosters comprehensive training, flexibility and self-management, promotes active learning, develops technical and social skills, and encourages engineers to solve problems in complex situations, among other relevant features [20].

CBE is conceived as an educational approach that is outcome-based and integrates mechanisms for the delivery of instruction and assessment methods designed to assess and evaluate student learning through the putting into practice of acquired knowledge, applications of that knowledge, and soft skills [159]. The new challenges demanded by I4.0 thus require a competent engineer, i.e., he or she must be able to have knowledge of the subject matter and a set of skills that put the acquired knowledge into practice, as well as certain attitudinal skills. The definition of competencies for I4.0 requires a universal understanding of what they are, namely the characteristics in terms of knowledge, skills and attitudes which enable the tasks entrusted to be performed to a satisfactory level [160]. There are many definitions of what a competency is. One of them is presented below [161]:

*A student's competency contains three elements: knowledge, skills, and attitudes/values. These components are integrated to perform a specific activity, with measurable results, which clearly indicate what the student is capable of doing. A competency comprises a set of resources and talents that an individual has to perform a specific task.*

It is possible to affirm that the design of the competencies required in I4.0 must take into account the combination of those learned knowledge, the skills that make it possible to put the knowledge into practice, and the attitudes necessary and recognized for companies to function adequately in I4.0. According to Armstrong and Taylor [162], competency is associated with a distinctive characteristic or trait that remains hidden (underlying) of an individual that results in or provides superior and effective performance. There are several ways to classify competencies. Martens [163] classifies them in the work context as follows:

- Generic competencies: these are related to work behaviors and attitudes specific to different areas of production, for example, the ability to work in teams, negotiation skills, planning, etc.
- Specific competencies: these are related to the technical aspects directly associated with the occupation and are not easily transferable to other work contexts (operation of specialized machinery and formulation of infrastructure projects, among others).

In education, there are also several classifications of competencies. Galdeano and Valiente [164], consider that competencies can be classified as follows:

1. Core competencies: these are the intellectual capacities necessary to support and promote the learning and knowledge of a specific profession; some examples of these are: cognitive, technical and methodological competencies; most of these competencies were developed in previous educational systems or levels.
2. Generic (transversal) competencies: This type of competencies are described as those attributes or characteristics that an alumnus or university graduate should possess with total independence of his or her career or profession, and are not designed with purely technical considerations, but should take into account the human aspects. These competencies collect generic considerations of all those abilities, skills, knowledge, potentialities and capacities that any university graduate should possess during his or her education and before finding a job.
3. Specific competencies: These types of competencies are described as those attributes or characteristics that students should have before graduation and should be built according to the experiences and practices of the students or graduates themselves. With the specific professional competencies, the aim is to start with the typical functions or role of the professional in society and the typical situations of the professional field in which graduates are generally incorporated, and then identify the professional competencies in terms of the actions, context or conditions for carrying them out and the quality criteria for their execution.

On the other hand, the transition from teacher-centered pedagogy to student-centered learning requires the adaptation of new active teaching strategies that allow [165]:

- To give students a greater role in their education.
- Encourage collaborative work.
- Organize teaching according to the competencies to be acquired.
- Stimulate the acquisition of autonomous and lifelong learning.

In this sense, the use of active methods related to the teaching and learning processes can support and promote those pedagogies that consider the student as the center of the educational process, since they favor in an active way, and from diverse experiences and real contexts, the construction and use of knowledge. These methods displace teachers from the center of knowledge and place them as facilitators of educational processes, whose main function is to stimulate students to reflect on the different contexts and to problematize situations within the environments in which they develop and where they construct and are the protagonists of their learning [166]. Some authors have discussed learning methodologies in depth [167,168] and agree that changes are required in the strategies employed by universities to design the expected profile of an engineer in the I4.0 era.

The methodologies considered active (strategies, techniques and methods) are used by teachers with the aim of transforming the processes and activities related to teaching into tasks in which students participate actively and where they are able to direct their own significant learning [169]. Currently, there are several classifications of learning methods where active methodologies occupy an essential place. Some of the active methodologies that exist and are regularly applied are listed below [170]: Cooperative Learning, Project-Based Learning, Learning Contract, Problem-Based Learning, Exposition/Lecture, Case Study, and Simulation and Game, among others.

According to Jimenez et al. [171], one of the active methodologies that adapts naturally to engineering education is Project Based Learning (PBL), since it is based on the construction of meanings and problem solving, where students abstract knowledge and extrapolate it to other fields in a dynamic way, that is, students learn while they create. Some important characteristics of PBL are presented below [172]: it fosters relationships between the academic world, the context of realities and competencies for work; it promotes that students are able to self-evaluate and reflect on things, and that they accept feedback and allow being evaluated by knowledgeable people; it encourages students to accept being evaluated on the basis of evidence of their learning; it helps to design and develop specific competencies and specific objectives according to current needs.

PBL is an active methodology that is applied in various universities, particularly in engineering programs, e.g., Lin et al. [173] investigated the effects of infusing an engineering design process using PBL in science, technology, engineering and mathematics (STEM) projects to develop and evaluate the cognitive structures of technology teachers in training for engineering design thinking. In [174], PBL is used for mechanical vibration control studies using Matlab 9.4/Simulink Software. In applications for the teaching of mechatronics, the APB has been applied in the field of Renewable Energies in the solution of projects that integrate various fields of knowledge [175]. In [176] the PBL is applied to the teaching of mechatronics in the subject of robotics.

PBL has been applied to support theoretical concepts in engineering curricula and to provide a learning experience that develops practical skills and competencies for I4.0 [177]. There are other concepts that attempt to relate I4.0 to education. Abele [178] defines the concept of a Learning Factory 4.0 *as a learning environment that includes four distinctive features: (a) authentic processes, which include multiple stations and comprise technical and organizational aspects, (b) an environment that is changeable and resembles a real value chain, (c) a physical product that is manufactured, and (d) a didactic concept that comprises formal, informal and non-formal learning, enabled by the actions of the learners in an in situ learning approach.*

The interconnectedness of a Learning Factory 4.0 (which is the fundamental idea of I4.0) is based on cyber-physical production systems (CPPS). Learning Factories 4.0 are intended to prepare learners for the challenges of I4.0. The implementation of these

Learning Factories 4.0 interconnected with technical vocational training centers can promote the development of technical skills related to the subject as well as multidisciplinary digital skills [179].

### 2.2.3. Technical Considerations for the Design of the Specific Competencies of the Mechatronic Engineer under the I4.0 Vision

Once known, in the previous sections, the generalities of I3.5, I4.0 and I5.0 and their methods and technologies, as well as the general aspects of CBE, active methodologies and mechatronic engineering, it is possible to propose technical considerations that make it possible to design specific competencies for the training of the mechatronic engineer.

Specific competencies are related to the technical aspects and considerations of a career or profession [180]. To approach the study of this type of competencies in faculties or schools, the starting point is the profile of the graduate that the study programs have in order to contrast it with the expectations in the professional field both regionally and nationally, as well as internationally, find similarities and differences, and arrive at the selection of the elements that could be recommended for the profession. The specific competencies are divided into two major classes: (1) The class associated with the training of disciplines that graduates must acquire, called academic disciplinary competencies and (2) Those competencies related to professional training that future graduates must possess: professional competencies [164]. Although the specific competencies are not universal, they can be designed based on the analysis of the key technological concepts of I4.0, such as IoT, simulation, CPS, DT, AI and technological reconditioning (retrofitting), among others. For the case of mechatronic engineers, the development of specific competencies can be carried out under the following technical considerations:

1. Similarly to mechatronics, I4.0 is considered as a paradigm that integrates various technologies whose purpose is to improve and optimize production systems under the operation of CPS. It can be stated that synergic integration represents a characteristic feature of I4.0. In this sense, the Mechatronic Engineer must be able to integrate disruptive technologies since these are the basis of I4.0. The most important feature of a CPS is the high integration, mainly of software and hardware resources, with the aim of carrying out various tasks of calculation, control, computation and communication, taking into consideration for the design of the same to the technological assets and their theories [68].

2. Mechatronics is rather an evolution of electromechanical systems and CPS (which represent the heart of I4.0) coming from an evolution of cyber-systems [151] or IT and software development. In this vision, the Mechatronic Engineer must further improve his knowledge and expertise in electromechanical technologies and gradually venture more into IT with the purpose of realizing technological integrations and more specialized applications, including Big Data, Cloud Computing and IoT, among others.

3. One of the computing disciplines that the Mechatronics Engineer must address in greater depth, both in his training and applications, is AI, since a large part of I4.0 is based on the development and operation of intelligent systems and the DTs that make up the CPS.

4. If one starts from the premise that I4.0 is actually a large-scale optimization of I3.0 [20], this concept should be considered necessary in the training and applications of the mechatronic engineer. It is worth mentioning that the applications of analytical methods such as: stochastic optimization and mathematical optimization, among others, are already well known, especially in the analysis and studies of large databases, and that they are applied to have an optimized planning and to have instantaneous time control of operations and processes. Optimization models considered as large-scale have various applications, such as in design, manufacturing and intelligent production [181]. All these optimization techniques should be valued and learned by mechatronic engineers.

5. Another key computational tool that should be considered by Mechatronics Engineering is computational simulation, since it is the basis of the DTs and the support of the CPS. However, for this tool to be applied efficiently in I4.0, it must move from applications in engineering and decision-making processes and from the phases and steps of design and engineering, towards simulation that takes into account various areas of knowledge (multidisciplinary) [125]. That is, to a simulation that integrates different models and methods from different areas to provide more robust solutions. Multidisciplinary simulation requires the use of large databases and high information processing, which implies knowing and mastering technologies that handle large amounts of data and cloud technologies.

6. IoT is a necessary tool that the mechatronic engineer must know and apply, since it allows connecting various digital devices and appliances and promotes interactions between humans and computers. This tool applied in conjunction with other disruptive technologies, such as data mining, autonomous learning and AI, can be used for specialized applications [116]. IoT facilitates communication and cooperation between CPSs. This technology is already essential in the operation of today's factories so the mechatronic engineer must be familiar with its management and operation.

7. Mechatronics engineers must transition from traditional design and manufacturing methodologies to digital design and manufacturing conceived within the I4.0 vision. Digital design and manufacturing technologies provide great support for product conception throughout the product lifecycle, which includes product sales and services [182]. Custom design and manufacturing using 3D printing are technologies that improve designs and accelerate production. The world's leading companies in design and manufacturing are already implementing digitalization as the basis of competitiveness, so it is already a pressing need for the Mechatronic Engineer to master the new methodologies.

8. Although design and manufacturing methodologies are very important in the training and applications of mechatronic engineers in the I4.0 vision, they must be complemented by introducing Reverse Engineering methods [20,171]. Technological reconditioning and to a large extent the maintenance of equipment, machines and production systems, and directly or indirectly use some method of reverse engineering to solve various problems. The information and information models generated by the application of reverse engineering can be used for the design of DT in the maintenance of production systems. Reverse engineering is a method of analysis that companies are requiring since it is not only used for technological reconditioning, but it is also applied for the improvement of products and processes.

9. Although industrial automation has been a field of action for mechatronic engineers in companies for decades, today, in order to meet the challenges of the I4.0, it is necessary to integrate new technologies and new equipment to industrial production systems with which optimization and systematic continuous improvement processes can be carried out. Similarly, it is required to know and apply different automation architectures that allow greater flexibility and modularity, and that can interoperate between different manufacturers to enable automated, optimized and efficient production systems, as well as to design viable individualized and low-cost solutions in production systems. This implies that mechatronic engineers must know the forms and operation of modern technologies so that they can automate processes in the vision of I4.0.

10. Mechatronic engineers must have knowledge and skills on technological reconditioning methods, both traditional (which consists of the replacement of parts and components in machines, processes and systems, to optimize different variables) and intelligent (which aims to adapt existing systems, equipment and devices at a low cost) [59], since several problems that currently arise, and that will arise in the future, will be related to technological upgrading. This implies that engineers must have knowledge of CPSs conceived under the I3.0 approach and of modern disruptive

technologies so that, under some technological retrofitting methodology, a traditional CPS can be transformed at a lower cost to one that can operate under the vision and needs of I4.0.

11. In today's production systems, maintenance is one of the most important tasks due to its contribution to the organization, and nowadays it is being considered more frequently in corporate objectives [183]. With the advancement of technology in companies and industries, maintenance methods and techniques are designed and developed to be able to adapt to the demands of new producers or manufacturers. With the emergence of I4.0, novel methods and maintenance techniques have been created with the purpose of meeting the new demands; all these novel techniques are conceived under the concept of Maintenance 4.0 [184], which is described as the application of I4.0 to operations and maintenance activities. The objective is simple: to maximize production uptime by eliminating unplanned reactive maintenance [185]. The mechatronic engineer must then have knowledge and develop practical skills in maintenance 4.0 and its implications in order to meet the challenges brought by I4.0. Maintenance 4.0 involves knowledge of DT and CPS, as well as the use of various disruptive technologies such as cloud computing, big data management, and augmented reality.

12. One of the technologies that supports both I4.0 and I5.0 is robotics. Mechatronic engineers must not only be able to integrate new robots into production systems conceived under the I4.0 vision, but they must also know about collaborative robotics (the basis of I5.0 [53]), which deals with the integration of machines and humans into processes.

13. Interoperability is a concept of utmost importance in I4.0, and is understood as the ability of two or more software components to cooperate despite differences in language, interface, and execution platform [134]. In order for interoperability to take place, standardization is necessary. In this sense, mechatronic engineers must be able to know and master the concept of interoperability and its implications in the design and operation of CPSs.

14. IT security in I4.0 is of utmost importance for the protection of information in companies and industries. Traditional architectures and structures that exist today to achieve cyber security have different security mechanisms and systems that provide services such as integrity, access control and confidentiality, among others [186]. Nowadays, computer security is tested by various events and specialized cyber attacks. It is necessary to consider the various methods and techniques with which it is possible to detect intrusions and respond to hackers. These methods must be used with other techniques to prevent intrusions in order to build more robust, efficient and effective defense systems. Cybersecurity tasks should be considered necessary and important by the mechatronics engineer because design, manufacturing and production information, among others, represent the heart of any company.

15. Ergonomic design of workplaces that enable human-machine interaction, design of training methods for human-machine symbiosis, personalized manufacturing, cognitive computing, and IoT-based smart spaces are some tasks and technologies that are contextualized in I5.0. The concept of operator 4.0 relates to the design of human-cyber-physical production systems [54] that aim to boost, improve, enhance, empower and optimize the capabilities of human-machine interactions. These tasks and needs should be kept in mind by the mechatronics engineer as collaborative robotics and close interaction and collaboration between operators and machines will become more and more prevalent. In I5.0, the technologies of DT, CPS and AI that consider humans in symbiosis with machines will be listed as the core technologies.

The incorporation of disruptive technologies and new design methods to the mechatronics environment does not necessarily imply more knowledge and application burden to engineers, but rather they boost and encourage teamwork among the different engineering disciplines that are responsible for designing, managing, operating, maintaining and innovating modern productive systems characterized by CPS. The 15 technical considerations described above can be used to assist in the design of specific competencies for mechatronic engineers according to the needs of the local or national industrial sector. Some of the specific competencies that can be derived from the above considerations are listed below:

1. CPS design.
2. CPS operation.
3. DT Design.
4. CPS maintenance.
5. Conversion and reconditioning of CPS.
6. Automation of intelligent production systems.
7. Design, manufacturing and digitalized production.
8. Operation of collaborative robots.
9. Cyber-security management in the CPS.
10. Technology integration in smart factories.
11. Optimization of industrial processes.
12. Design of intelligent production systems.
13. Design of man-machine systems.
14. Analysis of large databases.
15. Interdisciplinary simulation in intelligent production systems.

## 3. Final Considerations

The tasks of mechatronic engineers are essential for the implementation of I4.0 and I5.0, so it is necessary that their training is transformed according to the needs and requirements of the current industrial revolutions. Due to its integration nature, mechatronics is similar to I4.0, since precisely this new business vision tries to integrate disruptive technologies to the productive processes looking for improvements and optimization. However, it should be understood that I4.0 implies a real change in industrial processes and does not consist of simple improvements generated by the incursion of technologies. This fact has important implications for the training of engineers, since it is not only a matter of training them in knowledge and in the specific or integrated applications of disruptive technologies, but also requires profound changes in production methodologies and education.

CBE and active methodologies should be promoted in universities for the training of mechatronic engineers. These educational approaches and active learning methods are the basis for the transformation from traditional (teacher-centered) education to constructivist (learner-centered) education. Just as industrial processes must be fundamentally transformed due to the incursion of disruptive technologies, engineering education must do the same, i.e., it must also be fundamentally transformed if it is to graduate competent engineers capable of meeting the challenges of the new industrial revolutions. CBE and active learning methods can be the basis of the transformation required by engineering education, mainly in mechatronics engineering.

The training of mechatronic engineers should take into account not only I4.0 and I5.0, but also I3.5, since technological transitions have implications in the shaping of specific competencies. For example, technological re-engineering is a task that will occur more frequently in companies that do not have sufficient capital to upgrade their production systems. This method of re-design, together with the methodology of reverse engineering, should be promoted and taught in universities. Currently, engineering education gives more priority to direct design methodologies (a process that starts from a need and ends in a product or service) than to reverse engineering methods, despite the fact that a high percentage of technological development is based on information from existing technologies in order to improve them. Technological reconditioning is a pressing need in I4.0.

The design of specific competencies in mechatronic engineers should consider the industrial needs of each region, state or country. However, technical considerations that take into account the main technologies, tools and processes, such as CPS and DT, can be raised to assist in the design of specific competencies. The 15 considerations proposed in this work can be expanded and do not exhaust the field of study. The proposals can be very useful for those universities that are developing or improving a career in Mechatronics Engineering, since these proposals could help and guide the development of specific competencies according to the needs of the local environment.

## 4. Conclusions

In this paper, 15 technical considerations have been proposed that can be used for the shaping of specific competencies for the benefit of mechatronic engineers. The conclusions of this paper are as follows:

- Engineering education, especially Mechatronics Engineering, must be fundamentally transformed in order to train the professionals who will face the challenges of I4.0 and I5.0.
- I4.0 implies profound changes in production processes, in design and manufacturing methods, in the configuration of factories and in the roles that mechatronic engineers will play. In this sense, Mechatronics Engineering as such must reinvent itself to keep pace with the technological changes in companies.
- CBE and active methodologies can be the basis for engineering education to transform itself and for universities to train the engineers required by the new industrial revolutions.
- The core training of mechatronic engineers should consider CPS, DT and AI as basic concepts that should be developed throughout their careers and that should be integrated into the curricula.
- The role of the mechatronics engineer in the I4.0 is not technical mastery of disruptive technologies, but should rather play the role of integrator and administrator of engineering groups made up of different disciplines. Companies must rethink the central role of the mechatronics engineer, since they are currently entrusted with various tasks that are not of their profile (preferably integrator and technology manager). I4.0 requires a mechatronics engineer capable of managing human resources from other fields of knowledge in order to solve specialized problems.
- Industrial maintenance, technological reconditioning and reverse engineering should be promoted as subjects of study in universities for the training of the mechatronic engineer. These skills are often learned more in companies than in universities. Technological reconditioning is a pressing need due to the fact that we are in a stage of technological transition between two industrial revolutions and that many companies do not have the capital to invest in new technologies.
- The 15 technical considerations proposed in this article can be used as a guide for the formation of specific competencies through which mechatronic engineers will acquire the knowledge and applications specifically required by the work environment in which they will work. These considerations take into account the transitions between the industrial revolutions and the most representative technologies of the I4.0, such as CPS, DT, IoT, simulation and AI. The technical considerations can be viewed as generic since they propose activities that most companies apply or will apply in one way or another in I4.0.
- For the training of the mechatronic engineer in the context of I4.0 and I5.0, all the competencies (basic, generic and specific) must be designed with the rigor indicated by the different CBE proposals and models.

**Author Contributions:** Conceptualization and writing, E.J.L.; Analysis and writing, F.C.J.; State of the art, G.L.S.; Proofreading and editing, F.J.O.E.; Research, M.A.M.M.; Supervision F.M.; Analysis and editing, P.A.L.L. All authors have read and agreed to the published version of the manuscript.

**Funding:** This research received no external funding.

**Institutional Review Board Statement:** Not applicable.

**Informed Consent Statement:** Not applicable.

**Data Availability Statement:** Not applicable.

**Acknowledgments:** The authors of this work are grateful to their universities and technological institutes for the facilities provided to carry out this research.

**Conflicts of Interest:** The authors declare no conflict of interest.

# References


1. Hozdić, E.; Jurković, Z. Cybernetization of industrial product-service systems in network environment. In *New Technologies, Development and Application*, 1st ed.; Karabegović, I., Ed.; Springer: Cham, Switzerland, 2019; pp. 262–270.
2. Bauernhansl, T. Industry 4.0: Challenges and opportunities for the automation industry. In Proceedings of the 7th EFAC Assembly Technology Conference 2013, Davos, Switzerland, 18–19 January 2013.
3. Akpinar, M.E. Industry 4.0 and Applications in Manufacturing Industry. In *Industry 4.0 and Global Businesses*, 1st ed.; Yakut, E., Ed.; Emerald Publishing Limited: Bingley, UK, 2022; pp. 111–124.
4. Choi, T.-M.; Kumar, S.; Yue, X.; Chan, H.-L. Disruptive Technologies and Operations Management in the Industry 4.0 Era and Beyond. *Prod. Oper. Manag.* **2022**, *31*, 9–31. [CrossRef]
5. Görçün, O.F. *Industry 4.0*; Beta Publisher: Paris, France, 2018; pp. 1–199.
6. Rasor, R.; Göllner, D.; Bernijazov, R.; Kaiser, L.; Dumitrescu, R. Towards collaborative life cycle specification of digital twins in manufacturing value chains. *Procedia CIRP* **2021**, *98*, 229–234. [CrossRef]
7. Zezulka, F.; Marcon, P.; Vesely, I.; Sajdl, O. Industry 4.0—An Introduction in the phenomenon. *IFAC-PapersOnLine* **2016**, *49*, 8–12. [CrossRef]
8. Karabegović, I.; Karabegović, E.; Mahmić, M.; Husak, E. Implementation of Industry 4.0 and Industrial Robots in the Manufacturing Processes. In *New Technologies, Development and Application II. NT 2019*, 1st ed.; Lecture Notes in Networks and Systems; Karabegović, I., Ed.; Springer: Cham, Switzerland, 2020; Volume 76, pp. 3–14.
9. Lu, Y. Cyber Physical System (CPS)-Based Industry 4.0: A Survey. *J. Ind. Eng. Manag.* **2017**, *2*, 1750014. [CrossRef]
10. Askarpour, M.; Ghezzi, C.; Mandrioli, D.; Rossi, M.; Tsigkanos, C. Formal Methods in Designing Critical Cyber-Physical Systems. In *From Software Engineering to Formal Methods and Tools, and Back*, 1st ed.; Lecture Notes in Computer, Science; ter Beek, M., Fantechi, A., Semini, L., Eds.; Springer: Cham, Switzerland, 2019; Volume 11865, pp. 110–130.
11. Badkilaya, S.K.; Bhat, H.P. The Need for a System to Benefit the Implementation of Digital Twin, by Helping Visualize the Virtual Dynamics Remotely. In *Cyber-Physical Systems and Digital Twins. REV2019 2019*, 1st ed.; Lecture Notes in Networks and, Systems; Auer, M., Ram, B.K., Eds.; Springer: Cham, Switzerland, 2020; Volume 80, pp. 38–50.
12. Qin, J.; Liu, Y.; Grosvenor, R. Multi-source data analytics for AM energy consumption prediction. *Adv. Eng. Informat.* **2018**, *38*, 840–850. [CrossRef]
13. Lee, J.; Davari, H.; Singh, J.; Pandhare, V. Industrial Artificial Intelligence for industry 4.0-based manufacturing systems. *Manuf. Lett.* **2018**, *18*, 20–23. [CrossRef]
14. Doyle, M.; Kopacek, P. Adoption of Collaborative Robotics in Industry 5.0. An Irish industry case study. *IFAC-PapersOnLine* **2021**, *54*, 413–418. [CrossRef]
15. Bednar, P.M.; Welch, C. Socio-Technical Perspectives on Smart Working: Creating Meaningful and Sustainable Systems. *Inf. Syst. Front.* **2019**, *22*, 281–298. [CrossRef]
16. AlMaadeed, M.A.A.; Ponnamma, D. Role of Research and Higher Education on Industry 4.0, Material Science as an example. In Proceedings of the 2020 IEEE International Conference on Informatics, IoT, and Enabling Technologies (ICIoT), Doha, Qatar, 2–5 February 2020. [CrossRef]
17. Mogos, R.; Bodea, C.N.; Dascălu, M.I.; Safonkina, O.; Lazaraou, E.; Trifan, E.L.; Nemoianu, I.V. Technology enhanced learning for industry 4.0 engineering education. *Rev. Roum. Sci. Techn.-Électrotechn. Énerg.* **2018**, *63*, 429–435.
18. Muktiarni, M.; Widiaty, I.; Abdullah, A.G.; Ana, A.; Yulia, C. Digitalisation trend in education during industry 4.0. *J. Phys. Conf. Ser.* **2019**, *1402*, 077070. [CrossRef]
19. Hernandez-de-Menendez, M.; Morales-Menendez, R.; Escobar, C.A.; McGovern, M. Competencies for Industry 4.0. *Int. J. Interact. Des. Manuf.* **2020**, *14*, 1511–1524. [CrossRef]
20. Jimenez, E.; Ochoa, F.J.; Luna, G.; Muñoz, F.; Cuenca, F.; Maciel, M.A. Competency-based Education of the Mechatronics Engineer in the Transition from Manufacturing 3.0 to Industry 4.0. In Proceedings of the 2nd IFSA Winter Conference on Automation, Robotics & Communications for Industry 4.0 (ARCI' 2022), Andorra la Vella, Andorra, 2–3 February 2022; pp. 84–87.
21. Bischof, C.; de Oliveira, E. Production Engineering Competencies in the Industry 4.0 context: Perspectives on the Brazilian labor market. *Production* **2020**, *30*, 1–10. [CrossRef]
22. Ambigaipagan, N.; Fauziah, W.; Sivan, R. Industry 4.0 Competence Model for Malaysia Industry4WRD. In Proceedings of the 33rd International Business Information and Management Association IBIMA Conference, Granada, Spain, 10–11 April 2019.
23. Coles, M.; Werquin, P. *Qualifications Systems. Bridges to Lifelong Learning*, 1st ed.; OCDE: Paris, France, 2007.
24. Katrantzis, E.; Moulianitis, V.; Miatliuk, K. Conceptual Design Evaluation of Mechatronic Systems. In *Emerging Trends in Mechatronics*, 1st ed.; Azizi, A., Ed.; IntechOpen: London, UK, 2019; pp. 27–50.
25. Nagy, R. Literature review of contemporary industrial revolutions as decision support resources. *J. Agric. Inform.* **2022**, *13*, 19–25. [CrossRef]

26. Lambrechts, W.; Sinha, S.; Marwala, T. The Global South and Industry 4.0: Historical Development and Future Trajectories. In *The BRICS Order*, 1st ed.; Monyae, D., Ndzendze, B., Eds.; Palgrave Macmillan: Cham, Switzerland, 2021; pp. 249–281.
27. Nagy, J. Az ipar 4.0 fogalma és kritikus kérdései—Vállalati interjúk alapján. *Vez.-Bp. Manag. Rev.* **2019**, *50*, 14–26. [CrossRef]
28. Doyle, M.; Kopacek, P. Industry 5.0: Is the Manufacturing Industry on the Cusp of a New Revolution? In Proceedings of the International Symposium for Production Research 2019, ISPR2019, Vienna, Austria, 28–30 August 2019; Lecture Notes in Mechanical Engineering. Durakbasa, N., Gençyılmaz, M., Eds.; Springer: Cham, Switzerland, 2020; pp. 432–441.
29. Caglar, T.; Teker, S. Industrial revolutions and its effects on quality of life. In Proceedings of the Global Business Research Congress, Istanbul, Turkey, 30–31 May 2019; Volume 56, pp. 304–311.
30. Ratanlal, P. Industry revolution 5.0. *J. Manag. Res.* **2021**, *10*, 1–2. [CrossRef]
31. Dilmé, M. Fundamental Concepts of Industry 4.0. In *Digital Manufacturing for SMEs: An Introduction*, 1st ed.; Chaplin, J., Pagano, C., Fort, S., Eds.; University of Nottingham: Nottingham, UK, 2020; pp. 5–26.
32. Kagermann, H.; Wahlster, W.; Helbig, J. *Recommendations for Implementing the Strategic Initiative Industrie 4.0*; Final Report of the Industrie 4.0 Working Group; ACATECH National Academy of Science and Engineering: Frankfurt, Germany, 2013.
33. World Economic Forum (WEF). The Future of Jobs: Employment, Skills and Workforce Strategy for the Fourth Industrial Revolution (Geneva). 2016. Available online: https://www3.weforum.org/docs/WEF_Future_of_Jobs.pdf (accessed on 6 June 2022).
34. Oztemel, E.; Gursev, S. Literature review of Industry 4.0 and related technologies. *J. Intell Manuf.* **2020**, *31*, 127–182. [CrossRef]
35. Suleiman, Z.; Shaikholla, S.; Dikhanbayeva, D.; Shehab, E.; Turkyilmaz, A. Industry 4.0: Clustering of concepts and characteristics. *Cogent Eng.* **2022**, *91*, 2034264. [CrossRef]
36. Meissner, H.; Ilsen, R.; Aurich, J.C. Analysis of Control Architectures in the Context of Industry 4.0. *Procedia CIRP* **2017**, *62*, 165–169. [CrossRef]
37. Anshari, M. Workforce Mapping of Fourth Industrial Revolution: Optimization to Identity. *J. Phys. Conf. Ser.* **2020**, *1477*, 072023. [CrossRef]
38. Shafiq, S.I.; Sanin, C.; Toro, C.; Szczerbicki, E. Virtual engineering object (VEO): Toward experience-based design and manufacturing for industry 4.0. *Cybern. Syst.* **2015**, *46*, 35–50. [CrossRef]
39. Qin, J.; Liu, Y.; Grosvenor, R. A Categorical Framework of Manufacturing for Industry 4.0 and Beyond. *Procedia CIRP* **2016**, *52*, 173–178. [CrossRef]
40. Morrar, R.; Arman, H.; Mousa, S. The fourth industrial revolution (Industry 4.0): A social innovation perspective. *Technol. Innov. Manag. Rev.* **2017**, *7*, 12–20. [CrossRef]
41. Kagermann, H.; Wahlster, W.; Helbig, J. Umsetzungsempfehlungen für das Zukunftsprojekt Industry 4.0. Abschlussbericht des Arbeitskreises Industry 4.0. Germany's future as a production standard. In *Promotorengruppe Kommunikation der Forschungsunion Wirtschaft—Wissenschaft, Erste ausgabe*; Promotorengruppe Kommunikation, Forschungsunion Wirtschaft—Wissenschaft, Herausgeber: Frankfurt, Germany, 2013.
42. Rodríguez, G.; Bribiesca, G. Assessing Digital Transformation in Universities. *Future Internet* **2021**, *13*, 52. [CrossRef]
43. Akundi, A.; Euresti, D.; Luna, S.; Ankobiah, W.; Lopes, A.; Edinbarough, I. State of Industry 5.0-Analysis and Identification of Current Research. *Trends Appl. Syst. Innov.* **2022**, *5*, 27. [CrossRef]
44. Chander, B.; Pal, S.; De, D.; Buyya, R. Artificial Intelligence-based Internet of Things for Industry 5.0. In *Artificial Intelligence-based Internet of Things Systems. Internet of Things*, 1st ed.; Pal, S., De, D., Buyya, R., Eds.; Springer: Cham, Switzerland, 2022; pp. 3–45.
45. Gopalakrishna, S.; Haldorai, K.; Seok, W.; Gon, W. COVID-19 and hospitality 5.0: Redefining hospitality operations. *Int. J. Hosp. Manag.* **2021**, *94*, 102869.
46. Özdemir, V.; Hekim, N. Birth of Industry 5.0: Making Sense of Big Data with Artificial Intelligence: The Internet of Things and Next-GenerationTechnology Policy. *Int. J. Integr. Biol.* **2018**, *22*, 65–76. [CrossRef]
47. Kumar, R.M.; Quoc, P.; Prabadevi, B.; Deepa, N.; Kapal, D.; Thippa, G.; Rukhsana, R.; Madhusanka, L. Industry 5.0: A survey on enabling technologies and potential applications. *J. Ind. Inf. Integr.* **2022**, *26*, 100257.
48. Rada, M. Industry 5.0 Definition. 2020. Available online: https://michael-rada.medium.com/industry-5--0-definition-6a2f9922dc48#:~{}:text=INDUSTRY%205.0%20is%20future%2C%20but,and%20wasting%20including%20INDUSTRIAL%20UPCYCLING (accessed on 6 June 2022).
49. Longo, F.; Padovano, A.; Umbrello, S. Value-oriented and ethical technology engineering in industry 5.0: A human-centric perspective for the design of the factory of the future. *Appl. Sci.* **2020**, *10*, 4182. [CrossRef]
50. Xun, X.; Yuqian, L.; Vogel, B.; Wang, L. Industry 4.0 and Industry 5.0-Inception, conception and perception. *J. Manuf. Syst.* **2021**, *61*, 530–535.
51. Hernandez, G.M.; Habib, L.; Garcia, F.A.; Montemayor, F. Industry 4.0 and Engineering Education: An Analysis of Nine Technological Pillars Inclusion in Higher Educational Curriculum. In *Best Practices in Manufacturing Processes*, 1st ed.; García, J., Rivera, L., González, R., Leal, G., Chong, M., Eds.; Springer: Cham, Switzerland, 2019; pp. 525–543.
52. Melnik, S.; Magnotti, M.; Butts, C.; Putman, C.; Aqlan, F. A Concept Relationship Map for Industry 4.0. In Proceedings of the 5th NA International Conference on Industrial Engineering and Operations Management, Detroit, MI, USA, 10–14 August 2020.
53. Haleem, A.; Javaid, M. Industry 5.0 and its applications in orthopaedics. *J. Clin. Orthop. Trauma* **2019**, *10*, 807–808. [CrossRef]
54. Ruppert, T.; Jaskó, S.; Holczinger, T.; Abonyi, J. Enabling Technologies for Operator 4.0: A Survey. *Appl. Sci.* **2018**, *8*, 1650. [CrossRef]
55. Romero, D.; Stahre, J.; Wuest, T.; Noran, O.; Bernus, P.; Fast, A.; Gorecky, D. Towards an Operator 4.0 Typology: A Human-centric Perspective on the Fourth Industrial Revolution Technologies. In Proceedings of the CIE46, Tianjin, China, 29–31 October 2016.

56. Yadav, S.; Prakash, S. Blockchain critical success factors for sustainable supply chain. *Resour. Conserv. Recycl.* **2020**, *152*, 104505. [CrossRef]

57. Yu, T.; Simbara, F.A.; Chien, H. Developing a framework for Industry 3.5 to strengthen manufacturer performance. *Int. J. Logist. Appl. Res. Appl.* **2021**, 1–22. [CrossRef]

58. Chen, C.; Tzu, H.; Hong, G. A Conceptual Framework for "Industry 3.5" to Empower Intelligent Manufacturing and Case Studies. *Procedia Manuf.* **2017**, *11*, 2009–2017. [CrossRef]

59. Al-Maeeni, S.S.H.; Kuhnhen, C.; Engel, B.; Schiller, M. Smart retrofitting of machine tools in the context of industry 4.0. *Procedia CIRP* **2020**, *88*, 369–374. [CrossRef]

60. Guerreiro, B.V.; Lins, R.G.; Sun, J.; Schmitt, R. Definition of Smart Retrofitting: First Steps for a Company to Deploy Aspects of Industry 4.0. In *Advances in Manufacturing. Lecture Notes in Mechanical Engineering*, 1st ed.; Hamrol, A., Ciszak, O., Legutko, S., Jurczyk, M., Eds.; Springer: Cham, Switzerland, 2018; pp. 161–170.

61. Di Carlo, F.; Mazzuto, G.; Bevilacqua, M.; Ciarapica, F.E. Retrofitting a Process Plant in an Industry 4.0 Perspective for Improving Safety and Maintenance Performance. *Sustainability.* **2021**, *13*, 646. [CrossRef]

62. Ramanathan, L.; Nandhini, R.S. Cyber-Physical System-An Architectural Review. In Proceedings of the Information and Communication Technology for Competitive Strategies (ICTCS 2020), Jaipur, India, 14–16 September 2020; Lecture Notes in Networks and, Systems. Joshi, A., Mahmud, M., Ragel, R.G., Thakur, N.V., Eds.; Springer: Singapore, 2022; Volume 191, pp. 133–142.

63. Sanz, V.; Urquia, A. Cyber-physical system modeling with Modelica using message passing communication. *Simul. Modell. Pract. Theory* **2022**, *117*, 102501. [CrossRef]

64. Putnik, G.D.; Ferreira, L.; Lopes, N.; Putnik, Z. What is a Cyber-Physical System: Definitions and models spectrum. *FME Trans.* **2019**, *47*, 663–674. [CrossRef]

65. Mladineo, M.; Veza, I.; Gjeldum, N. Solving partner selection problem in cyber-physical production networks using the HUMANT algorithm. *Int. J. Prod. Res.* **2017**, *55*, 2506–2521. [CrossRef]

66. Edward Ashford, E.; Arunkumarr, S. *Introduction to Embedded Systems: A Cyber-Physical Systems Approach*, 2nd ed.; The MIT Press: Cambridge, MA, USA, 2016.

67. DeSmit, Z.; Elhabashy, A.E.; Lee, J.; Wells, L.; Camelio, J.A. Cyber-Physical Vulnerability Assessment in Manufacturing Systems. *Procedia Manuf.* **2016**, *5*, 1060–1074. [CrossRef]

68. Zacchia, Y.; DInnocenzo, A.; Smarra, F.; Malavolta, I.; Benedetto, M.D. State of the art of cyber-physical systems security: An automatic control perspective. *J. Syst. Software.* **2019**, *149*, 174–216. [CrossRef]

69. Kumar, A.; Sreenath, N. Cyber Physical Systems: Analyses, challenges and possible solutions. *Internet Things Cyber-Phys. Syst.* **2021**, *1*, 22–33.

70. Kopetz, H. Cyber-Physical Systems Are Different. In *Simplicity Is Complex: Foundations of Cyber-Physical System Design*, 1st ed.; Kopetz, H., Ed.; Springer: Cham, Switzerland, 2019; pp. 69–74.

71. Shyr, W.J.; Juan, H.C.; Tsai, C.Y.; Chang, Y.J. Application of Cyber-Physical System Technology on Material Color Discrimination. *Electronics* **2022**, *11*, 920. [CrossRef]

72. Wang, Q.; Zhou, X.; Yang, G.; Yang, Y. Behavior Modeling of Cyber-physical System Based on Discrete Hybrid Automata. In Proceedings of the 2013 IEEE 16th International Conference on Computational Science and Engineering, Sydney, NSW, Australia, 3–5 December 2013. [CrossRef]

73. Barišić, A.; Ruchkin, I.; Savić, D.; Mohamed, M.A.; Al-Ali, R.; Li, L.W.; Mkaouar, H.; Eslampanah, R.; Challenger, M.; Blouin, D.; et al. Multi-Paradigm Modeling for Cyber-Physical Systems: A Systematic Mapping Review. *J. Syst. Softw.* **2022**, *183*, 11108. [CrossRef]

74. Klötzer, C.; Pflaum, A. Cyber-physical systems as the technical foundation for problem solutions in manufacturing, logistics and supply chain management. In Proceedings of the 5th International Conference on the Internet of Things (IOT), IEEE 2015, Seoul, Korea, 26–28 October 2015; pp. 12–19.

75. Chen, H. Applications of Cyber-Physical System: A Literature Review. *J. Ind. Eng. Manag.* **2017**, *2*, 1–28. [CrossRef]

76. Fitzgerald, J.; Gorm, P.; Pierce, K. Multi-modelling and Co-simulation in the Engineering of Cyber-Physical Systems: Towards the Digital Twin. In *From Software Engineering to Formal Methods and Tools, and Back*, 1st ed.; ter Beek, M.H., Fantechi, A., Semini, L., Eds.; Springer: Cham, Switzerland, 2019; Volume 11865, pp. 40–55.

77. Nota, G.; Matonti, G.; Bisogno, M.; Nastasia, S. The contribution of cyber-physical production systems to activity-based costing in manufacturing. An Interventional Research Approach. *Int. J. Eng. Bus. Manag.* **2020**, *12*, 1750012. [CrossRef]

78. Lin, W.D.; Low, Y.H.; Chong, Y.T.; Teo, C.L. Integrated Cyber Physical Simulation Modelling Environment for Manufacturing 4.0. In Proceedings of the 2018 IEEE IEEM, Bangkok, Thailand, 16–19 December 2018; pp. 1861–1865.

79. Palkina, E.S.; Zhuravleva, N.A.; Panychev, A.Y. New Approach to Transportation Service Pricing Based on the Stakeholder Model of Corporate Governance. *Mediterr. J. Soc. Sci.* **2015**, *6*, 299–308. [CrossRef]

80. Kupriyanovsky, V.P.; Namnot, E.D.; Sinyakov, S.A. Cyber-Physical Systems as the Basis of the Digital Economy. *Int. J. Open Inf. Technol.* **2016**, *4*, 18–25.

81. Zhilenkov, A.; Chernyi, S.; Emelianov, V. Application of Artificial Intelligence Technologies to Assess the Quality of Structures. *Energies* **2021**, *14*, 8040. [CrossRef]

82. Ali, S.; Hafeez, Y.; Bilal, M.; Saeed, S.; Sup, K. Towards Aspect Based Components Integration Framework for Cyber-Physical System. *Comput. Mater. Contin. CMC* **2022**, *70*, 655–668. [CrossRef]

83. Jay, L.; Bagheri, B.; Hung, K. A Cyber-Physical Systems architecture for Industry 4.0-based manufacturing systems. *Manuf. Lett.* **2015**, *3*, 18–23.

84. Shafiq, S.I.; Sanin, C.; Szczerbicki, E.; Toro, C. Virtual engineering object/virtual engineering process: A specialized form of cyber physical system for Industrie 4.0. *Procedia Comput. Sci.* **2015**, *60*, 1146–1155. [CrossRef]

85. Lee, J.; Noh, S.D.; Kim, H.J.; Kang, Y.S. Implementation of Cyber-Physical Production Systems for Quality Prediction and Operation Control in Metal Casting. *Sensors* **2018**, *18*, 1428. [CrossRef] [PubMed]

86. Prajwal, A.T.; Muddukrishna, B.S.; Vasantharaju, S.G. Pharma 4.0-Impact of the Internet of Things on Health Care. *Int. J. App. Pharm.* **2020**, *12*, 64–69.

87. Arden, N.A.; Fisher, A.C.; Tyner, K.; Yu, L.X.; Lee, S.L.; Kopcha, M. Industry 4.0 for pharmaceutical manufacturing: Preparing for the smart factories of the future. *Int. J. Pharm.* **2021**, *602*, 120554. [CrossRef] [PubMed]

88. Ouranidis, A.; Vavilis, T.; Mandala, E.; Davidopoulou, C.; Stamoula, E.; Markopoulou, C.K.; Karagianni, A.; Kachrimanis, K. mRNA Therapeutic Modalities Design, Formulation and Manufacturing under Pharma 4.0 Principles. *Biomedicines* **2022**, *10*, 50. [CrossRef]

89. Ouranidis, A.; Davidopoulou, C.; Tashi, R.K.; Kachrimanis, K. Pharma 4.0 Continuous mRNA Drug Products Manufacturing. *Pharmaceutics* **2021**, *13*, 1371. [CrossRef]

90. Ouranidis, A.; Davidopoulou, C.; Kachrimanis, K. Integrating Elastic Tensor and PC-SAFT Modeling with Systems-Based Pharma 4.0 Simulation, to Predict Process Operations and Product Specifications of Ternary Nanocrystalline Suspensions. *Pharmaceutics* **2021**, *13*, 1771. [CrossRef]

91. Canuto, G.; Kaminski, P.C. From Embedded Systems (ES) to Cyber Physical Systems (CPS): An Analysis of Transitory Stage of Automotive Manufacturing in the Industry 4.0 Scenario. SAE Technical Paper Series. In Proceedings of the 25th SAE BRASIL International Congress and Display 2016, Sao Paulo, Brazil, 25–27 October 2016; pp. 1–6. [CrossRef]

92. Babiceanu, R.; Seker, R. Big Data and virtualization for manufacturing cyber-physical systems: A survey of the current status and future outlook. *Comput. Ind.* **2016**, *81*, 128–137. [CrossRef]

93. Suh, S.C.; Tanik, U.J.; Carbone, J.N.; Eroglu, A. *Applied Cyber-Physical Systems*, 1st ed.; Springer Science & Business Media: Berlin/Heidelberg, Germany, 2014.

94. Parvin, S.; Hussain, F.K.; Hussain, O.K.; Thein, T.; Park, J.S. Multi-cyber framework for availability enhancement of cyber physical systems. *Computing* **2013**, *95*, 927–948. [CrossRef]

95. Broy, M.; Schmidt, A. Challenges in engineering cyber-physical systems. *Computer* **2014**, *47*, 70–72. [CrossRef]

96. Tan, Y.; Goddard, S.; Perez, L.C. A prototype architecture for cyber-physical systems. *ACM SIGBED Rev.* **2008**, *5*, 1–2. [CrossRef]

97. Stark, R.; Damerau, T. Digital Twin. In *CIRP Encyclopedia of Production Engineering*, 1st ed.; Chatti, S., Tolio, T., Eds.; Springer: Berlin/Heidelberg, Germany, 2019; pp. 1–8.

98. Jian, D.; Tian, M.; Qing, Z.; Zhen, L.; Ji, Y. Design and application of digital twin system for the blade-rotor test rig. *J. Intell. Manuf.* **2021**, 1–17. [CrossRef]

99. Tao, F.; Zhang, M.; Nee, A.Y.C. *Digital Twin Driven Smart Manufacturing*, 1st ed.; Academic Press: London, UK, 2019.

100. Barricelli, B.; Casiraghi, E.; Fogli, D. A Survey on Digital Twin: Definitions, Characteristics, Applications, and Design Implications. *IEEE Access* **2019**, *7*, 167653–167671. [CrossRef]

101. Semeraro, C.; Lezoche, M.; Panetto, H.; Dassisti, M. Digital twin paradigm: A systematic literature review. *Comput. Ind.* **2021**, *130*, 103469. [CrossRef]

102. Pei, H.; Ki, K.; Maartje, S. Ethical Issues of Digital Twins for Personalized Health Care Service: Preliminary Mapping Study. *J. Med. Internet Res.* **2022**, *24*, e33081.

103. Friederich, J.; Francis, D.P.; Sanja Lazarova, S.; Mohamed, N. A framework for data-driven digital twins of smart manufacturing systems. *Comput. Ind.* **2022**, *136*, 103586. [CrossRef]

104. Neethirajan, S.; Kemp, B. Digital Twins in Livestock Farming. *Animals* **2021**, *11*, 1008. [CrossRef]

105. Tao, F.; Liu, W.; Liu, J.; Liu, X.; Liu, Q.; Qu, T.; Hu, T.; Zhang, Z.; Xiang, F.; Xu, W.; et al. Digital Twin and its Potential Application Exploration. *Int. J. Comput. Integr. Manuf.* **2018**, *24*, 1–18.

106. Tao, F.; Zhang, H.; Liu, A.; Nee, A.Y.C. Digital Twin in Industry: State-of-the-art. *IEEE Trans. Industr. Inform.* **2018**, *15*, 2405–2415. [CrossRef]

107. Malakuti, S.; Schlake, J.; Ganz, C.; Harper, K.E.; Petersen, H. Digital Twin: An Enabler for New Business Models. In Proceedings of the Automation Congress 2019, Hangzhou, China, 22–24 November 2019.

108. Lattanzi, L.; Raffaeli, R.; Peruzzini, M.; Pellicciari, M. Digital twin for smart manufacturing: A review of concepts towards a practical industrial implementation. *Int. J. Comput. Integr. Manuf.* **2021**, *34*, 567–597. [CrossRef]

109. Moyne, J.; Qamsane, Y.; Balta, E.C.; Kovalenko, I.; Faris, J.; Barton, K.; Tilbury, D.M. A Requirements Driven Digital Twin Framework: Specification and Opportunities. *IEEE Access* **2020**, *8*, 107781–107801. [CrossRef]

110. Eyre, J.; Freeman, C. Immersive applications of industrial digital twins. The Industrial Track of EuroVR 2018. In Proceedings of the 15th Annual EuroVR Conference, London, UK, 22–23 October 2018; VTT Technical Research Centre of Finland, VTT Technology: London, UK, 2018; pp. 11–20.

111. Tao, F.; Qi, Q.; Wang, L.; Nee, A.Y.C. Digital Twins and Cyber-Physical Systems Toward Smart Manufacturing and Industry 4.0: Correlation and Comparison. *Engineering* **2019**, *5*, 653–661. [CrossRef]

112. Abiola, A.; Chimay, J.; Omobolanle, O. Towards next generation cyber-physical systems and digital twins for construction. Special issue: Next Generation ICT—How distant is ubiquitous computing? *Inf. Technol. Constr.* **2021**, *26*, 505–525.

113. Tao, F.; Cheng, J.; Qi, Q.; Zhang, M.; Zhang, H.; Sui, F. Digital twin-driven product design, manufacturing and service with big data. *Int. J. Adv. Manuf. Technol.* **2018**, *94*, 3563–3576. [CrossRef]

114. Zeb, S.; Mahmood, A.; Hassan, S.; Jalil, M.D.; Gidlund, M.; Guizani, M. Industrial digital twins at the nexus of NextG wireless networks and computational intelligence: A survey. *J. Netw. Comput. Appl.* **2022**, *200*, 103309. [CrossRef]

115. Atzori, L.; Iera, A.; Morabito, G. The Internet of Things: A survey. *Comput. Netw.* **2010**, *54*, 2787–2805. [CrossRef]

116. Jain, P.; Aggarwal, P.K.; Chaudhary, P.; Makar, K.; Mehta, J.; Garg, R. Convergence of IoT and CPS in Robotics. In *Emergence of Cyber Physical System and IoT in Smart Automation and Robotics. Advances in Science, Technology & Innovation*, 1st ed.; Singh, K.K., Nayyar, A., Tanwar, S., Abouhawwash, M., Eds.; Springer: Cham, Switzerland, 2021; pp. 15–30.

117. Munirathinam, S. Chapter Six—Industry 4.0: Industrial Internet of Things (IIOT). In *Advances in Computers*, 1st ed.; Pethuru, R., Preetha, E., Eds.; Elsevier: Amsterdam, The Netherlands, 2020; Volume 117, pp. 129–164.

118. Al, A.; Guizani, M.; Mohammadi, M.; Aledhari, M.; Ayyash, M. Internet of things: A survey on enabling technologies, protocols, and applications. *IEEE Commun. Surv. Tutor* **2015**, *17*, 2347–2376.

119. Franco, J.; Aris, A.; Canberk, B.; Uluagac, A.S. A Survey of Honeypots and Honeynets for Internet of Things, Industrial Internet of Things, and Cyber-Physical Systems. *IEEE Commun. Surv. Tutor* **2021**, *23*, 2351–2383. [CrossRef]

120. Banks, J.; Carson II, J.S.; Nelson, B.L.; Nicol, D.M. *Discrete Event System Simulation*, 5th ed.; Pearson Education: Madrid, Spain, 2009.

121. Rodič, B. Industry 4.0 and the New Simulation Modelling Paradigm. *Organizacija* **2017**, *50*, 193–207. [CrossRef]

122. Weyer, S.; Meyer, T.; Ohmer, M.; Gorecky, D.; Zühlke, D. Future Modeling and Simulation of CPS-based Factories: An Example from the Automotive Industry. *IFAC-PapersOnLine* **2016**, *49*, 97–102. [CrossRef]

123. Suzuki, A.; Masutomi, K.; Ono, I.; Ishii, H.; Onoda, T. CPS-Sim: Co-Simulation for Cyber-Physical Systems with Accurate Time Synchronization. *IFAC-PapersOnLine* **2018**, *51*, 70–75. [CrossRef]

124. Posada, J.; Toro, C.; Barandiaran, I.; Oyarzun, D.; Stricker, D.; Amicis, R.; Pinto, E.; Eisert, P.; Dollner, J.; Vallarino, I. Visual Computing as Key Enabling Technology for Industry 4.0 & Industrial Internet. *IEEE Comput. Graph. Appl.* **2015**, *35*, 26–40. [PubMed]

125. Krishnamurthi, R.; Kumar, A. Modeling and Simulation for Industry 4.0. In *Smart Production, Sharp Business and Sustainable Development. Advances in Science, Technology & Innovation*, 1st ed.; Nayyar, A., Kumar, A., Eds.; Springer: Cham, Switzerland, 2020; pp. 127–141.

126. Gunal, M.M. Simulation and the Fourth Industrial Revolution. In *Simulation for Industry 4.0. Springer Series in Advanced Manufacturing*, 1st ed.; Gunal, M., Ed.; Springer: Cham, Switzerland, 2019; pp. 1–17.

127. Souza, M.L.H.; da Costa, A.; de Oliveira, G.; da Rosa, R. A survey on decision-making based on system reliability in the context of Industry 4.0. *J. Manuf. Syst.* **2020**, *56*, 133–156. [CrossRef]

128. Ermağan, İ. Worldwide Artificial Intelligence Studies with a Comparative Perspective: How Ready is Turkey for This Revolution? In *Artificial Intelligence Systems and the Internet of Things in the Digital Era. EAMMIS 2021*, 1st ed.; Lecture Notes in Networks and, Systems; Musleh, A.M., Razzaque, A., Kamal, M.M., Eds.; Springer: Cham, Switzerland, 2021; Volume 239, pp. 500–512.

129. Kamble, R.; Shah, D. Applications of artificial intelligence in human life. *Int. J. Res. Granthaalayah* **2018**, *6*, 178–188. [CrossRef]

130. Chu, Y.; Chen, Q.; Fan, Z.; He, J.-J. Tunable V-Cavity Lasers Integrated With a Cyclic Echelle Grating for Distributed Routing Networks. *IEEE Photon. Technol. Lett.* **2019**, *31*, 943–946. [CrossRef]

131. Radanliev, P.; De Roure, D.; Kleek, M.; Santos, O.; Ani, U. Artificial intelligence in cyber physical systems. *AI Soc.* **2021**, *36*, 783–796. [CrossRef]

132. Lv, Z.; Xie, S. Artificial intelligence in the digital twins: State of the art, challenges, and future research topics. *Digit. Twin* **2022**, *1*, 1–23. [CrossRef]

133. Silva, R.; Jia, X.; Lee, J.; Sun, K.; Walter, A.; Barata, J. Industrial Artificial Intelligence in Industry 4.0—Systematic Review, Challenges and Outlook. *IEEE Access* **2020**, *8*, 220121–220139.

134. Wegner, P. Interoperability. *ACM Comput. Surv.* **1996**, *28*, 285–287. [CrossRef]

135. IEEE. *IEEE Standard Computer Dictionary: A Compilation of IEEE Standard Computer Glossaries*; IEEE Standard: Piscataway, NJ, USA, 1991; Volume 610, pp. 1–217.

136. Gonzalez, I.; Calderon, A.J.; Figueiredo, J.; Sousa, J.M.C. A Literature Survey on Open Platform Communications (OPC), Applied to Advanced Industrial Environments. *Electronics* **2019**, *8*, 510. [CrossRef]

137. Chen, D.; Vernadat, F.B. *Enterprise Interoperability: A Standardisation View*, 1st ed.; Springer: Berlin/Heidelberg, Germany, 2003; pp. 273–282.

138. Zeid, A.; Sundaram, S.; Moghaddam, M.; Kamarthi, S.; Marion, T. Interoperability in Smart Manufacturing: Research Challenges. *Machines* **2019**, *7*, 21. [CrossRef]

139. Yang, L. Industry 4.0: A survey on technologies, applications and open research issues. *J. Ind. Inf. Integr.* **2017**, *6*, 1–10.

140. Coşkun, S.; Kayıkcı, Y.; Gençay, E. Adapting Engineering Education to Industry 4.0 Vision. *Technologies* **2019**, *7*, 10. [CrossRef]

141. Sakhapov, R.; Absalyamova, S. Fourth industrial revolution and the paradigm change in engineering education. *MATEC Web Conf.* **2018**, *245*, 12003. [CrossRef]

142. Motyl, B.; Baronio, G.; Uberti, S.; Speranza, D.; Filippi, S. How will Change the Future Engineers' Skills in the Industry 4.0 Framework? A Questionnaire Survey. *Procedia Manuf.* **2017**, *11*, 1501–1509. [CrossRef]

143. Popkova, E.G.; Ragulina, Y.V.; Bogoviz, A.V. Fundamental Differences of Transition to Industry 4.0 from Previous Industrial Revolutions. In *Industry 4.0: Industrial Revolution of the 21st Century. Studies in Systems, Decision and Control*, 1st ed.; Popkova, E., Ragulina, Y., Bogoviz, A., Eds.; Springer: Cham, Switzerland, 2019; Volume 169, pp. 21–29.

144. McKee, S.; Gauch, D. *Implications of Industry 4.0 on Skills Development. Anticipating and Preparing for Emerging Skills and Jobs. Education in the Asia-Pacific Region: Issues, Concerns and Prospects*, 1st ed.; Panth, B., Maclean, R., Eds.; Springer: Cham, Switzerland; Singapore, 2020; Volume 55, pp. 279–288.

145. Kaur, R.; Awasthi, A.; Grzybowska, K. Evaluation of Key Skills Supporting Industry 4.0-A Review of Literature and Practice. In *Sustainable Logistics and Production in Industry 4.0. EcoProduction (Environmental Issues in Logistics and Manufacturing)*, 1st ed.; Grzybowska, K., Awasthi, A., Sawhney, R., Eds.; Springer: Cham, Switzerland, 2020; pp. 19–29.

146. Jakovljevic, Z.; Nedeljkovic, D. Cyber Physical Systems in Manufacturing Engineers Education. In *Machine and Industrial Design in Mechanical Engineering. KOD 2021. Mechanisms and Machine Science*, 1st ed.; Rackov, M., Mitrović, R., Čavić, M., Eds.; Springer: Cham, Switzerland, 2022; Volume 109, pp. 735–743.

147. Harshama, F.; Tomizuka, M.; Fukuda, T. MechatronicsWhat is it, why, and how?—An editorial. *IEEE/ASME Trans. Mechatron.* **1996**, *1*, 1–4. [CrossRef]

148. Bradley, D.; Russell, D.; Hehenberger, P.; Azorin, J.; Watt, S.; Milne, C. From Mechatronics to the Cloud. In *Reinventing Mechatronics*, 1st ed.; Xiu, Y., Bradley, D., Russell, D., Moore, P., Eds.; Springer: Cham, Switzerland, 2020; pp. 17–33.

149. Hehenberger, P.; Vogel, B.; Bradley, D.A.; Eynard, B.; Tomiyama, T.; Achiche, S. Design, modelling, simulation and integration of cyber physical systems: Methods and applications. *Comput. Ind.* **2016**, *82*, 273–289. [CrossRef]

150. Yan, X.T.; Bradley, D. Reinventing Mechatronics. In *Reinventing Mechatronics*, 1st ed.; Xiu, Y., Bradley, D., Russell, D., Moore, P., Eds.; Springer: Cham, Switzerland, 2020; pp. 1–10.

151. Guérineau, B.; Bricogne, M.; Durupt, A.; Rivest, L. Mechatronics vs. cyber physical systems: Towards a conceptual framework for a suitable design methodology. In Proceedings of the 2016 11th France-Japan & 9th Europe-Asia Congress on Mechatronics (MECATRONICS)/17th International Conference on Research and Education in Mechatronics (REM), Compiegne, France, 15–17 June 2016; pp. 314–320.

152. Craig, K. Mechatronic Capstone Design Course. *Mechatron. Appl. Int. J.* **2021**, *2*, 47–60. [CrossRef]

153. Russell, D. Reinventing Mechatronics—Final Thoughts. In *Reinventing Mechatronics*, 1st ed.; Xiu, Y., Bradley, D., Russell, D., Moore, P., Eds.; Springer: Cham, Switzerland, 2020; pp. 179–186.

154. Efendi, R.; Jama, J.; Yulastri, A. Development of Competency Based Learning Model in Learning Computer Networks. *J. Phys. Conf. Ser.* **2019**, *1387*, 012109. [CrossRef]

155. Combéfis, S.; de Moffarts, G. Reinventing Evaluations with Competency Based Assessments: A Practical Experiment with Future Computer Science Engineers. In Proceedings of the 2020 IEEE Frontiers in Education Conference (FIE), Uppsala, Sweden, 21–24 October 2020; pp. 1–5.

156. Caratozzolo, P.; Membrillo, J. Evaluation of Challenge Based Learning Experiences in Engineering Programs: The Case of the Tecnologico de Monterrey, Mexico. In *Visions and Concepts for Education 4.0. ICBL 2020. Advances in Intelligent Systems and Computing*, 1st ed.; Auer, M.E., Centea, D., Eds.; Springer: Cham, Switzerland, 2021; Volume 1314, pp. 419–428.

157. Acakpovi, A.; Nutassey, K. Adoption of competency based education in TVET Institutions in Ghana: A case study of Mechanical Engineering Department, Accra Polytechnic. *Int. J. Voc. Tech. Educ.* **2015**, *7*, 64–69.

158. Mace, K.L.; Welch, C.E. The future of health professions education: Considerations for competency-based education in athletic training. *Athl. Train. Educ. J.* **2019**, *14*, 215–222. [CrossRef]

159. Gervais, J. The operational definition of competency-based education. *J. Competency-Based Educ.* **2016**, *1*, 98–106. [CrossRef]

160. Filipowicz, G. Zarządzanie kompetencjami. In *Perspektywa Firmowa i Osobista Warszawa*, 1st ed.; Wolters Kluwer SA: Alphen aan den Rijn, The Netherlands, 2014.

161. Ramirez, R.A.; Morales, R.; Iqbal, H.; Parra, R. Engineering Education 4.0: -proposal for a new Curricula. In Proceedings of the 2018 IEEE Global Engineering Education Conference (EDUCON), Santa Cruz de Tenerife, Spain, 17–20 April 2018.

162. Armstrong, M.; Taylor, S. *Armstrong's Handbook of Human Resource Management Practice*, 13th ed.; Kogan Page: London, UK, 2014.

163. Mertens, L. *Labor Competition: Systems, Emergence and Models*, 1st ed.; Cinterfor/OIT: Montevideo, Uruguay, 1996.

164. Galdeano, C.; Valiente, A. Professional competencies. *Educ. Quím Univ. Nac. Autónoma De México* **2010**, *21*, 28–32.

165. Crisol, E. Using Active Methodologies: The students' view. *Procedia Soc. Behav. Sci.* **2017**, *237*, 672–677.

166. Navas, N.A. Active methodologies and the nurturing of students' autonomy, Semina: Ciências Sociais e Humanas. *Londrina* **2011**, *32*, 25–40.

167. Baena, F.; Guarin, A.; Mora, J.; Sauza, J.; Retat, S. Learning factory: The path to industry 4.0. *Procedia Manuf.* **2017**, *9*, 73–80. [CrossRef]

168. Prinz, C.; Morlock, F.; Freith, S.; Kreggenfeld, N.; Kreimeier, D.; Kuhlenkötter, B. Learning factory modules for smart factories in industrie 4.0. *Procedia CiRp* **2016**, *54*, 113–118. [CrossRef]

169. Labrador, J. *Active methodologies*, 1st ed.; Polytechnic University of Valencia: Valencia, Spain, 2008.

170. De Miguel, M. Teaching Modalities Focused on the Development of Competences: Orientations to Promote Methodological Change in the European Higher Education Area. Project EA2005–0118, 1st ed.Ediciones Universidad de Oviedo: Oviedo, Spain, 2005.

171. López, E.J.; Flores, M.A.; Sandoval, G.L.; Velázquez, B.L.; Vázquez, J.J.D.; Velasquez, L.A.G. Reverse Engineering and Straight-forward Design as Tools to Improve the Teaching of Mechanical Engineering. In *Industry Integrated Engineering and Computing Education*, 1st ed.; Abdulwahed, M., Bouras, A., Veillard, L., Eds.; Springer: Cham, Switzerland, 2019; pp. 93–118.

172. Navarro, L.; Jiménez, E.; Bojórquez, I.; Ramírez, G. Competencies and teaching strategies: Experiences at the La Salle University Norwest. In *Educación Handbook T-I*, 1st ed.; Ecorfan: Guanajuato, Mexico, 2013; pp. 101–113.

173. Lin, K.Y.; Wu, Y.T.; Hsu, Y.T.; Williams, P.J. Effects of infusing the engineering design process into STEM project-based learning to develop preservice technology teachers' engineering design thinking. *Int. J. STEM Educ.* **2021**, *1*, 1. [CrossRef]

174. Liu, Y.; Whitaker, S.; Hayes, C.; Logsdon, J.; McAfee, L.; Parker, R. Establishment of an experimental-computational framework for promoting Project-based learning for vibrations and controls education. *Int. J. Mech. Eng. Educ.* **2022**, *50*, 158–175. [CrossRef]

175. Pandian, S.R. Intelligent Mechatronic Technologies for Green Energy Systems. In Proceedings of the 2010 ASEE Gulf-Southwest Annual Conference, McNeese State University, Lake Charles, LA, USA, 24–26 March 2010; pp. 1–12.

176. Atef, A. Project-Based Learning in Mechatronics Engineering: Modelling and development of an autonomous wheeled mobile robot for firefighting. *Eurasia Proc. Educ. Soc. Sci.* **2016**, *4*, 198–204.

177. Zarte, M.; Pechmann, A. Implementing an Energy Management System in a Learning Factory—A Project-Based Learning Approach. *Procedia Manuf.* **2020**, *45*, 72–77. [CrossRef]

178. Abele, E. Learning factory. In *CIRP Encyclopedia of Production Engineering*; Springer: Berlin/Heidelberg, Germany, 2016; pp. 1–5.

179. Roll, M.; Ifenthaler, D. Learning Factories 4.0 in technical vocational schools: Can they foster competence development? *Empir. Res. Voc. Ed. Train.* **2021**, *13*, 20. [CrossRef]

180. Tirado, L.; Estrada, J.; Ortiz, R.; Solano, H.; González, J.; Alfonso, D.; Restrepo, G.; Delgado, J.; Ortiz, D. Professional competencies: A strategy for the successful performance of industrial engineers. *Rev. Fac. Ing. Univ. Antioq.* **2007**, *40*, 123–139.

181. Granrath, L. Large Scale Optimization Is Needed for Industry 4.0 and Society 5.0. In *Optimization in Large Scale Problems. Springer Optimization and Its Applications*, 1st ed.; Fathi, M., Khakifirooz, M., Pardalos, P., Eds.; Springer: Cham, Switzerland, 2019; Volume 152, pp. 3–6.

182. Qin, S.F.; Cheng, K. Future Digital Design and Manufacturing: Embracing Industry 4.0 and Beyond. *Chin. J. Mech. Eng.* **2017**, *30*, 1047–1049. [CrossRef]

183. Sharma, A.; Yadava, G.S.; Deshmukh, S.G. A literature review and future perspectives on maintenance optimization. *J. Qual. Maint. Eng.* **2011**, *17*, 5–25. [CrossRef]

184. Al, B.; Algabroun, H.; Jonsson, M. Maintenance 4.0 to fulfill the demands of Industry 4.0 and Factory of the Future. *Int. J. Eng. Res.* **2018**, *8*, 20–31.

185. Almagor, D.; Lavid, D.; Nowitz, A.; Vesely, E. *Maintenance 4.0 Implementation Handbook*, 1st ed.; Reliabilityweb Inc.: Fort Myers, FL, USA, 2020.

186. Thames, L.; Schaefer, D. Cybersecurity for Industry 4.0 and Advanced Manufacturing Environments with Ensemble Intelligence. In *Cybersecurity for Industry 4.0. Springer Series in Advanced Manufacturing*, 1st ed.; Thames, L., Schaefer, D., Eds.; Springer: Cham, Switzerland, 2017; pp. 243–265.

MDPI

*Article*

# Additively Manufactured Robot Gripper Blades for Automated Cell Production Processes

Ferdinand Biermann [1], Stefan Gräfe [1], Thomas Bergs [1,2] and Robert H. Schmitt [1,2,*]

[1]  Fraunhofer Institute for Production Technology IPT, 52074 Aachen, Germany
[2]  Laboratory for Machine Tools and Production Engineering (WZL), RWTH Aachen University, 52074 Aachen, Germany
*  Correspondence: robert.schmitt@ipt.fraunhofer.de

**Abstract:** The automation of cell production processes demands strict requirements with regard to sterility, reliability, and flexibility. Robots work in such environments as transporting devices for a huge variety of disposables, e.g., cell plates, tubes, cassettes, and other objects. Therefore, the blades of their grippers must be designed to hold all of these different materials in a stable, gentle manner, and in defined positions, which means that the blades require complex geometries. Furthermore, they should have as few edges as possible, so as to be easy to clean. In this report, we demonstrate how these requirements can be met by producing stainless steel robot grippers by additive manufacturing.

**Keywords:** laboratory processes automation; additive manufacturing; robotic production; laser powder bed fusion

**Citation:** Biermann, F.; Gräfe, S.; Bergs, T.; Schmitt, R.H. Additively Manufactured Robot Gripper Blades for Automated Cell Production Processes. *Processes* **2022**, *10*, 2080. https://doi.org/10.3390/pr10102080

Academic Editor: Sergey Y. Yurish

Received: 13 August 2022
Accepted: 11 October 2022
Published: 14 October 2022

**Publisher's Note:** MDPI stays neutral with regard to jurisdictional claims in published maps and institutional affiliations.

## 1. Introduction

Cell-based therapies are gaining a growing interest in research and industry as they promise efficient therapies for diseases that are hardly or not at all curable today. In order to decrease their production cost as well as to increase quality and product safety, research and industry are putting more and more effort in fully automating these therapies. Examples are the projects StemCellFactory [1], iCellFactory [2], and AutoCRAT, all of which involve cell production platforms that cultivate, modify, and analyze stem cells fully automatically. In the AutoCRAT project, the platforms AUTOSTEM [3] and StemCellDiscovery [4] are being enhanced in order to enable novel cell production and analysis processes.

Many of such production platforms consist of an arrangement of several automated biotechnological devices, which perform the different processing or analysis steps, and a centrally positioned robot, which transfers all needed objects between the stations. These objects can be all kinds of cell plates, tubes, flasks, pipettes, and other components, all of which have different shapes and sizes. In a manual lab, handling of such objects is no problem, as they are developed to be handled by humans. A human hand has flexibility, adaptivity, and a variety of sensory inputs that cannot be attained by most industrial robot grippers today. To the best of the authors' knowledge, it is not possible to design robot grippers and gripper blades that can handle all of the existing objects used in biotechnological labs. For reliable handling, robot gripper blades need to be custom designed to reliably hold the objects required for the specific production process [5]. Even when the object variety is minimized, several object types with different geometries remain in most cases. In order to solve these issues, one or several of the following three aspects need to be implemented:

- Shape the gripper blades in a way to handle different object sizes and geometries. Depending on the shape and quantity of the different object types, the grippers become more complex and difficult to manufacture.
- Use objects that have standardized surfaces for robot handling. Those are, at least for now, not commercially available for all standard sizes and shapes. Therefore,

custom-made objects or adapter pieces that can be fixed onto the objects need to be developed. Both increase cost and presumably add effort in material preparation.

- Use several grippers or gripper blades, which the robot can change between handling actions. However, this increases technical complexity as well as processing time.

To save production time and cost, the first solution should be preferred. The effort for designing and manufacturing complex grippers needs to be invested only once. Hereafter, they amortize themselves.

Another challenge of processing in biotechnological environments is cleanliness and sterility. In order to avoid unwanted dead particles and, even more critically, living organisms and organic structures, strict requirements are defined for such environments in standards and guidelines, the most important of which are the good manufacturing practice (GMP) guidelines [6–8]. They demand that air and surfaces be kept antiseptic, especially those that are close to the biologic product, and that processes, materials, movements, and air flows protect the product from contaminants as best as possible. This also implies that all relevant surfaces must be easy to clean and decontaminate; therefore, they must be resistant against the required cleaning agents and sterilization gases, and have a geometry that allows fast and reliable particle removal. With regard to the latter, surfaces should have as few holes or gaps as possible, rounded corners, and a low roughness, which is, according to the state of the art, Ra $\leq$ 0.8 µm [9].

The ideal material for robot gripper blades in biotechnological applications is stainless steel, as it provides the stability and durability to be used reliably for a long time period, and can withstand most chemical cleaning and sterilizing agents. As of now, the most common way to produce such tools is by subtractive manufacturing, i.e., milling. Modern milling machines have up to five axes, are driven by computerized numerical control (CNC), and are able to produce a huge variety of complex geometries. However, they reach their limits when it comes to undercuts, hollow inclusions, complex radii and curvatures, or fine geometries such as rips or grids [10].

Additive manufacturing (AM) is a novel method for producing metal objects. Currently, several AM technologies are in development, for example, laser metal deposition (LMD) with wire or powder, or laser powder bed fusion (LPBF).

The process principle of LPBF is shown in Figure 1. Laser radiation is used to create a weld track in a powder bed by melting the powder to solid material. Several weld tracks next to each other result in a layer. Hereafter, the powder bed is moved down by the thickness of one layer, and a new layer of powder from the powder reservoir is placed on the previously welded tracks by a recoater. Several melted layers on top of each other result in a solid metal body. In this way, a 3D-printed metal component is created layer by layer from a sliced computer-aided design (CAD) file on top of a build plate [11–14]. Support structures are added to the part for the purpose of attaching the part to the base plate, for stabilization of overhangs, and for heat dissipation. The support structure needs to be removed afterwards, e.g., by milling or grinding processes, or by manual material removal [13,15].

The layered structure of the components allows much greater geometric design freedom compared to conventional manufacturing technologies [12]. This also makes it possible to round corners and edges easily. In milling, most radii are formed by producing a sharp edge first and then shaping the radius with a special tool. If the object has many radii or a complex curvature, it is difficult to clamp in a stable and defined position. In additive manufacturing, the round edge can be produced immediately, which makes such geometries much simpler to manufacture. However, this may make the addition of support structures necessary.

Another advantage of the opportunities AM provides in geometric design is weight reduction. Many gripper devices, despite being able to lift heavy objects, have limits on the weight of the gripper blades. The reason for this is the momentum of the blades on their bearings, which are more sensitive when opening or closing the gripper. One way to reduce weight is by removing unneeded material wherever possible, e.g., by decreasing thickness and using ribs to maintain stability. Another option is to create hollow sections inside of the component. However, this is not possible for every additive manufacturing technique. In LPBF, for example, every spot that is not welded remains covered with powder; consequently, every hollow section in the finished part is filled with powder. If it is not possible to remove the powder, weight can only be reduced slightly.

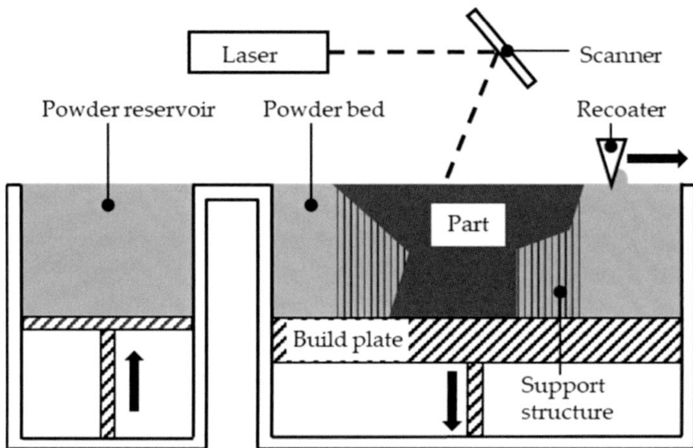


**Figure 1.** Process principle of laser powder bed fusion.

The roughness of the part varies depending on the orientation of the surface during the build-up. Top-skin (upward-facing surfaces during build process) and side surfaces have lower roughness and better quality compared to down-skin surfaces (downward-facing surfaces during build process) [16,17]. The printing tolerance of the parts can reach $\pm 0.1$ mm; however, it can increase, e.g., due to thermal distortion to $\pm 0.5$ mm or more [14]. Post-machining processes increase surface roughness and accuracy. Common post-processes are sand or glass-bead blasting, milling (also fine milling), grinding, and polishing.

In our research, we have used the additive metal manufacturing technology LPBF to produce robot gripper blades for the automated handling of objects in biotechnology. The gripper blade design enables the gripper to hold several materials with different geometries at different positions. Furthermore, we produced several design proposals that are shaped to improve stability and simplify cleaning.

## 2. Materials and Methods

The gripper blades discussed in this article are designed to be used in the cell production platform StemCellDiscovery [4], mounted on a robot (VS-087, Denso, Kariya, Japan) with servo grippers (SG-00014, P.T.M. Präzisionstechnik GmbH, Gröbenzell, Germany). In order to perform several cell cultivation and analysis processes, several vessels (Table 1) need to be held and transported by the robot. The variety of these vessels has been reduced to a minimum, but some geometries in different orientations still need to be held. These are 50 mL falcon tubes, 5 mL corning tubes, micro titer plates (MTP), pipette holder plates for the liquid handling unit (Microlab STAR, Hamilton Company, Reno, NV, USA), and test cartridges for an endotoxin tester (Endosafe® nexgen-PTS™, Charles River Laboratories, Wilmington, MA, USA) of the platform, as shown in Table 1 and Figure 2.

**Table 1.** Disposables used in the StemCellDiscovery which need to be transported by the robot.

| No. | Name | Type and Manufacturer | Gripping From |
|---|---|---|---|
| A | 50 mL falcon tube | 50 mL Centrifuge tubes, 525-0156, VWR, Radnor, PA, USA | side, top |
| B | 5 mL corning tube | Vial Cryogenic Classic 5.0 mL, LW3338, Alpha Laboratories, Hampshire, UK | side |
| C | micro titer plate (MTP) | VWR Tissue Culture Plates 6 wells, Surface treated, Sterile, VWR, Radnor, PA, USA | side, top |
| D | pipette holder | High Volume Tips (1000 µL), Hamilton Company, Reno, NV, USA | side, top |
| E | endotoxin test cartridge | Rapid Test Cartridges for LAL & Beta-D-Glucan Detection Assays, Charles River Laboratories, Wilmington, MA, USA | side |

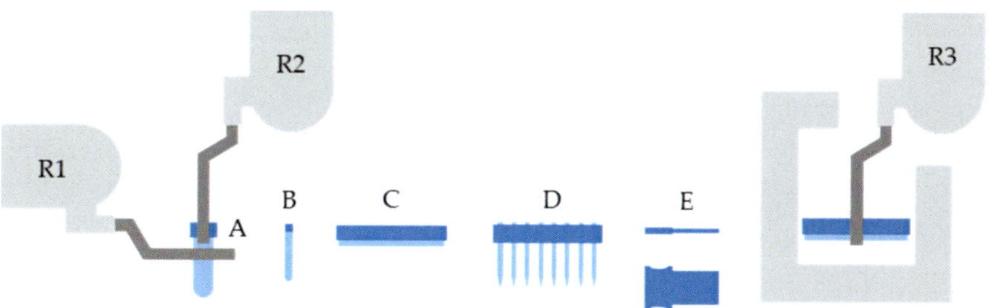


**Figure 2.** Robot grippers holding the following objects from the side (R1) or from the top (R2): 50 mL falcon tubes (A), 5 mL corning tubes (B), micro titer plates (MTP) (C), pipette holders (D), and Charles River Endotoxin test cartridges (E, shown from side and top view). (R3) shows an example of a difficult position for the grippers to reach.

Furthermore, the gripper blades may not exceed a certain size, and must be shaped to fit in all areas of the platform and to avoid collisions. For example, an MTP needs to be placed onto a microscope that barely leaves space for the robot to move. A schematic visualization of that area is shown in Figure 2, R3.

Based on these requirements, the gripper blades were modeled in 3D CAD (Figure 3). On the left side of Figure 3, the first version of the gripper blades is shown, which was designed for conventional production. This design is suitable for manufacturing by milling and die-sinking electrical discharge machining (EDM). It contains geometries for holding round objects, such as tubes, as well as flat objects, such as plates, and special objects, such as test cartridges. Furthermore, the design has an angled geometry, a mounting part to fix them to the robot gripper, and holes to insert rubber parts which improve the grip's ability to hold objects. However, its shape shows many even surfaces and edges, as such a design is easy and efficient to mill. On the right side of Figure 3, the gripper blades are improved by redesigning them in CAD, so that they can be additively manufactured using LPBF. Here, edged geometries are rounded wherever possible, making the gripper much easier to clean and much lighter, without losing stability. This design would be very challenging for milling, but is suitable for additive manufacturing. The outer dimensions of the gripper blades are shown in Figure 4.

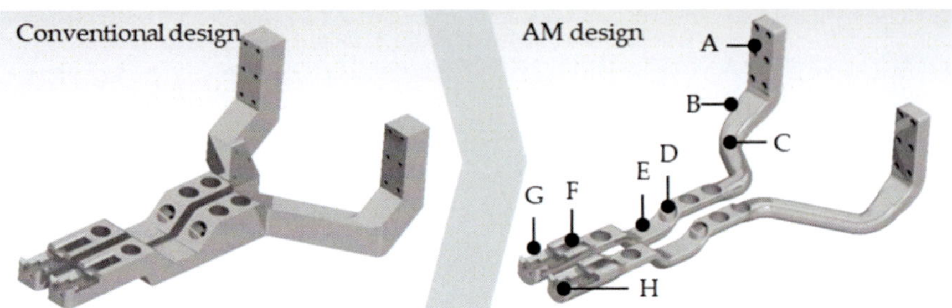



**Figure 3.** Conventional (**left**) and AM design (**right**) of gripper blades with mounting part (A), radii for simple cleaning (B), angled shape to reach places with limited space (C), mounting holes for rubber pieces to increase grip (D), diamond shape to hold tubes (E), pocket and circular geometry for fitting endotoxin testing cartridge (F), rounded shape to hold tubes horizontally (G), and edges to safely hold tubes and plates from the top (H).

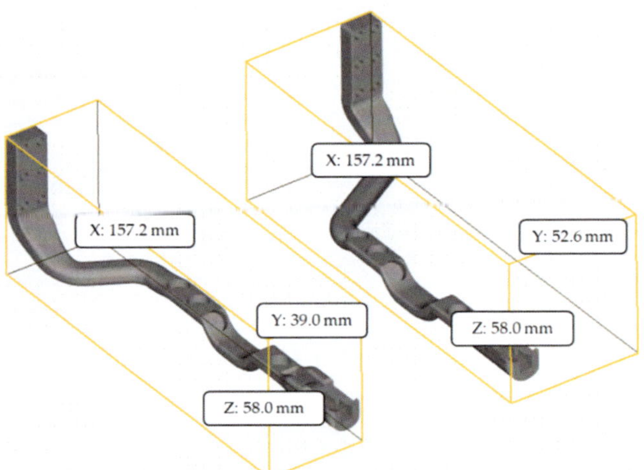



**Figure 4.** Outer dimensions of the gripper blades.

Based on this 3D data, the gripper blades were produced with the LPBF process out of stainless steel SS316L powder with a particle size of 10–45 μm (LPW Technology Ltd., Widnes, UK). The machine tool used was an EOS M290 (EOS GmbH Electro Optical Systems, Krailing, Germany). In order to achieve a good quality of the specific geometrical elements (Figure 3, D–H), these elements were placed face up on the build plate. The orientation and the support structures used are shown in Figure 5A. The block support used (Figure 5A, blue) includes cone supports, which create a strong connection between gripper and base plate and increase the thermal flow, therefore reducing the distortion of the grippers. Block support can be removed easily; thus, the part does not need to be machined completely after printing. After defining placement and support structures of the part, process parameters were set. In order to achieve a high material density, process parameters with an energy density of 57.7 J/mm$^3$ were chosen. The laser power was 214.2 W, the scanning speed was 928.1 mm/s, and the hatch distance was 100 μm. The build-up was carried out with a layer thickness of 0.04 mm, and the process required 8.5 h to produce one set of gripper blades.

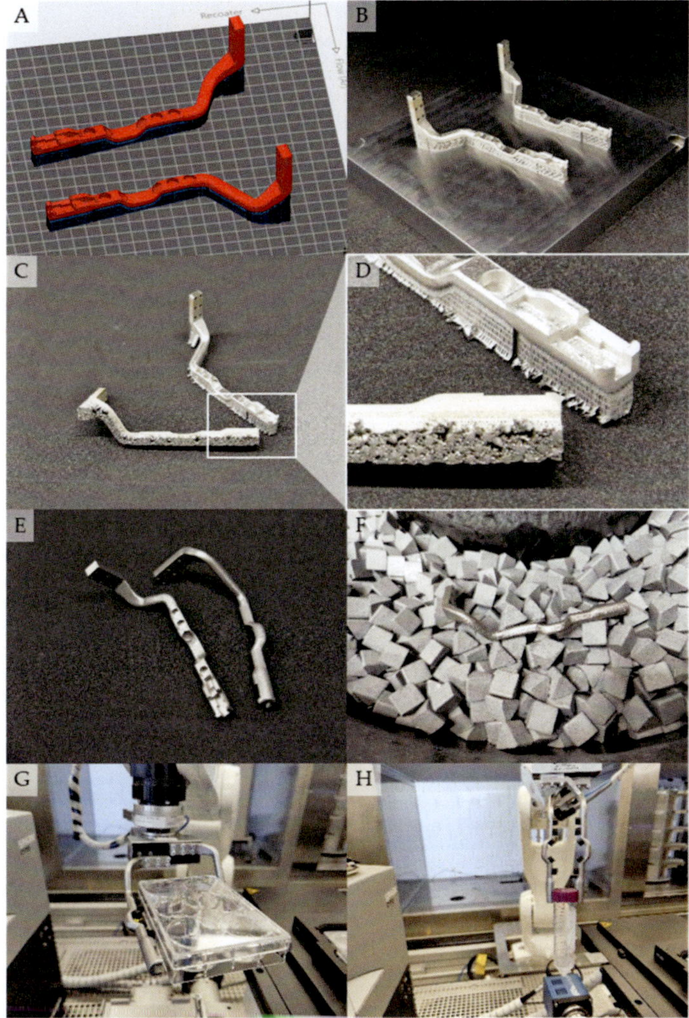

**Figure 5.** Stages of the gripper blade production: (**A**) placement of the 3D model (red) onto the base plate (grey) and addition of support structure (blue and black), (**B**) gripper blades after LPBF production), (**C**) blades after removal from base plate, (**D**) magnified view of the support structures, (**E**) blades after support structure removal and glass bead blasting, (**F**) blades in the tumbling machine, (**G**) blades assembled onto the robot, holding a micro titer plate, (**H**) blades holding a 50 mL falcon tube.

After production (Figure 5B), the gripper blades were sawed from the base plate (Figure 5C) and the support structures were removed by hand. Afterwards, the corresponding surfaces were ground. Then, the parts were sand and glass-bead blasted, and, finally, tumbled to smoothen their surface (Figure 5D). After production, glass bead blasting, and tumbling, the surface roughness was measured with an Alicona Infinite Focus G5 (Alicona Imaging GmbH, Raaba, Austria). Finally, fine structures, such as small holes, were added or revised manually, where necessary.

The gripper blades were equipped with rubber pieces in order to increase their grip, and then installed onto the robot gripper of the StemCellDiscovery and tested for handling the intended objects (Figure 5E,F).

## 3. Results

It took 8.5 h to produce the gripper blades with LPBF. Processing and slicing of the 3D CAD data depends on the computers used and the operator's skills. For the given sample, the required preparation time was 60 min. After the printing process, support structures were removed and the surfaces were blasted and tumbled for four hours, which took, in total, another five hours. The times for manufacturing the gripper blades shown in Figure 3 (left) with milling were discussed with experts in that field, and were calculated as 16 h for machine preparation and programming, 20 h for machining, and 1.5 h for deburring. All of these times are shown in Figure 6 for comparison.

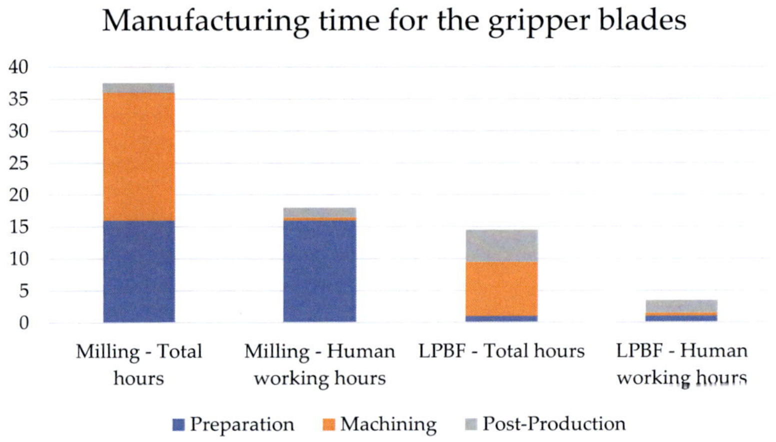

**Figure 6.** Comparison of the manufacturing time for the gripper blades with milling and LPBF in total and in human labor time.

After printing, only the mounting holes had to be postprocessed by drilling to remove small material residues. In addition, the parts showed a light deformation after being loosened from the base plate, which was corrected by manual bending of the parts. Apart from this, the part showed all geometrical features which had been designed in CAD. Even fine geometries, such as the hooks at the tips of the gripper blades, had been produced in sufficient detail.

The surface of the gripper after LPBF was very rough. The measured roughness of the surface after LPBF production was Ra = 5.9 µm. Sand and glass-pearl blasting reduced it to Ra = 3.4 µm. Tumbling reduced the roughness to Ra = 2.3 µm.

When the volume of the conventional design was compared with the presented design in CAD (Figure 3), weight is reduced by 58% in the additively manufactured part. This was confirmed by weighing the real parts. The conventional design parts had a mass of 228 g and 234 g, while the presented design parts weighed 96 g and 99 g, resulting in weight reductions of 58% in both.

## 4. Discussion

The robot gripper blades presented here are designed to be able to hold several object types in different orientations.

The parts have a smooth geometry and very few edges, both of which limit space for contaminants to remain and simplify cleaning. However, even glass-pearl blasting and tumbling could not reduce the surface roughness down to the required value of Ra $\leq$ 0.8 µm. Here, further methods for surface polishing need to be investigated, for which different technologies for mechanical, chemical, or laser-based polishing exist and can be applied [18–22].

The deformation that the parts showed after being removed from the base plate results from thermal stresses during the printing process. These can be prevented by adapting the 3D model before printing, by bending it in the other direction, so that the thermal stresses then form it into the desired geometry.

The only other necessary post-processing was removing small residues from the mounting holes. These holes are intended for M3 screws and are very narrow (i.e., a diameter of 3 mm) to ensure the correct position of the gripper blade on the gripper. In other applications, where the through hole can be bigger in relation to the intended screw size, the holes will probably not need to be postprocessed.

Compared to a model being designed for milling, the design presented here shows a great weight reduction. It would theoretically be possible to reduce volume and weight even further without affecting the LPBF production difficulty or speed; however, this may lead to less smooth surfaces.

The comparison of the manufacturing times with LPBF and milling displayed in Figure 6 shows that the total milling process takes more than twice as long as LPBF. For both methods, great amounts of the manufacturing time consist of machining time, during which processes such as LPBF, milling, or tumbling can run unsupervised. Therefore, the human labor time required is much lower in both cases, as shown in Figure 6. It can also be seen that the human labor time for LPBF is much lower than for milling, which is mainly due to its lower preparation time.

If this or a similar geometry were manufactured by milling, it would need to be milled from a large metal block, 94% of which would be wasted. Another option would be to split it into several parts that would need to be bolted or welded thereafter. This would require additional shapes for mounting, and would lower the geometric accuracy due to additional manufacturing tolerances of the assembling step. In addition, bolted parts can loosen over time, reducing the gripper blades' reliability or even making them completely unusable.

In both cases, more preparation would be necessary before milling, e.g., CNC programming and loading the corresponding milling tools into the machine. Furthermore, it is likely that some clamps would need to be modified or produced beforehand to hold the part during milling. For milling, the part would need to be processed from several sides, requiring manual reclamping steps. Some geometries, such as the radii (Figure 3, B), would be challenging to manufacture, while others, such as the cartridge pockets (Figure 3, F), would not be possible at all in this form. After milling, however, few or no post-processing steps would be necessary.

A third option for producing such parts would be injection molding or sintering. Both manufacturing techniques require extensive preparations for creating the molds, and are only economic for large part quantities. Moreover, here, some post-processing for removing residues from gate and feeder and milling surfaces with low tolerances are necessary.

For the purpose of manufacturing parts with a geometry similar to those presented here, additive manufacturing techniques such as LPBF appear to be the most efficient. The preparation effort is low, and although the manufacturing time may be long, this method does not require manual interventions and can run automatically and unsupervised. Further advantages of LPBF are a great freedom in geometry and much lower waste of material. Especially for prototyping or custom-made gripper blades, this method shows great promise. If, for example, the geometry of a tube, flask, or plate changes or a new object needs to be handled, new adapted 3D-printed gripper blades can be manufactured easily without much preparation or programming effort. Radii and curvatures are much simpler to realize with LPBF, which simplifies cleaning. In addition, undercuts and pockets can be manufactured easily, which provides many more options for geometries to grip and hold objects.

Two major weakness of LPBF compared to milling are manufacturing tolerances and surface roughness. Where needed, these surfaces should be smoothed after printing, e.g., by means of grinding, polishing, sandblasting, or glass-bead blasting. If most surfaces of the parts have such requirements, manufacturing by milling may be more efficient.

Furthermore, milling or injection molding is likely to be cheaper when it comes to mass production. On the other hand, the research in additive manufacturing of metal parts is developing quickly, and may one day become the gold standard for manufacturing all kinds of robot grippers in biotechnological applications.

Especially for biotechnological applications of robot gripper blades, LPBF is a suitable and efficient manufacturing process. Due to strict guidelines, stainless steel is the ideal material for gripper blades, and a curved and rounded geometry simplifies cleaning and decontamination. Biotechnological processes can require handling a high variety of different objects, some of which are not even designed for robotic handling. The freedom in geometric design possible with LPBF enables a user to create one set of gripper blades that is able to hold all required objects reliably. This has been proven with successful tests of the gripper blades presented in this paper, which were used in the StemCellDiscovery to hold and transport all required objects. Further advantages are lightweight designing, flexible adaption, and unsupervised production.

In summary, the LPBF production method shows great potential for manufacturing complex geometries such as robot gripper blades in automated biotechnological environments. Once remaining issues, such as the surface structure, are solved, LPBF could become the ideal manufacturing method for such devices.

**Author Contributions:** Conceptualization, F.B.; writing—original draft preparation, F.B. and S.G.; writing—review and editing, F.B. and S.G.; supervision, T.B. and R.H.S.; project administration, R.H.S.; funding acquisition, T.B. and R.H.S. All authors have read and agreed to the published version of the manuscript.

**Funding:** The project AutoCRAT has received funding from the European Union's Horizon 2020 research and innovation program under grant agreement No 874671. The materials presented and views expressed here are the responsibility of the author(s) only. The EU Commission takes no responsibility for any use made of the information set out.

**Institutional Review Board Statement:** Not applicable.

**Informed Consent Statement:** Not applicable.

**Data Availability Statement:** Not applicable.

**Conflicts of Interest:** The authors declare no conflict of interest.

# References


1. Elanzew, A.; Nießing, B.; Langendoerfer, D.; Rippel, O.; Piotrowski, T.; Schenk, F.; Kulik, M.; Peitz, M.; Breitkreuz, Y.; Jung, S.; et al. The StemCellFactory: A Modular System Integration for Automated Generation and Expansion of Human Induced Pluripotent Stem Cells. *Front. Bioeng. Biotechnol.* **2020**, *8*, 580352. [CrossRef]
2. Brecher, C.; Herfs, W.; Malik, A.; Valest, C.; Wein, S.; Wanek, P.; Zenke, M. Adaptive Automatisierung für die Zellproduktion: Mechatronische Rekonfigurierbarkeit von Produktionsanlagen. *GIT Labor-Fachz.* **2016**, *60*, 27–39.
3. Ochs, J.; Hanga, M.P.; Shaw, G.; Duffy, N.; Kulik, M.; Tissin, N.; Reibert, D.; Biermann, F.; Moutsatsou, P.; Ratnayake, S.; et al. Needle to needle robot-assisted manufacture of cell therapy products. *Bioeng. Transl. Med.* **2022**, *7*, e10387. [CrossRef]
4. Ochs, J.; Biermann, F.; Piotrowski, T.; Erkens, F.; Nießing, B.; Herbst, L.; König, N.; Schmitt, R.H. Fully Automated Cultivation of Adipose-Derived Stem Cells in the StemCellDiscovery—A Robotic Laboratory for Small-Scale, High-Throughput Cell Production Including Deep Learning-Based Confluence Estimation. *Processes* **2021**, *9*, 575. [CrossRef]
5. Biermann, F.; Mathews, J.; Nießing, B.; König, N.; Schmitt, R. Automating Laboratory Processes by Connecting Biotech and Robotic Devices—An Overview of the Current Challenges, Existing Solutions and Ongoing Developments. *Processes* **2021**, *9*, 966. [CrossRef]
6. European Commission. EudraLex—Volume 4—Good Manufacturing Practice (GMP) Guidelines. Available online: https://ec.europa.eu/health/documents/eudralex/vol-4 (accessed on 3 May 2021).
7. 13.040.35 (DIN EN ISO 14644-1:2015); Reinräume und Zugehörige Reinraumbereiche—Teil 1: Klassifizierung der Luftreinheit anhand der Partikelkonzentration. DIN Deutsches Institut für Normung e.V.: Berlin, Germany, 2016.
8. 13.040.35 (DIN EN 17141:2021); Reinräume und Zugehörige Reinraumbereiche—Biokontaminationskontrolle. DIN Deutsches Institut für Normung e.V.: Berlin, Germany, 2021.
9. Bobe, U.; Wildbrett, G. Anforderungen an Werkstoffe und Werkstoffoberflächen bezüglich Reinigbarkeit und Beständigkeit. *Chem. Ing. Tech.* **2006**, *78*, 1615–1622. [CrossRef]

10.  Pereira, T.; Kennedy, J.V.; Potgieter, J. A comparison of traditional manufacturing vs. additive manufacturing, the best method for the job. *Procedia Manuf.* **2019**, *30*, 11–18. [CrossRef]
11.  Lachmayer, R.; Lippert, R.B. (Eds.) *Additive Manufacturing Quantifiziert: Visionäre Anwendungen und Stand der Technik*; Springer Vieweg: Berlin/Heidelberg, Germany, 2017; ISBN 9783662541128.
12.  Gebhardt, A. *Additive Fertigungsverfahren: Additive Manufacturing und 3D-Drucken für Prototyping—Tooling—Produktion*, 5th ed.; neu bearbeitete und erweiterte Auflage; Hanser: München, Germany, 2016; ISBN 3446444017.
13.  Sander, J. *Selektives Laserschmelzen Hochfester Werkzeugstähle*; Saechsische Landesbibliothek-Staats-und Universitaetsbibliothek Dresden; Technische Universität Dresden: Dresden, Germany, 2018.
14.  Scherer, T.M. *Beanspruchungs-und Fertigungsgerechte Gestaltung Additiv Gefertigter Zerspanungswerkzeuge*; Shaker Verlag: Düren, Germany, 2020.
15.  VDI-Gesellschaft Produktion und Logistik. *Additive Fertigungsverfahren: Grundlagen, Begriffe, Verfahrensbeschreibungen: Basics, Definitions, Processes = Additive Manufacturing Processes, Rapid Manufacturing*; VDI-Richtlinien VDI 3405: Berlin, Germany, 2014.
16.  Strano, G.; Hao, L.; Everson, R.M.; Evans, K.E. Surface roughness analysis, modelling and prediction in selective laser melting. *J. Mater. Process. Technol.* **2013**, *213*, 589–597. [CrossRef]
17.  Fox, J.C.; Moylan, S.P.; Lane, B.M. Effect of Process Parameters on the Surface Roughness of Overhanging Structures in Laser Powder Bed Fusion Additive Manufacturing. *Procedia CIRP* **2016**, *45*, 131–134. [CrossRef]
18.  Lamikiz, A.; Sánchez, J.A.; López de Lacalle, L.N.; Arana, J.L. Laser polishing of parts built up by selective laser sintering. *Int. J. Mach. Tools Manuf.* **2007**, *47*, 2040–2050. [CrossRef]
19.  Tyagi, P.; Goulet, T.; Riso, C.; Garcia-Moreno, F. Reducing surface roughness by chemical polishing of additively manufactured 3D printed 316 stainless steel components. *Int. J. Adv. Manuf. Technol.* **2019**, *100*, 2895–2900. [CrossRef]
20.  Boschetto, A.; Bottini, L.; Macera, L.; Veniali, F. Post-Processing of Complex SLM Parts by Barrel Finishing. *Appl. Sci.* **2020**, *10*, 1382. [CrossRef]
21.  Ginestra, P.; Ceretti, E.; Lobo, D.; Lowther, M.; Cruchley, S.; Kuehne, S.; Villapun, V.; Cox, S.; Grover, L.; Shepherd, D.; et al. Post Processing of 3D Printed Metal Scaffolds: A Preliminary Study of Antimicrobial Efficiency. *Procedia Manuf.* **2020**, *47*, 1106–1112. [CrossRef]
22.  Kaynak, Y.; Tascioglu, E. Post-processing effects on the surface characteristics of Inconel 718 alloy fabricated by selective laser melting additive manufacturing. *Prog. Addit. Manuf.* **2020**, *5*, 221–234. [CrossRef]

 *processes*

*Article*

# Multiple Mobile Robots Coordination in Shared Workspace for Task Makespan Minimization

Jarosław Rudy [1], Radosław Idzikowski [1], Elzbieta Roszkowska [2,*] and Konrad Kluwak [1]

[1] Department of Control Systems and Mechatronics, Wrocław University of Science and Technology, 50-370 Wrocław, Poland
[2] Department of Cybernetics and Robotics, Wrocław University of Science and Technology, 50-372 Wrocław, Poland
* Correspondence: elzbieta.roszkowska@pwr.edu.pl

**Abstract:** In this paper we consider a system of multiple mobile robots (MMRS) and the process of their concurrent motion in a shared two-dimensional workspace. The goal is to plan the robot movement along given fixed paths so as to minimize the completion time of all the robots while ensuring that they never collide. Thus, the considered problem combines the problems of robot schedule optimization with collision and deadlock avoidance. The problem formulation is presented and its equivalent reformulation that does not depend explicitly on the geometry of the robot paths is proposed. An event-based solution representation is proposed, allowing for a discrete optimization approach. Two types of possible deadlocks are identified and deadlock avoidance procedures are discussed. We proposed two types of solving methods. First, we implemented two metaheuristics: the local-search-based taboo search as well as the population-based artificial bee colony. Next, we implemented 14 simple constructive algorithms, employing dispatch rules such as first-in first-out, shortest distance remaining first, and longest distance remaining first, among others. A set of problem instances for different numbers of robots is created and provided as a benchmark. The effectiveness of the solving methods is then evaluated by simulation using the generated instances. Both deterministic and lognormal-distributed uncertain robot travel times are considered. The results prove that the taboo search metaheuristic obtained the best results for both deterministic and uncertain cases, with only artificial bee colony and a few constructive algorithms managing to remain competitive. Detailed results as well as ideas to further improve proposed methods are discussed.

**Keywords:** discrete optimization; multiple mobile robots systems; multiple resources; uncertainty; simulation; metaheuristics

**Citation:** Rudy, J.; Idzikowski, R.; Roszkowska, E.; Kluwak, K. Multiple Mobile Robots Coordination in Shared Workspace for Task Makespan Minimization. *Processes* **2022**, *10*, 2087. https://doi.org/10.3390/pr10102087

Academic Editor: Sergey Y. Yurish

Received: 28 August 2022
Accepted: 12 October 2022
Published: 15 October 2022

**Publisher's Note:** MDPI stays neutral with regard to jurisdictional claims in published maps and institutional affiliations.

## 1. Introduction

The use of mobile robots in many processes is steadily increasing, driven by advances in robotics, improving autonomous vehicle design, as well as emergence of Industry 4.0 and similar paradigms. This includes the multiple mobile robots system (MMRS), which has a wide range of possible applications in many processes, starting from in-factory transport and drone-based vehicle inspection, through search and rescue operations, and forestry and agriculture to extraction of minerals and space exploration. In some MMRSs, the tasks are separate, but in most cases the tasks are connected in some way and the robots work as one group towards a common goal. In such situations, the entire operation is considered complete only when all of the tasks of all robots have been completed.

There are a few crucial issues that determine the practical usefulness of an MMRS. The first is the time taken to complete all the tasks, called the *makespan*. In some cases it is important to finish all the task before some given deadline. Even with no such requirement, the shorter the makespan the better since it decreases the operational costs or increases the effectiveness of the overall operation (which can directly translate to more lives saved during a search and rescue operation).

The second issue is to ensure that the robots never collide with each other. In some systems this can be solved by letting the robots deviate from their planned paths and bypass each other. Otherwise, the robot can simply wait for another one to pass before proceeding. Both solutions result in delays in task execution, affecting the makespan. Thus, it is important to avoid collision while also minimizing the resulting delays.

The final issue is the possibility of deadlocks occurring, which is especially dangerous as deadlocks prevent the task from being completed at all. Moreover, in a small two-dimensional workspace, deadlock of even two robots can quickly cause all of the robots to become stuck indefinitely, and the bigger the deadlock is, the more difficult it is to resolve. It is thus crucial to either prevent deadlocks or resolve them effectively once they occur.

Thus, any practical solving method for MMRS should ensure collision- and deadlock-free solution—usually through definition and properly carried out acquisition of resources by the robots—while also reducing the makespan as much as possible. All of the aforementioned problems are non-trivial, with detection of deadlocks and makespan optimization for most practical scheduling problems both being considered NP-hard. Naturally, those problems become more complex when tackled at once, while also affecting each other.

This paper is related to earlier research shown in papers [1–4], where control systems for an MMRS with robots moving in a shared two-dimensional workspace were considered. These works propose solutions that provide collision- and deadlock-free robot motion control for any arbitrary dispatch rule. However, the problem of the system efficiency optimization through robot motion scheduling is not considered. In the current paper, we aim to focus on this problem and propose a number of solving methods for MMRS to optimize the makespan while avoiding collisions and deadlocks among robots. We then test the effectiveness of these methods during a simulation using a larger number of problem instances we provide publicly. Moreover, we transform the original continuous-valued 2-D path-defined problem into a discrete optimization formulation that does not use the concept of 2-D paths. In the light of this, the main contributions of our paper can be summed up as follows:

1. We formulate a discrete mathematical model for the problem of makespan optimization for a robot movement scheduling system with possibility of deadlocks.
2. We propose a solution representation together with an evaluation method and avoidance procedure for two types of deadlocks.
3. We propose two metaheuristic solving methods for the considered problem.
4. We propose a publicly available set of problem instances to be used as a benchmark for this problem.
5. We research the effectiveness of the proposed methods against a number of simple constructive algorithms for both deterministic and uncertain cases.

The remainder of this paper is organized as follows. In Section 2, we present a short overview of related literature. In Section 3, we formulate the problem and present its convenient reformulation. In Section 4, we discuss solution representation and its deadlock-free evaluation. In Section 5, we propose several solving algorithms. Section 6 contains the results of a computer experiment. Finally, Section 7 contains conclusions.

## 2. Related work

We will start our overview with the aforementioned paper [1]. The authors combine the discrete event system introduced in [2] with a continuous time system, resulting in a hybrid control system. The discrete representation of the MMRS is obtained through the division of the robot paths into sectors, and the supervisory control, providing collision and deadlock free robot motion is expressed in the Petri net formalism. Petri nets were also employed by Kloetzer et al. [5], but with different assumptions: a robot could choose several alternative trajectories instead of single fixed path. Division into sectors occurred, but it was applied not to paths, but to the workspace itself, with the constraint that only a specific number of robots was allowed in the same sector at once.

A problem similar to that considered in [1], but with a specific approach to path division into sectors, was studied in papers [3,4]. Namely, the workspace was divided into a grid of squares, whose size was equal to or larger than the robot disk diameter. In such a system, the squares are considered as resources, and a robot always occupies (i.e., its disk overlaps) from one to four grid squares at a time. Then the robots' paths were partitioned into sectors such that the subset of resources occupied by a robot in any given point of any given sector was constant. Similar to [2], the problem of the paper was to ensure the correct concurrent robot motion rather than the optimization of the system efficiency.

Next, Gakuhari et al. [6] consider an MMRS with non-holonomic robots in an environment with obstacles. Global path planning is executed periodically, allowing adaption to environment changes and deadlock-free navigation, supplemented with the $A^*$ algorithm. The effectiveness of the navigation method is tested on several environment examples through simulation. Čáp et al. [7] apply an asynchronous decentralized approach called reverse prioritizing planning for coordination of robot movement, while also considering two types of conflicts. Experiments using complex examples with many robots showed that the method obtains a solution twice as fast as synchronous centralized approach. Ferrera et al. [8,9] also considered a decentralized robot navigation systems, but assumed semi-fixed paths: in general, robots try to stay on the main path, but employ "evasive maneuvers" when the robots get too close to each other. The authors also consider high density of robots.

We also note that metaheuristic algorithms are also used for some variants of MMRS. For example, Bhattacharjee et al. [10] employed artificial bee colony (ABC) in an environment with obstacles, while ensuring enough distance from obstacles was kept during motion. A similar ABC-based approach for online path planning and collision avoidance was considered by Liang and Lee [11]. As a last ABC-based example, Mansury et al. [12] considered paths for soccer robots based on Ferguson splines and compare it against genetic algorithm and particle swarm optimization variants. On the other hand, Kumar et al. [13] employed taboo search (TS) method for optimizing the movement of a single robot in complex environment. While interesting, only a single-robot single-environment case was presented. An interesting example of using the TS metaheuristic is a paper by Balan and Luo [14], where due to unknown or unstable environment a robot path is given as a number of waypoints. The robot has to travel to each subsequent waypoint in the shortest possible manner with the help of map building.

Lygeros et al. [15] considered an automated highway system for safe navigation of autonomous vehicles. The authors used hybrid controllers and game theory to guarantee safety, while also providing an upper bound on the achievable highway throughput. Tomlin et al. [16] considered a multiagent hybrid system for control of aircraft and resolution of of conflicts in air traffic. The system allows aircraft to have high flexibility by choosing their own trajectories and altitudes, but ensures no conflicts through safe zones. The authors provide examples, but limited to a few aircraft at once. Pecora et al. [17] considered the problem of vehicle trajectory planning. The system assumes minimal and maximal vehicle speed as well as requirements about floor space usage and deadlines of tasks. The employed approached of "trajectory envelopes" allows to obtain alternative execution patterns for the vehicles, but due to the constraints the obtained solutions are not always feasible. Moreover, the running time of the algorithm turned out to be exponential in the number of the vehicles. A similar general approach with trajectory envelopes was proposed by Andreasson et al. [18]. The approach was verified for a scenario of five forklifts in a factory setting. Grover et al. [19] performed an extensive analysis on characteristics of deadlock for MMRS and provided a provably correct decentralized algorithm for deadlock resolution and collision avoidance. However, only two- and three-robot examples were used to verify the approach.

Akella and Hutchinson [20] considered a makespan minimization for collision-free trajectory coordination of the MMRS. They proposed a mixed integer linear programming formulation for up to 20 robots and remarked that the problem remains NP-hard even with fixed trajectories. However, only a maximum of 80 collision zones per scenario were considered. Wang and Gombolay [21] proposed an interesting approach to robot coordination with specific time and location constraints using machine learning and graph attention networks. This allowed obtaining competitive results with greatly reduced computation time for scenarios of up to five robots and 100 tasks. A more comprehensive review of recent approaches for various types of MMRSs (including ground, underwater, and aerial robots) can be found in a paper by Lin et al. [22].

We will now move onto other, non-robot-related works employing multiple resources and resolving conflicts and deadlocks. Such problems are essentially discrete optimization problems, where decision has to be made regarding the order of allocation of resources to tasks. As such, they can be viewed as related to resource-constrained project scheduling (RCPS) or various scheduling and job scheduling problems. For example, De Frene et al. considered heuristics for a RCPS with spatial resources for construction projects [23]. A procedure is used to transform priority lists into appropriate precedence lists to avoid deadlocks. The authors consider a large number of dispatch rules and verify their effectiveness. Similarly, Prashant Reddy et al. [24] tackle a multi-mode multi-RCPS with preemption. The authors make use of Petri nets to model the problem and identify deadlocks as well as propose a genetic algorithm approach.

Considering job scheduling problems, Lawley et al. [25] tackled a flexible manufacturing system with buffer space and modeled it using discrete event system approach. Both job shop and flow shop settings were considered. The authors performed safety analysis to guarantee deadlock-free execution with polynomial complexity on constraints execution. The results also confirmed that first-come first-served dispatch rule is not enough in practice. Similarly, Golmakani et al. [26] considered a scheduling problem for flexible manufacturing cells with assumption of transportation, buffers, and precedence constraint. Automata theory is used to obtain deadlock-free solution for three classes of manufacturing cells, each verified through medium-sized numerical example.

Very interestingly, Sun et al. [27] considered a mix of job shop scheduling and mobile robots with the goal of minimizing makespan. The authors propose two approaches for deadlock avoidance and perform numerical tests. Result showed that the number of jobs has more impact on the makespan than the number of operations and that a fixed entrance strategy can significantly reduce computation time with little effect on makespan. A similar approach, considering the use of automated guided vehicles in a manufacturing systems was presented in [28]. The authors proposed taboo search and genetic algorithm metaheuristics, while using an enhanced mixed-integer programming as a reference point. However, the authors did not explicitly take path shapes into consideration and the considered scenarios were of moderate size.

Finally, regarding scheduling in networks and computer systems, Sun et al. [29] considered scheduling for high-performance computing with multiple resources. The authors compare two scheduling paradigms: list scheduling and pack scheduling and propose a method of transforming the problem into single-resource equivalent. Pack scheduling is shown to perform better in practice despite its worse theoretical properties. Gopalan and Kang [30] considered the problem of allocation of multiple resources for real-time operation systems and proposes a multiple resource allocation and scheduling (MURALS) algorithm to solve the problem.

To our best knowledge, the problem of deadlock avoidance in systems of resource sharing processes and the problem of optimal scheduling of their operations have been considered separately so far. In this paper, in the context of systems of mobile robots sharing their motion space, we consider both of these problems simultaneously, which is a significant novelty. Moreover, the above-discussed literature review shows that existing works on mobile robot coordination are limited to small numbers of robots, testing scenarios, and algorithms used. Thus, researchers and practitioners could benefit from more extensive research on effectiveness of algorithms, which we address in this paper.

## 3. Problem Formulation

In this section we will formulate the considered problem of robots coordination in a shared movement space for makespan minimization. Then we will reformulate the model to a form that does not uses the concepts of 2-D paths. Concerning notation, $\mathbb{R}_{>0}$ and $\mathbb{R}_{\geq 0}$ denote the set of positive and non-negative real numbers, respectively. Similarly, $\mathbb{N}_0$ and $\mathbb{N}_+$ denote the set of natural numbers with and without zero, respectively.

### 3.1. Base Problem

Let $\mathcal{A} = \{1, 2, \cdots, n\}$ be a set of $n \in \mathbb{N}_+$ autonomous robots. Let $\phi_a, v_a \in \mathbb{R}_{>0}$ be the radius and speed of robot $a$, $a \in \mathcal{A}$. For simplicity, we assume that robots are disk-shaped, but in general each robot $a$ can be of any shape as long as its body is contained in a circle with the center at $(x_a, y_a)$ and radius $\phi_a$, assuming $(x_a, y_a)$ is the current location of robot $a$. Next, let $\mathcal{P}_a$ be a path of robot $a \in \mathcal{A}$. $\mathcal{P}_a$ is composed of a finite sequence of elements, each being either a line segment or circle arc. Robot paths are thus always finite. Cycles are possible, as long as they have a finite number of repetitions. The length of path $\mathcal{P}_a$ is denoted $L_a \in \mathbb{R}_{>0}$. An example of base instance with three robots is shown in Figure 1.

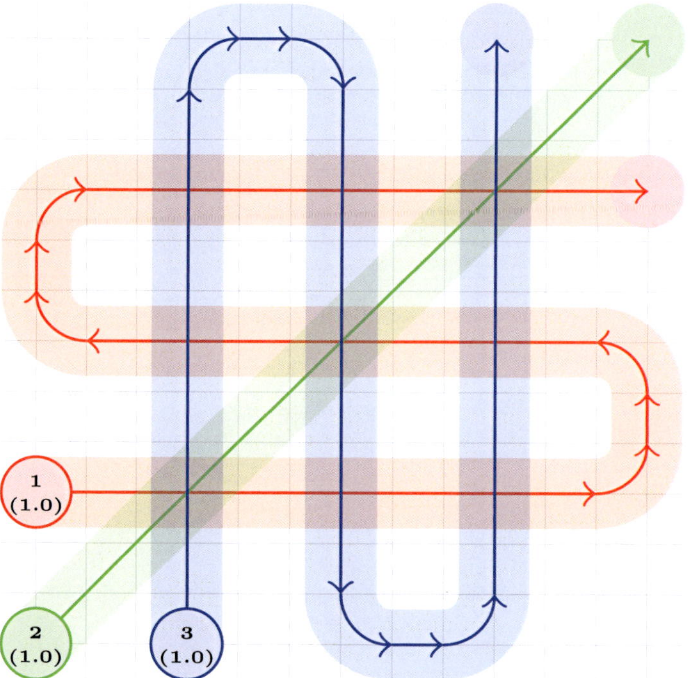


**Figure 1.** Example base problem instances with 3 robots. Each robot is portrayed with a different color. Circles denote robot starting positions. Circle labels show robot number and speed (in parentheses).

The goal is to coordinate the movement of the robots, such as to minimize the time at which all robots complete traveling their respective paths under the following conditions:

1. At any given time the speed of robot $a$ is either $v_a$ or 0 (i.e., robots are always either stopped or moving at their regular speed, acceleration is instantaneous).
2. Robots travel only "forward" along their paths—they cannot change direction and travel in reverse.

3. Robots cannot overlap each other, i.e., for any given time and any two robots $a$ and $b \neq a$ the Euclidean distance between their positions $(x_a, y_a)$ and $(x_b, y_b)$ has to be higher than the sum of their radii:

$$\sqrt{(x_a - x_b)^2 + (y_a - y_b)^2} > \phi_a + \phi_b. \tag{1}$$

As a result, for each robot we want to define a movement schedule, e.g., a set of time intervals in which robot is supposed to move (we assume the robot is stopped at other times). Let $C_a(\pi)$ be the time at which robot $a$ completes traveling its path according to some problem solution $\pi$ (e.g., movement schedule). Then the makespan $C_{\max}(\pi)$ is given as:

$$C_{\max}(\pi) = \max\{C_1, C_2, \ldots, C_n\}. \tag{2}$$

and our goal is to minimize it:

$$C_{\max}(\pi^*) = \min_{\pi \in \Pi_{\text{feas}}} C_{\max}(\pi). \tag{3}$$

where $\pi^*$ is the optimal solution and $\Pi_{\text{feas}}$ is the set of all feasible solutions.

*3.2. Derived Problem*

To solve the problem defined in Section 3, one needs to know when two robots are about to get too close to each other. The original problem formulation is cumbersome for this purpose. Thus, here we will transform the original problem into more convenient formulation. As example, we will use a simple instance shown in Figure 2.

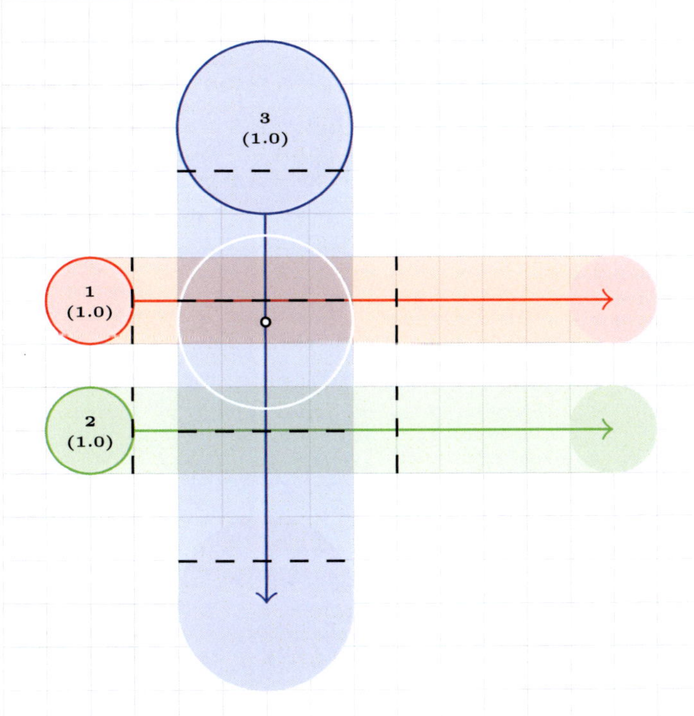


**Figure 2.** Exemplary instance for 3 robots. The white dot and circle denote a position and disk range somewhere along the path of robot 3. The dashed lines represent the borders of sectors (which will be explained further on).

In order to ensure the robots will never overlap each other, we divide each robot path into *sectors*. We will also introduce a set of *resources*. To travel a sector, a robot will require a (possibly empty) set of resources, which it will hold until leaving the sector. The procedure to define sectors and resources is as follows.

Let us consider robot $a$. For each point $p$ on the path of $a$ we define set $\mathcal{L}_p^a \subseteq \mathcal{A} \setminus \{a\}$ of robots that might overlap with $a$ when it is located at point $p$. More specifically, for point $p \in \mathcal{P}_a$ the set $\mathcal{L}_p^a$ is defined as:

$$\mathcal{L}_p^a \{ b \in \mathcal{A} \setminus \{a\} : \exists_{p' \in \mathcal{P}_b}\, d(p,p') \leq \phi_a + \phi_b \}, \tag{4}$$

where $d(p,p')$ is the Euclidean distance between points $p$ and $p'$. To illustrate it, let us consider the instance from Figure 2. The colored areas represent areas of the plane that are at some point covered by given robot (red for 1, green for 2, blue for 3), with mixed colors representing multiple robots occupancy. Moreover, the white dot represents some point on path of robot 3, while the area inside the white circle represents the occupancy of robot 3 when it is located at that point. We can see that the point itself overlaps with the path of robot 1 only, but the circle also overlaps with the path of robot 2, thus $\mathcal{L}_p^3 = \{1,2\}$.

Now we will move onto defining sectors. First, let $p_a(d) \in \mathcal{P}_a$ with $d \in [0, L_a]$ be a point on the path of robot $a$ that $a$ reaches after traveling distance $d$ from the starting point of $\mathcal{P}_a$ without stopping. Note that, even if $\mathcal{P}_a$ passes through the same point $p$ multiple times, it will result in different $d$. Thus, $d$ uniquely determines a point $p \in \mathcal{P}_a$.

With this we can define robot sectors. The idea is that a sector is the longest continuous section of the robot path that has constant value of $\mathcal{L}$ and that sectors are disjoint. Sector $i$-th will be denoted $\mathcal{S}_i$, where $i \in \{1,2,\ldots,\sum_{a=1}^n S_a\}$ and $S_a$ is the number of sectors of robot $a$ (which will be determined in a moment). The first sector overall is defined as

$$\{ p_1(d) \in \mathcal{P}_1 : d \in [0, d_1^1] \}, \tag{5}$$

where $d_1^1$ is the furthest that robot 1 can travel from its starting position before value $\mathcal{L}_{p(d)}^1$ changes or value $L_1$ is reached, whichever comes first. In other words $d_1^1 \leq L_1$ is the largest value such that for all $d \in [0, d_1^1]$ it holds that $\mathcal{L}_{p(d)}^1 = \mathcal{L}_{p(0)}^1$. The length of sector 1 is $l_1 = d_1^1$.

Next, we define the subsequent sectors of robot 1 based on the previous sector. In general, the $j$-th sector is defined as:

$$\{ p_1(d) \in \mathcal{P}_1 : d \in (d_1^{i-1}, d_1^i] \}, \tag{6}$$

where $d_1^i$ is the furthest we can travel from $p(d_1^{i-1})$ before value $\mathcal{L}_{p(d)}^1$ becomes different than $\mathcal{L}_{p(d_1^{i-1})}^1$ (or we reach $L_1$). The length of the $i$-th sector is $l_i = d_1^i - d_1^{i-1}$. If $d_1^i = L_1$, then we defined all sectors of robot 1.

Next, we move onto subsequent robots. In general, the first sector of robot $a$ will be the $\sum_{b=1}^{a-1} S_b + 1$-th sector overall, defined as:

$$\{ p_a(d) \in \mathcal{P}_a : d \in [0, d_a^1] \}, \tag{7}$$

where the meaning of $d_a^1$ is analogous to $d_1^1$. The length of this sector is equal to $d_a^1$. Similarly, we define the $i$-th sector of robot $a$ as:

$$\{ p_a(d) \in \mathcal{P}_a : d \in (d_a^{i-1}, d_a^i] \}, \tag{8}$$

which has a length of $d_a^i - d_a^{i-1}$. The total numbers of sectors is denoted as $S = \sum_{a=1}^n S_a$. We thus distinguish between the $i$-th sector overall ($i \in \{1,2,\ldots,S\}$) and the $i$-the sector of robot $a$ ($i \in \{1,2,\ldots,S_a\}$), and which one we mean at the time will be clear from the context.

As example of sectors, let us consider instance from Figure 2. Robot 1 has three sectors:

1. $\mathcal{S}_1$—before it starts to overlap the path of robot 3;

2.     $\mathcal{S}_2$—when it overlaps with the path of robot 3;
3.     $\mathcal{S}_3$—after it stops overlapping with the path of robot 3.

Robot 2 has the same setting with sectors $\mathcal{S}_4$, $\mathcal{S}_5$, $\mathcal{S}_6$. Finally, robot 3 has five sectors:

1.     $\mathcal{S}_7$—before it starts to overlap the path of robot 1;
2.     $\mathcal{S}_8$—while it overlaps the path of robot 1, but not 2;
3.     $\mathcal{S}_9$—while it overlaps paths of both robot 1 and 2;
4.     $\mathcal{S}_{10}$—while it overlaps with the path of robot 2 only;
5.     $\mathcal{S}_{11}$—after it stops overlapping path of either robot 1 or 2.

Thus, for instance from Figure 2 we obtain 11 sectors. The sectors along with their lengths are shown in Table 1 (columns 2 and 3).

**Table 1.** Derived instance based on instance from Figure 2 ($\phi_a$ and $v_a$ are the same as original).

| Robot $a$ | Sector $i$ | Length $l_i$ | Conflicts $\mathcal{C}_i$ | Resources $\mathcal{R}_i$ |
|---|---|---|---|---|
| | 1 | 0.975 | | |
| 1 | 2 | 6.025 | $\{8,9\}$ | $\{1,2\}$ |
| | 3 | 5.000 | | |
| | 4 | 0.975 | | |
| 2 | 5 | 6.025 | $\{9,10\}$ | $\{3,4\}$ |
| | 6 | 5.000 | | |
| | 7 | 0.975 | | |
| | 8 | 3.000 | $\{2\}$ | $\{1\}$ |
| 3 | 9 | 3.025 | $\{2,5\}$ | $\{2,3\}$ |
| | 10 | 3.000 | $\{5\}$ | $\{4\}$ |
| | 11 | 1.000 | | |

Next, for each pair of sectors we define a conflict relation. Sectors $\mathcal{S}_i$ and $\mathcal{S}_j$ of robots $a$ and $b$, respectively, are in conflict if and only if $a \neq b$ and

$$\exists_{p \in \mathcal{S}_i, p' \in \mathcal{S}_j}\, d(p, p') \leq \phi_a + \phi_b. \tag{9}$$

In other words, sectors $\mathcal{S}_i$ and $\mathcal{S}_j$ are in conflict if they belong to different robots and the disks of those robots might overlap when they are in those respective sectors. The relation is symmetric: if $\mathcal{S}_i$ is conflicted with $\mathcal{S}_j$, then $\mathcal{S}_j$ is conflicted with $\mathcal{S}_i$. The set of sector numbers of all sectors conflicted with $\mathcal{S}_i$ is denoted $\mathcal{C}_i$. Once again we will illustrate it with example for instance from Figure 2, which has eight conflicts, as follows:

1.     Sector 2 (robot 1) is conflicted with sectors 8 and 9 (robot 3).
2.     Sector 5 (robot 2) is conflicted with sectors 9 and 10 (robot 3).
3.     Sector 8 (robot 3) is conflicted with sector 2 (robot 1).
4.     Sector 9 (robot 3) is conflicted with sectors 2 (robot 1) and 5 (robot 2).
5.     Sector 10 (robot 3) is conflicted with sector 5 (robot 2).

The contents of $\mathcal{C}_i$ for all sectors are shown in Table 1 (column 4), except when $\mathcal{C}_i = \{\}$.

Finally, we define resources. The idea is simple: a resource is an ordered pair $(i, j)$ where $j > i$ such that sectors $\mathcal{S}_i$ and $\mathcal{S}_j$ are in conflict. The number of resources is half the number of conflicts. For convenience, we number the resources with subsequent natural numbers according to natural sorting order i.e., $(1, 3)$ would go after $(1, 2)$ but before $(2, 1)$. Let us denote the total number of resources by $R$, in this case $R = 4$. Finally, we define set $\mathcal{R}_i$ which will contain the numbers of resources required to enter sector $\mathcal{S}_i$.

To illustrate, in our example we have only four resources: $1 = (2, 8)$, $2 = (2, 9)$, $3 = (5, 9)$, and $4 = (5, 10)$, and thus $R = 4$. With regards to sets $\mathcal{R}_i$, let us consider $i = 9$. Sector 9 is conflicted with sectors 2 and 5, so to enter sector 9 we require resources $(2, 9)$ (resource 2) and $(5, 9)$ (resource 3); thus, $\mathcal{R}_9 = \{2, 3\}$. The contents of $\mathcal{R}_i$ for all sectors are shown in Table 1 (column 5), except when $\mathcal{R}_i = \{\}$.

With regards to the above formulation, a practical issue arises. Namely, how to obtain values $d_a^i$. One solution is an analytical approach, where those values are determined exactly. Its advantage is accuracy, but the method is not obvious, especially for more complex shapes than line segments and circle arcs. Another approach is to transform the continuous path $\mathcal{P}_a$ into a finite set of discrete points. This approach is easier to use for generic path shapes, but has lower accuracy. In this paper, we have adopted the latter, discrete approach. Specifically, for path $\mathcal{P}_a$ we consider points $p_a(ks)$, where $k \in \mathbb{N}_0$ and $s$ is step size, as well as point $p_a(L_a)$. With sufficiently small step size, for example

$$s = \frac{\min\{\phi_1, \phi_2, \ldots, \phi_n\}}{20}, \tag{10}$$

we can model the base instance with enough accuracy.

With this we can transform any instance of the base problem into equivalent formulation that is independent of the shapes of 2-D paths. Throughout the rest of the paper we will mainly deal with such "derived" problem, referred to as simply "the problem".

## 4. Solution Representation and Feasibility

As mentioned in Section 3, the direct solution to our problem could be, for example, for each robot $a$ a set of time intervals in which $a$ moves. However, such representation is cumbersome as (1) it is still a continuous optimization problem, admitting infinite and uncountable number of solutions, (2) is obviously admits non-optimal solutions, and (3) the number of time intervals is not necessarily $\mathcal{O}(S)$. For these reasons, we will propose an alternative solution representation that will reduce our problem to discrete optimization.

Let us consider resource $(i, j)$. That resource is required for some robot $a$ to enter sector $\mathcal{S}_i$ as well as for some robot $b \neq a$ to enter sector $\mathcal{S}_j$. We know that robots hold all resources they need while they are inside a sector. Thus, it is enough to only decide when a robot is allowed to enter the sector—once it has entered, it can safely move until it has reached the end of the sector.

Let us notice that each sector is entered only once, thus the $(i, j)$ resource will be required once by $a$ and once by $b$—no other robot will ever require it. Thus, for such resource $(i, j)$, we need only decide whether robot $a$ or $b$ should acquire resource $(i, j)$ first. Since all resources work in this way, our solution reduces to a binary vector $\pi = (1, 2, \ldots, R)$, where $\pi(r)$ is the $r$-th element of $\pi$. Let us assume that resource $r$ is required by robots $a$ and $b > a$. If $\pi(r) = 0$ then $a$ is to acquire resource $r$ first, then $b$. Otherwise ($\pi(r) = 1$), $b$ is to acquire $r$ before $a$. Such a representation results in $2^R$ possible solutions. In our exemplary instance, a possible solution could be:

$$\pi = (0, 0, 1, 0). \tag{11}$$

For resources 1 and 2 the competing robots are 1 and 3. Since $\pi(1) = \pi(2) = 0$, then robot 1 is supposed to get access before robot 3. For resources 3 and 4 the competing robots are 2 and 3. Since $\pi(3) = 1$ and $\pi(4) = 0$, then robot 2 will be the second to access resource 3, but the first to access resource 4.

The above setup would work well if each robot required only one resource at a time. However, it is possible that robot will require several resources, each with its own two-element queue. With multiple queues the concept of "being first" is not trivial. We have considered three possibilities:

1.  Robot $a$ is first if it is first in all relevant queues.
2.  Robot $a$ is first if it is first in any relevant queue.
3.  We compute robot priority by summing robot positions in all relevant queues and divide it by number of such queues. For example, if robot $a$ needs three resources with queues $(a, b)$, $(b, a)$, and $(a, c)$ then the priority is $(1 + 2 + 1)/3 = 4/3$. We then compute this priority for all other robots in relevant queues (robots $b$ and $c$ in this example). Robot $a$ is first if its priority is lowest among considered robots.

Options 1 and 2 are simple, but they cause problems when the number of resource is high as the robot would almost never (option 1) or almost always (option 2) be considered first. This would either cause frequent deadlocks or would defeat the purpose of solution $\pi$ enforcing moving order. Thus, we have chosen option 3.

With this, we have formulated the task of robot coordination in continuous 2-D workspace as a discrete optimization problem with decision variables in a form of a single binary vector. As such, our formulation could be seen as a variant of a scheduling problem, for example a job shop scheduling problem, except that a job could be processed on multiple machines at once.

### 4.1. Feasibility and Deadlock Avoidance

As with other problems concerned with acquiring resources, there is possibility of deadlocks occurring. A deadlock occurs when there is non-empty set of robots that cannot proceed (are stopped indefinitely), despite not having completed their task yet. Deadlocks result in infeasible solutions, thus they need to be avoided or resolved. In the case of our problem there are two types of deadlocks, type I and type II. We will describe those as well as methods of avoiding deadlocks of type I and resolving deadlocks of type II.

Type I is a classic deadlock caused by resource holding by robots while waiting for another resource and the formation of resource awaiting cycles. A simple example would be with two robots $a$ and $b$. First, $a$ acquires resource $r_1$ and $b$ acquires resource $r_2$. Next, $a$ wants to acquire $r_2$ , while $b$ wants to acquire $r_1$. As a result, robot $a$ is holding resource $r_1$ while waiting for $r_2$, and robot $b$ is holding resource $r_2$ while waiting for $r_1$. Thus, no robot will proceed as the required resource is unavailable and no resource will be released. We will now show a method for avoiding such deadlocks, based on the method shown in [2].

Let us start by introducing some useful notation. Let $x$ be a $n$-element vector, describing the "sector state" of the system, with its $a$-th element $x_a \in \{0, 1, \ldots, S_a, S_a + 1\}$ describing the state of robot $a$. If $a$ is in its $i$-th sector, then $x_a = i$. It should be pointed out that this numbering is not the same as the $1, 2, \ldots, S$ overall numbering for sectors. Two special values 0 and $S_a + 1$ are used for when the robot has not started its task yet and for when it has completed its task, respectively. The initial state of the system is $x_0 = (0, 0, \ldots, 0)$. Similarly, the final state is $x_F = (S_1 + 1, S_2 + 1, \ldots, S_n + 1)$.

As explained earlier, we are not concerned what happens when a robot moves along a given sector, as it has no effect on resources. We are only concerned when a robot changes sectors, described as an *event*. Event $e_a$ means that robot $a$ moves to the next sector, changing system state from $x = (x_1, x_2, \ldots, x_a, \ldots, x_n)$ to $x' = (x_1, x_2, \ldots, x_a + 1, \ldots, x_n)$. Ideally, event $e_a$ can occur in state $x$ if and only if:

1. Robot $a$ has not completed its task i.e., $x_a \leq S_a$.
2. All resources required for $a$ to enter its next sector are free. Equivalently, this means that in state $x$ no robot is in sector that is conflicted with sector robot $a$ wants to enter.
3. State $x'$ to which event $e_a$ leads is *safe*.

In general, state $x$ is safe if the final state $x_F$ is reachable from it. However, the problem of determining whether a given state is safe is considered NP-hard [31]. Due to this, we will not fully check the safety of $x$. Instead, we will check the sufficient condition of $x$ being safe. This might lead to use treating some safe states as unsafe, but this is much easier to determine.

First, let us define non-conflicting sectors. A sector $i$ is non-conflicting if and only if $C_i = \{\}$. Next, the nearest non-conflicting sector for robot $a$ is the first sector, starting from $a$'s current sector, which is non-conflicting. For this purpose, robots before start of their task ($x_a = 0$) and after completing their task ($x_a = S_a + 1$) are considered to be in their nearest non-conflicting sectors.

With this, for a state $x$ to be safe, it is sufficient to show there is a "safe sequence" of robots $z = (z_1, z_2, \ldots, z_n)$, such that $z_1$ moves first to its nearest non-conflicting sector, while remaining robots do not change sectors. Next, $z_2$ moves and so on up to $z_n$. The resulting state $x'$ is safe as all robots are in non-conflicting sectors, meaning all resources are free. Thus, $x_F$ is reachable from $x'$. Since $x'$ is reachable from $x$, then $x_F$ is reachable from $x$, meaning that $x$ is safe as well. The procedure for finding out a safe robot sequence for state $x$ is as follows:

1. $\mathcal{A}^* \leftarrow \mathcal{A}$.
2. For each robot $a \in \mathcal{A}$ determine its nearest non-conflicting sector $b_a$ based on current state $x$.
3. Find the first robot $a$ in $\mathcal{A}^*$, such that $a$ can move to $b_a$ i.e., either (1) $b_a = x_a$ or (2) no robot in state $x$ is in sector that is in conflict with any sector $x_a$ through $b_a$.

    (a) If $a$ does not exist, then algorithm stops and $x$ is not safe.
    (b) Else $\mathcal{A}^* \leftarrow \mathcal{A}^* \setminus \{a\}$.

4. If $|\mathcal{A}^*| > 0$, then go to step 3, otherwise, state $x$ is safe.

The above method solves the issues of type I deadlocks: we avoid them by avoiding entering unsafe states. By doing so, it will be always possible to move all robots to their nearest non-conflicting sectors, which require no resources, meaning all resources will become available.

Next we will consider type II deadlocks. Those deadlocks are never caused by resource unavailability alone, but also by the order of robots in solution $\pi$. We will illustrate it with simple example. First, we assume there are three robots and no robot has completed its task yet. We consider possible events from state $x$: $e_1$, $e_2$, and $e_3$. Let us assume only event $e_2$ is possible i.e., $e_1$ or $e_3$ would lead to unsafe states. In that situation, only robot 2 can move. Next, we assume that robot 2 has to acquire resource $r$ for which it competes with robot 1 and neither of them acquired this resource yet. Finally, let us assume that $\pi(r) = 0$, i.e., robot 1 is "planned" to acquire resource $r$ before robot 2. With this, a deadlock is reached. If robots 1 or 3 move, then we reach unsafe state, resulting in type I deadlock. However, robot 2 cannot move either, because it would need to acquire $r$, while $\pi$ will force it to wait until robot 1 acquires and releases it first. In this case, no robot will proceed, which we describe as type II deadlock.

We can also illustrate it with an instance from Table 1, except we will need to modify robot speeds. Let us assume the solution is $\pi = (1, 1, 1, 0)$. Thus robot 3 (blue) given high-enough speed will travel through sectors $\mathcal{S}_7$ and $\mathcal{S}_8$ and will enter $\mathcal{S}_9$. However, we choose the speeds in such a way that before robot 3 enters $\mathcal{S}_{10}$, robot 2 (green) will reach the end of $\mathcal{S}_4$ and will try to enter $\mathcal{S}_5$. This is, however, impossible, as robot 2 will require resources 3 and 4 for this, but resource 3 is unavailable due to robot 3 being still in sector $\mathcal{S}_9$. Robot 2 thus has to wait. However, robot 3 will not move, as to enter $\mathcal{S}_{10}$ it requires resource 4, but $\pi(4) = 0$, thus out of robots 2 and 3, the one with the lower number should obtain that resource first. We see that type II deadlock is not caused by physical resources, but by the solution $\pi$ (or a mix of both).

We considered two approaches to dealing with type II deadlocks. The first option is to detect such type II deadlocks and consider solution $\pi$ for which it occurred as infeasible (we will explain the detection as a part of the solution evaluation shortly). However, such an approach is cumbersome for the solving algorithms. For algorithms considering a single solution (e.g., constructive heuristics), if such solution is infeasible, then the algorithm does not work. For algorithms considering multiple candidate solutions (e.g., metaheuristics) infeasible solutions are also problematic as they force the algorithm to evaluate meaningless solutions. Metaheuristics also require initial solution, which should also be feasible.

Thus, we have adopted a second approach to resolving type II deadlocks: when such a deadlock is detected, we temporarily ignore the order imposed by $\pi$. Instead, we consider events from $e_1$ to $e_n$ and choose the first event $e_a$ which is feasible (i.e., $a$ has not completed its task yet, all required resources are available and resulting state is safe). This turns a potentially infeasible solution into a feasible one.

### 4.2. Solution Evaluation

The last issue related to solution is its evaluation, i.e., a method of transforming a solution $\pi$ into actual robot movement schedule in order to obtain task completion times $C_a$ and makespan $C_{\max}$. The procedure works by simulating robot movements and is as follows.

Initially, the system state is $x_0$ and time is $t = 0$. The procedure then enters a main loop, which continues until all robots complete their task i.e., until state $x_F$ is reached. In each loop iteration, each robot is assessed to see if it will be able to move in this iteration. For each robot $a$ there are the following possibilities:

1. If $a$ has already completed its task (i.e., $x_a = S_a + 1$), then $a$ is ignored and cannot move. We set $t_a \leftarrow \infty$.
2. Otherwise, if $a$ is not at the end of its current sector, then $a$ can move and we set $t_a$ to the time it will take $a$ to reach the end of its current sector.
3. Otherwise, $a$ has to be at the end of the current sector. We check if it is possible for $a$ to enter a new sector (all needed resources are available, the resulting new state is safe and $a$ is considered first in its resource queues). If the access is denied, then the robot cannot move and we set $t_a \leftarrow \infty$. If the access is granted, then the robot moves to the next sector and:

    (a) If $a$ completed its task, then we set $C_a \leftarrow t$ and $t_a \leftarrow \infty$.
    (b) Otherwise, the situation is similar to case 2, $a$ can move until the end of the newly entered sector, so we set $t_a$ to the time it will take $a$ to reach the end of it.

There is one additional possibility to consider. It might happen that robot $a$ cannot move (e.g., resources are not available), but then robot $b > a$ changes sectors, freeing the resources, enabling $a$ to move. Thus, the above robot assessment procedure is repeated, until no new robots were designated to move (no new values $t_a \neq \infty$ are assigned).

After the assessment is done, we have a set of values $(t_1, t_2, \ldots, t_n)$, with each value indicating either that $a$ cannot move ($t_a = \infty$) or that $a$ can safely move for time $t_a$ inside its sector. Let $\Delta t = \min(t_1, t_2, \ldots, t_n)$. Two possibilities can occur.

1. $\Delta t \neq \infty$ (i.e., at least one robot can advance). In that case each movable ($t_a \neq \infty$) robot $a$ advances for time $\Delta t$ (i.e., by distance $\Delta t \times v_a$). We record in the schedule that $a$ moves in time interval from $t$ to $t + \Delta t$. After all movable robots have advanced, we update the simulation time $t \leftarrow t + \Delta t$.
2. $\Delta t = \infty$ (i.e., no robot can advance). This indicates that a type II deadlock occurred. In this case, we repeat the assessment procedure once more, but this time we ignore solution $\pi$ and the queues, thus at least one robot will be able to move, reducing this to case 1.

After the procedure completes, we obtain for each robot its movement schedule (intervals during which it advances), task completion times $C_a$, and compute the makespan.

At this point, a careful reader might ask why a different solution representation was not used. Namely, a solution might have multiple resources between the same pair of robots, i.e., robots $a$ and $b$ compete for both $\pi(r_1)$ and $\pi(r_2)$. In such a case one could simply assume $\pi(r_1) = \pi(r_2)$, significantly reducing the solution space. However, there exist instances for which such "matching" policy is not optimal. For example, consider instance from Figure 3. Here there are only two resources, but robot 2 acquires them at the same time, while robot 1 acquires them separately. For this instance the possible solutions are $\pi_1 = (0,0)$, $\pi_2 = (0,1)$, $\pi_3 = (1,0)$, and $\pi_4 = (1,1)$. A simple brute force algorithm results in $C_{\max}(\pi_1) = 16.98$, $C_{\max}(\pi_2) = 11.00$, and $C_{\max}(\pi_3) = C_{\max}(\pi_4) = 17.96$. Thus, $\pi^* = \pi_2$. In general, the idea of such "matching" between some values of $\pi$ is interesting, but determining which elements of $\pi$ can be safely matched is non-trivial.

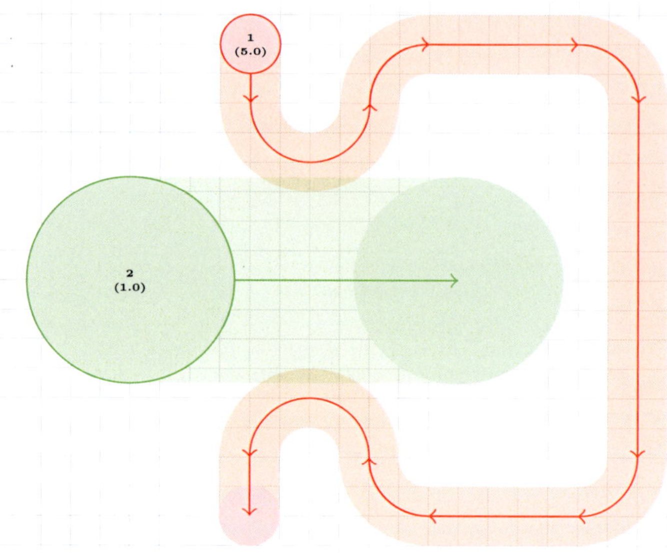


**Figure 3.** Counterexample instance for solution representation.

## 5. Solving Algorithms

In this section we describe the algorithms used to solve the aforementioned problem. We considered two types of algorithms: constructive algorithms and metaheuristics.

### 5.1. Constructive Algorithms

Constructive algorithms work by building a single solution according to some *policy*, often called a dispatch rule, and are popular in many applications, including scheduling [23], particularly in computer networks, high-performance computing, and operating systems [29,32]. The general procedure is similar to solution evaluation shown in Section 4.2, except that algorithm starts with empty $\pi$, which is constructed on-the-fly as the algorithm makes decisions according to its policy. As a result, such algorithms are simple and have short running time, but are usually not optimal, hence they are also heuristics. Below we list all 14 implemented constructive algorithms.

#### 5.1.1. Lower Number First

One of the simplest and fastest algorithms is lower number first, or LNF, done simply by setting $\pi = (0, 0, \ldots, 0)$. This means that for every resource $r$ that has robots $a$ and $b > a$ competing for it, the policy is for $a$ to acquire $r$ first. Of course, as explained earlier, this rule might be temporarily overlooked if type II deadlock occurs. This algorithm can be seen as deterministic version of random algorithms, since we can simply re-order the robots, in which case LNF might return a different solution.

#### 5.1.2. First-In First-Out

First-in first-out, or FIFO (also known as first-come first-served (FCFS), earliest arrival first (EAF), etc., depending on context), is a very basic and popular algorithm in many optimization, scheduling, and dispatching problems. For this algorithm, the policy is that the first robot to reach the beginning of the sector will be prioritized to enter it first. In practice, we keep a queue, initially empty, for every resource $r$. When robot $a$ wants to enter a conflicting sector, it adds itself to queues for all required sectors. Next, we compute robot priority as an average of its position in all the relevant queues (similar to what was shown in Section 4) and $a$ will be considered "first in" only when its priority is lowest among all robots queued alongside him in the relevant queues. Of course, $a$ also needs to clear the regular conditions for sector entry. Thus, $a$ will proceed only when (1) required resources are free, (2) resulting state is safe, and (3) $a$ is considered "first in". In such a case $a$ enters the new sector, removes itself from all relevant queues, and modifies $\pi$ accordingly. After this, similar to solution evaluation, the procedure terminates and $\pi$ will hold the constructed solution.

### 5.1.3. Shortest Distance First

Shortest distance traveled first (SDTF) is an algorithm that prioritizes robots that have traveled the least distance from their starting positions. In our implementation, when robot $a$ wants to enter a sector, the distance traveled so far by $a$ is computed and then compared against the distance traveled by other robots who are competing over the resources $a$ wants to acquire. It should be noted that when a robot enters a sector, it is removed from the queues of all resources it has just acquired. Thus, robot $a$ is compared only against robots that have yet to reach this point—robots that have already passed this point are not compared against $a$. We will also consider another variant of SDFT: our objective, the makespan, is based on time rather than distance. Thus, in the second variant the distance traveled is divided by robot speed and the resulting algorithms is called shortest time traveled first (STTF). Note that we count only the time spent moving, as total time spent so far is identical for all robots.

Next, we define two more algorithms that are similar to SDFT and STTF, but that consider remaining distance (time) rather than traveled. The resulting algorithms are shortest distance remaining first (SDRF) and shortest time remaining first (STRF). Finally, we introduce two algorithms that count the total robot path instead of splitting it into traveled-so-far and yet-to-travel part. Those algorithms are called shortest overall distance first (SODF) and shortest overall time first (SOTF).

### 5.1.4. Longest Distance Variants

Finally, we also define complementary algorithms to the ones defined above, which differ in that they prioritize longest distances instead of shortest. Thus, this leads to definition of the following six additional algorithms:

1.  Longest distance traveled first (LDTF) and longest time traveled first (LTTF).
2.  Longest distance remaining first (LDRF) and longest time remaining first (LTRF).
3.  Longest overall distance first (LODF) and longest overall time first (LOTF).

### 5.2. Metaheuristics

The constructive algorithms presented in the previous subsection are fast and simple, but they are not enough in all cases. A proof of that is supplied by the counterexample instance presented in Figure 4. For this very simple two-robot instance all 14 constructive algorithms obtained makespan of either 28.26 or 31.26. However, a simple brute force algorithm proved that the optimal makespan in this case is 23.85. This means that the constructive algorithms were 17% and 31% away from the optimum. Thus, alternative solving methods are required. Since exact methods are too time-consuming, we will propose two metaheuristic algorithms, described below. It should also be noted that both proposed metaheuristics could be easily modified into parallel algorithms to vastly decrease computation times while run in many-core or distributed computing environments. For the artificial bee colony, this could be achieved by assigning a number of bees to a given thread or network node. Similarly, for the taboo search, a number of solutions from the neighborhood could be assigned to a single thread/node.

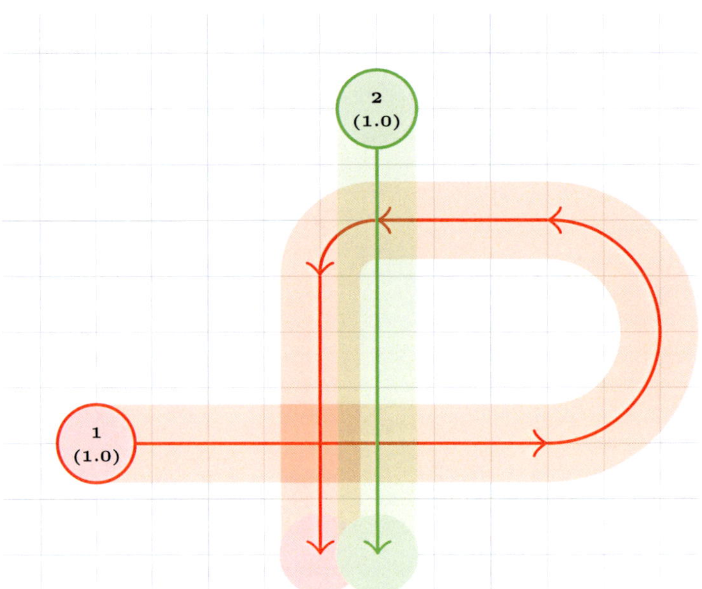



**Figure 4.** Counterexample instance for constructive algorithms.

### 5.2.1. Artificial Bee Colony

Artificial bee colony, or ABC, is a popular population-based probabilistic metaheuristic and was previously used for robot path planning and collision avoidance (see papers [10–12,33] for details). Moreover, ABC is both effective and simple to implement, with only a few tuning parameters required.

ABC uses food sources as solutions. Initially a population of $p$ food sources is created and each food source $k$ has its counter $c_k$ set to 0. This results in $p$ different solutions $\pi_k$ for $k \in \{1, 2, \ldots, p\}$. These solutions are created randomly. Then main algorithm loop starts with each iteration consisting of three phases. In phase 1, a worker bee is sent to each food source and tries to improve it. In our case, for a given solution $\pi_k$ we transform it into solution $\pi'_k$ by flipping a random bit $r$, i.e., $\pi'_k(r)$ is set to $\neg \pi_k(r)$. The resulting solution $\pi'_k$ is evaluated. If $C_{\max}(\pi'_k)$ is better than $C_{\max}(\pi_k)$, then $\pi'_k$ replaces $\pi_k$ and $c_k$ is set to zero. Otherwise, $\pi_k$ stays and $c_k$ increases by 1. In essence, this phase implements the standard valley descent local search procedure with multiple solutions.

Phase 2 is similar to phase 1 with one major difference. In phase 1, the $k$-th improvement is carried out for the $k$-th food source. In phase 2, the food source for the $k$-th improvement is random, but with assumption that better food sources have higher chance of being "tended to". As a result, statistically, good food sources will be improved multiple times in a single iteration, while bad food sources will receive less attention. In practice, many methods exist for such randomization, in our case we have employed the roulette-wheel selection via stochastic acceptance shown in [34], which guarantees food source choice in time $\mathcal{O}(1)$.

Finally, in phase 3 each food source $k$ is checked for the value of its counter $c_k$. If $c_k$ is higher than some limit value *maxCount* then $\pi_k$ is deemed non-promising and is replaced with random solution and $c_k$ is set to zero again. With this, phase 3 allows abandoning food sources that fail to improve the solution quality, leading to discovering of new food sources.

Regardless of phase, each time a new solution $\pi$ is generated, its quality is compared to best found solution so far $\pi_{\text{best}}$ and if $\pi$ is better, it replaces $\pi_{\text{best}}$. The algorithm stops after a fixed number of iterations *maxIter* is reached with $p$, *maxCount*, and *maxIter* as parameters. The overview of ABC method is shown in Algorithm 1.

---

> **Algorithm 1:** Artificial bee colony (method outline).

```
1  function ArtificialBeeColony(maxIter, p, maxCount):
2      P ← GenerateInitPopulation(p);
3      Evaluate(P);
4      while it < maxIter do
5          EmployedBeesPhase(P);
6          OnlookersBeesPhase(P);
7          ScoutBeesPhase(P, maxCount);
8          π' ← arg min_{π∈P} C_max(π);
9          if f_max(π') < C_max(π*) then
10             π_best ← π';
11         it ← it + 1;
12     return π_best;
```

### 5.2.2. Taboo Search

Taboo search (TS) is a local search metaheuristic. It is usually more complex to formulate than ABC, but TS can be made to be fully deterministic algorithm. Moreover, the authors have previously employed this metaheuristic for several discrete optimization problems (check, for example, papers [35,36]), where TS has proved its efficiency.

TS is an iterative algorithm as well. Its outline is shown in Algorithm 2. First, an initial solution $\pi_{\text{init}}$ is chosen. In our case it is the FIFO constructive algorithm, but other options are possible as well. Then, in each iteration for a current solution $\pi$ a neighborhood $\mathcal{N}(\pi)$ is generated, where each neighbor $\pi_r$ is created from $\pi$ by flipping the $r$-th bit in $\pi$ (thus, the neighborhood has $R$ elements). After all neighbors are generated and evaluated, the best of those neighbors $\pi_{\mathcal{N}}$ replaces $\pi$, regardless whether $\pi_{\mathcal{N}}$ improves $\pi$ or not.

In order to help the algorithm to escape local optima, a taboo list is used. When $\pi_{\mathcal{N}}$ is chosen to replace $\pi$ the move that led to this becomes forbidden for number of iterations defined by cadence parameter $c$. We have used an implementation that allows all taboo list operations in time $\mathcal{O}(1)$. Additionally, we employ aspiration criterion, that allows to ignore the taboo list if $\pi_r$ is better than the best known solution so far.

---

> **Algorithm 2:** Taboo search (method outline).

```
1  function TabooSearch(maxIter, c, π_init):
2      π ← π_init;
3      π_best ← π;
4      while it < maxIter do
5          π_N ← nul;
6          foreach π_r ∈ N(π) do
7              if not IsSolutionForbidden(π_r) or C_max(π_r) < C_max(π_best) then
8                  if π_N = nul or C_max(π_r) < C_max(π_N) then
9                      π_N ← π_r;
10         π ← π_N;
11         MakeMoveForbiddenForCadence(c);
12         if C_max(π) < C_max(π_best) then
13             π_best ← π;
14         it ← it + 1;
15     return π_best;
```

As with ABC, each time a new solution is accepted, it is compared against best known solution $\pi_{\text{best}}$ and replaces it if it is better. The algorithm finishes when the number of iterations *maxIter* is reached. Values $c$, *maxIter*, and $\pi_{\text{init}}$ are the parameters of the method.

## 6. Computer Experiment

In this section, we present the results of a computer experiment on a set of problem instances. We will start by describing the instance generation procedure, including the uncertain data assumptions, as well as the quality of measure used to compare algorithms. Then we will present the results and discuss them.

### 6.1. Instance Generation

The instance generation procedure is as follows. First, we assume a bounding rectangle from point $(0,0)$ to $(50,50)$. For each of $n$ robots the starting position is chosen randomly outside of the bounding rectangle (we keep a margin of 2 away from the rectangle). The robot radius is drawn from uniform distribution with range $[0.5,2]$ i.e., $\phi_a \sim \mathcal{U}(0.5,2)$. Similarly, for robot speed we have $v_a \sim \mathcal{U}(0.5,2.5)$. The robot path consists of $60 - 3n$ segments (allowing for shorter paths for more robots). We take care that when an arc follows a line segment, the arc starts angled correctly to the line segment (i.e., no sharp turns for robots). Due to this, if a line segment would follow another line segment, it would have the same angle, thus such line segments can be merged, therefore the path alternates between line segments and arcs. For each line its length is drawn from $\mathcal{U}(2,37)$ (maximum is around 75% of the bounding rectangle width). A safety mechanism is employed: if a path would lead the robot out of the bounding rectangle, the segment is rejected and a new one is attempted. After 20 unsuccessful attempts, a 180 degrees turn-around arc is used instead.

The above considers only the deterministic case. For the uncertain case, we proceed as follows. For a given instance $I$ we obtain a modified instance $I(Y)$ by replacing all sector lengths $l_i$ with values $l_i y_i$, where $y_i$ is a realization of random variable $Y$. In other words, $y_i$ is a scaling parameter of the original sector lengths. Index $i$ is used to denote that $y_i$ and $y_j$ are different realizations of $Y$. Additionally, since we assume robot speed is constant, changes to travel distances translate directly to changes to travel time, modeling uncertainty in robot movement. As for random variable $Y$, we would like to have the expected value of $l_i Y$ be equal to $l_i$ (i.e., the expected sector length is equal to its nominal value), so the expected value of $Y$ should be 1. We also assume standard deviation parameter $\sigma_Y$. Thus:

$$\mathbb{E}[Y] = \mu_Y = 1, \tag{12}$$

$$\mathbb{V}[Y] = \sigma_Y^2, \tag{13}$$

where $\mathbb{E}$ and $\mathbb{V}$ are expectation and variance operators, respectively.

Since sector lengths have to be positive, we choose to model $Y$ using log-normal distribution, i.e., $Y \sim LogNormal(\mu, \sigma^2)$. In order for $Y$ to have expected value and standard deviation assumed in (12) and (13), we set $\mu$ and $\sigma$ to:

$$\mu = \ln\left(\frac{\mu_Y^2}{\sqrt{\mu_Y^2 + \sigma_Y^2}}\right) = \ln\left(\frac{1}{\sqrt{1 + \sigma_Y^2}}\right), \tag{14}$$

$$\sigma^2 = \ln\left(1 + \frac{\sigma_Y^2}{\mu_Y^2}\right) = \ln\left(1 + \sigma_Y^2\right). \tag{15}$$

Using the above instance generator, we have prepared a benchmark of 50 instances, 10 for each size $n \in \{2,3,4,5,6\}$. Instances are numbered with subsequent natural numbers i.e., instances for $n = 2$ are numbered 1 through 10, for $n = 3$ are numbered 11 through 20 and so on. Number of sectors ranges from 38 to 693, while number of resources ranges from 21 to 4007. An example of generated instance for four robots is visualized in Figure 5. Let us note that the number of resources identified for this, relatively small, instance with only four robots was equal to 419, yielding a solution space with $2^{419}$ solutions. This is roughly equivalent to the solution space of traveling salesman problem with approximately 84 cities.

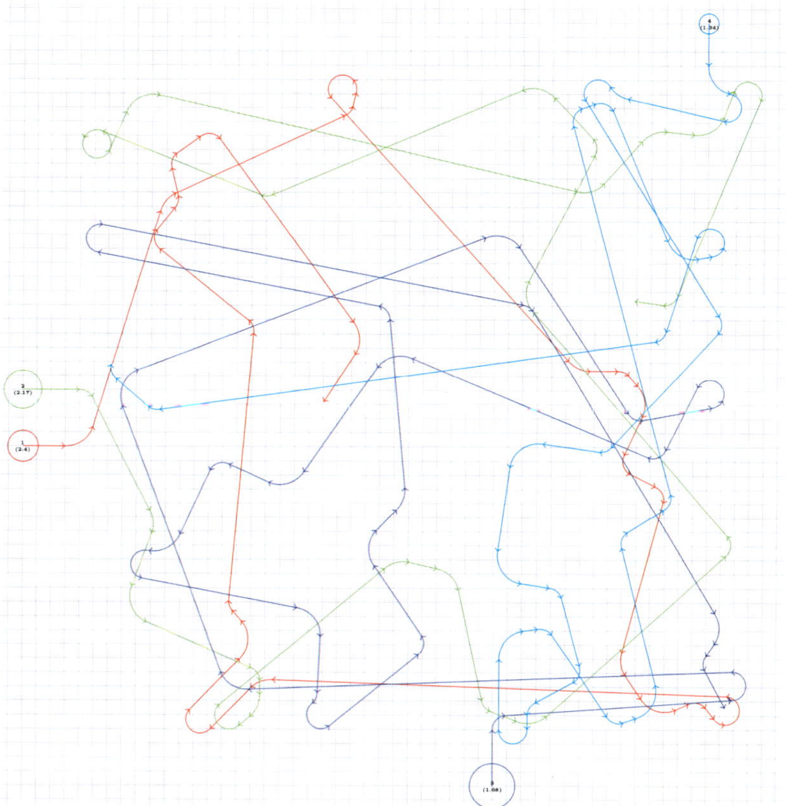

**Figure 5.** Visualization of benchmark instance 21 ($n = 4$) with 236 sectors and 419 resources.

*6.2. Measure of Quality*

Concerning the measure of quality, we will use percentage relative deviation, or PRD, defined as follows. Let $\mathcal{M} = \{LNF, FIFO, \ldots, ABC, TS\}$ be the set of solving methods (algorithms) we want to compare, described in Section 5. Let $\pi_m^I$ be solution obtained by method $m \in \mathcal{M}$ for problem instance $I$. Then for solving method $m$ and instance $I$ the value $PRD_m^I$ is equal to:

$$PRD_m^I = \frac{C_{\max}(\pi_m^I)}{\min_{z \in \mathcal{M}} C_{\max}(\pi_z^I)}. \tag{16}$$

In other words, we compare given method $m$ against a reference method $z \in \mathcal{M}$, which is the one which obtained the best (lowest) makespan for that instance $I$. Thus, $C_{\max}(\pi_z^I)$ serves as the best-known solution for instance $I$ and $z$ may change between different instances. For example, if $C_{\max}(\pi_z^I) = 250$ and $C_{\max}(\pi_m^I) = 350$ then $PRD_m^I = 1.4$, meaning that method $m$ obtained 40% worse result than best-known method $z$. Thus, *PRD* allows us to normalize the results, removing the bias caused by different instances and different problem sizes resulting in different ranges of observed makespans.

For the case with uncertain data we proceed slightly different. For a given instance $I$ we create a sample of 100 uncertain instances denoted $I(Y)_1$ through $I(Y)_{100}$. All instances draw from the same distribution, but each will be slightly different. Next, for solving method $m$, $m$ is used to solve the original (deterministic) instance $I$ with nominal values, obtaining solution $\pi_m^I$. Finally, we compute the value of $C_{\max}(\pi_m^{I(Y)})$ as the arithmetic mean:

$$C_{\max}(\pi_m^{I(Y)}) = \frac{1}{100}\sum_{i=1}^{100} C_{\max}(\pi_m^{I(Y)_i}),\tag{17}$$

thus, $C_{\max}(\pi_m^{I(Y)})$ estimates the expected value of makespan obtained by method $m$. Values $C_{\max}(\pi_m^{I(Y)})$ and $C_{\max}(\pi_z^{I(Y)})$ are then used to define $PRD_m^{I(Y)}$, similarly to (16).

### 6.3. Implementation and Equipment Details

All algorithms were implemented in Julia programming language (version 1.7.2) and run on a server with the AMD Ryzen Threadripper 3990X CPU (2.9 MHz, 64-cores) and 64 GB of RAM under Linux operating systems. Regarding the parameters of the ABC and TS methods, we performed a calibration using a portion of the instance described in Section 6.1, namely 30 instances with $n \in \{2, 3, 4\}$. For the TS method, four parameters were considered: (1) number of iterations, (2) initial solution, (3) cadence type, and (4) cadence size. For the number of iterations we have considered $maxIter \in \{60, 80, 100, 120\}$. For initial solution we considered $\pi_{\text{init}} \in \{\text{FIFO}, \text{LDRF}, \text{STTF}, \text{LNF}\}$. Three cadence types were considered. First, constant cadence with four values $c \in \{10, 100, 1000, 1000\}$. Second, linear cadence $c = \beta \times maxIter$ with four values $\beta \in \{0.2, 0.4, 0.6, 0.8\}$. Lastly, square root variant $c = \gamma \times \sqrt{maxIter}$ with four values $\gamma \in \{0.5, 1.0, 2.0, 4.0\}$. The preliminary research indicated that the best variant was $maxIter = 120$, $\pi_m athrminit = \text{LDRF}$, $c = \beta \times maxIter$, and $\beta = 0.4$. It should be noted that out of the four parameters only $maxIter$ has significant (and straightforward) effect on the running time. We have also consciously refrained from using higher number of iterations due to the size of the neighborhood.

Similar, preliminary research was done for the ABC method. Here, we have also assumed an upper bound for the computation time determined by the iteration limit and population size $maxIter \times p$. We have considered values $maxIter \in \{60, 80, 100, 120\}$ and $p \in \{5, 10, 15, 20\}$. Concerning the parameter $maxCount$, we have tested the same values as cadence $c$ for the TS method. The research indicated that the best results were obtained for $maxIter = 120$, $p = 20$ and linear $maxCount$ with $\beta = 0.4$.

### 6.4. Results

In our experiment we have assumed six different uncertainty scenarios. The first scenario is for deterministic data, i.e., $\sigma = 0$. The remaining five scenarios are for progressively increasing robot travel times uncertainty with $\sigma \in \{0.01, 0.05, 0.1, 0.15, 0.2\}$. Cases $\sigma = 0.01$, $\sigma = 0.1$, and $\sigma = 0.2$ can be taken to represents slightly, moderately, and largely uncertain data, respectively. First, we show exemplary results for two instances in Tables 2 and 3.

For instance, from Table 2 (which already has 132 resources), we observe that the TS and ABC methods provide the best results when the uncertainty is small ($\sigma < 0.05$), while the FIFO, SDTF, and STTF algorithms start to outperform them when uncertainty becomes more significant ($\delta \geq 0.05$). However, we still observe that TS and ABC outperform all other tested algorithms, often very significantly (50% to 90% differences). We also note that as $\delta$ increases, the quality of the performance of TS and ABC generally drops, while that of constructive algorithms increases. This is an expected result as constructive rule-based algorithms are usually viewed as more robust. Next, we will discuss the results for instance from Table 3 (which has nearly 20 times more resources). Here, the TS method managed to outperform all other algorithms for all $\delta$ values. On the other hand, ABC is consistently outperformed by three algorithms, but still remains ahead of the remaining ones. Interestingly, this time around the performance of ABC increases with the increase of $\delta$. From the constructive algorithms, FIFO and LDRF obtained the best results. We also note that the general performance gap between algorithms increased compared to instance from Table 2. This is easily seen by comparing the values of LOTF in both tables.

**Table 2.** *PRD* results for instance no. 13 ($n = 3$) with 121 sectors and 132 resources (best three results for each scenario type in bold).

| Algorithm | $\sigma = 0$ | $\sigma = 0.01$ | $\sigma = 0.05$ | $\sigma = 0.1$ | $\sigma = 0.15$ | $\sigma = 0.2$ |
|---|---|---|---|---|---|---|
| TS | **1.018** | **1.000** | **1.000** | **1.021** | 1.068 | 1.041 |
| ABC | **1.000** | **1.026** | **1.029** | **1.018** | 1.063 | 1.141 |
| LNF | 1.604 | 1.581 | 1.463 | 1.414 | 1.396 | 1.415 |
| FIFO | 1.444 | 1.097 | 1.043 | 1.035 | **1.032** | **1.032** |
| SDTF | 1.287 | 1.076 | 1.037 | 1.023 | **1.021** | **1.025** |
| STTF | 1.310 | **1.073** | **1.017** | **1.000** | **1.000** | **1.000** |
| LDTF | 1.630 | 1.229 | 1.167 | 1.158 | 1.162 | 1.183 |
| LTTF | 1.630 | 1.229 | 1.167 | 1.158 | 1.162 | 1.183 |
| SDRF | 1.630 | 1.229 | 1.167 | 1.158 | 1.162 | 1.183 |
| STRF | 1.630 | 1.229 | 1.167 | 1.158 | 1.162 | 1.183 |
| LDRF | **1.047** | 1.301 | 1.209 | 1.198 | 1.192 | 1.196 |
| LTRF | 1.174 | 1.102 | 1.186 | 1.201 | 1.162 | 1.193 |
| SODF | 1.900 | 1.384 | 1.313 | 1.286 | 1.284 | 1.279 |
| SOTF | 1.630 | 1.229 | 1.167 | 1.158 | 1.162 | 1.183 |
| LODF | 1.842 | 1.581 | 1.463 | 1.414 | 1.396 | 1.415 |
| LOTF | 1.760 | 1.554 | 1.395 | 1.320 | 1.244 | 1.274 |

**Table 3.** *PRD* results for instance no. 48 ($n = 6$) with 546 sectors and 2415 resources (best three results for each scenario type in bold).

| Algorithm | $\sigma = 0$ | $\sigma = 0.01$ | $\sigma = 0.05$ | $\sigma = 0.1$ | $\sigma = 0.15$ | $\sigma = 0.2$ |
|---|---|---|---|---|---|---|
| TS | **1.000** | **1.000** | **1.000** | **1.000** | **1.000** | **1.000** |
| ABC | 1.619 | 1.542 | 1.614 | 1.455 | 1.447 | 1.499 |
| LNF | 2.053 | 1.897 | 1.836 | 1.726 | 1.699 | 1.649 |
| FIFO | 1.679 | **1.355** | **1.314** | **1.245** | **1.234** | **1.197** |
| SDTF | 1.900 | 1.602 | 1.542 | 1.457 | 1.442 | 1.402 |
| STTF | 1.889 | 1.640 | 1.590 | 1.508 | 1.494 | 1.454 |
| LDTF | 1.615 | 1.963 | 1.903 | 1.799 | 1.777 | 1.713 |
| LTTF | 2.196 | 1.821 | 1.764 | 1.679 | 1.670 | 1.633 |
| SDRF | **1.327** | 1.988 | 1.927 | 1.826 | 1.807 | 1.756 |
| STRF | 1.864 | 1.987 | 1.926 | 1.825 | 1.804 | 1.752 |
| LDRF | 1.840 | **1.248** | **1.209** | **1.146** | **1.130** | **1.101** |
| LTRF | **1.258** | 1.495 | 1.449 | 1.370 | 1.358 | 1.321 |
| SODF | 1.528 | 1.993 | 1.932 | 1.831 | 1.811 | 1.760 |
| SOTF | 1.864 | 1.987 | 1.926 | 1.825 | 1.804 | 1.752 |
| LODF | 1.664 | 1.791 | 1.737 | 1.645 | 1.627 | 1.580 |
| LOTF | 1.677 | 1.918 | 1.860 | 1.761 | 1.744 | 1.693 |

We will now move onto aggregated (average) results grouped by number of robots $n$, which are shown in Table 4. For the sake of brevity, we have restricted ourselves to showing metaheuristics and three best constructive algorithms for each $n$. In general, we observe that effectiveness of both TS and ABC decreases with increase in $\sigma$. For the best constructive algorithms the relations is opposite—those algorithms improve under uncertainty. TS is also the best algorithm overall, being outperformed only a few times when $n$ is small and $\delta$ is high. ABC is significantly worse, being outperformed by TS and FIFO except for small $n$ and $\delta$. We also observe that FIFO remains fairly effective in almost all cases. SDTF, STTF, LDRF, and LTRF are consequently the next in the effectiveness order. However, all of those constructive algorithms are still often 20% to 70% worse than TS metaheuristic and this effect increases as $n$ increases.

**Table 4.** Aggregated average *PRD* results for selected algorithms, grouped by number of robots (best result for each scenario type in bold).

| $n$ | Algorithm | $\sigma = 0$ | $\sigma = 0.01$ | $\sigma = 0.05$ | $\sigma = 0.1$ | $\sigma = 0.15$ | $\sigma = 0.2$ |
|---|---|---|---|---|---|---|---|
| | TS | **1.000** | 1.094 | 1.094 | 1.090 | 1.087 | 1.083 |
| | ABC | **1.000** | **1.000** | **1.009** | 1.017 | 1.023 | 1.014 |
| 2 | FIFO | 1.020 | 1.020 | 1.021 | 1.020 | 1.020 | 1.017 |
| | SDTF | 1.018 | 1.021 | 1.019 | **1.016** | **1.015** | **1.011** |
| | LTRF | 1.033 | 1.021 | 1.021 | 1.020 | 1.020 | 1.017 |
| | TS | **1.015** | **1.013** | **1.053** | 1.089 | 1.103 | 1.104 |
| | ABC | 1.075 | 1.086 | 1.112 | 1.093 | 1.190 | 1.141 |
| 3 | LNF | 1.504 | 1.496 | 1.475 | 1.466 | 1.434 | 1.420 |
| | FIFO | 1.184 | 1.094 | 1.074 | **1.070** | **1.050** | **1.045** |
| | LDRF | 1.239 | 1.197 | 1.183 | 1.157 | 1.127 | 1.113 |
| | LTRF | 1.237 | 1.230 | 1.227 | 1.202 | 1.175 | 1.159 |
| | TS | **1.000** | **1.022** | **1.069** | **1.110** | **1.146** | **1.156** |
| | ABC | 1.260 | 1.320 | 1.300 | 1.325 | 1.324 | 1.384 |
| 4 | FIFO | 1.552 | 1.334 | 1.279 | 1.236 | 1.218 | 1.200 |
| | STTF | 1.682 | 1.286 | 1.242 | 1.204 | 1.187 | 1.179 |
| | LTRF | 1.361 | 1.457 | 1.401 | 1.352 | 1.335 | 1.317 |
| | TS | **1.025** | **1.136** | **1.204** | **1.258** | **1.301** | 1.311 |
| | ABC | 1.540 | 1.586 | 1.638 | 1.657 | 1.698 | 1.578 |
| 5 | FIFO | 1.386 | 1.358 | 1.338 | 1.313 | 1.312 | **1.298** |
| | LDRF | 1.643 | 1.529 | 1.503 | 1.463 | 1.444 | 1.417 |
| | LTRF | 1.498 | 1.676 | 1.648 | 1.614 | 1.601 | 1.584 |
| | TS | **1.000** | **1.014** | **1.031** | **1.038** | **1.048** | **1.070** |
| | ABC | 1.640 | 1.635 | 1.625 | 1.590 | 1.578 | 1.584 |
| 6 | FIFO | 1.437 | 1.297 | 1.253 | 1.214 | 1.192 | 1.182 |
| | STTF | 1.735 | 1.314 | 1.282 | 1.255 | 1.250 | 1.247 |
| | LTRF | 1.546 | 1.349 | 1.308 | 1.270 | 1.258 | 1.249 |

Finally, the aggregated results for all benchmark instances are shown in Table 5. Once again, TS's average performance is superior to all other tested algorithms. ABC remains competitive for deterministic case, but is outclassed by several algorithms (most notably FIFO) for uncertain cases. We also note that the implemented metaheuristics are fairly robust: TS and ABC suffered only 13.6% and 2.8% drop in PRD between the $\sigma = 0$ and $\sigma = 0.2$ cases, respectively. As previously, unlike the metaheuristics, the constructive algorithms become more competitive as uncertainty in the data increases. This is most visible for STTF, which gains 26.4% in PRD between the $\sigma = 0$ and $\sigma = 0.2$ cases, although it is still slightly behind FIFO. On the other hand, LNF improves the least by only 3.6%. Out of all the constructive algorithms, FIFO, STTF, and LDRF obtained the best results for all $\sigma$ scenarios. As for the worst algorithms, LODF, SOTF, and LTTF performed the worst for the $\sigma = 0$ case, while SOTF, STRF, LODF, SODF, and LNF performed the worst for $\sigma = 0.2$.

**Table 5.** Aggregated average *PRD* results for all benchmark instances (three best results for each scenario type in bold).

| Algorithm | $\sigma = 0$ | $\sigma = 0.01$ | $\sigma = 0.05$ | $\sigma = 0.1$ | $\sigma = 0.15$ | $\sigma = 0.2$ |
|---|---|---|---|---|---|---|
| TS | **1.008** | **1.056** | **1.090** | **1.117** | **1.137** | **1.145** |
| ABC | **1.303** | 1.325 | 1.337 | 1.337 | 1.363 | 1.340 |
| LNF | 1.586 | 1.620 | 1.590 | 1.563 | 1.545 | 1.531 |
| FIFO | **1.316** | **1.221** | **1.193** | **1.170** | **1.158** | **1.148** |
| SDTF | 1.502 | 1.316 | 1.292 | 1.268 | 1.254 | 1.242 |
| STTF | 1.509 | **1.261** | **1.236** | **1.213** | **1.201** | **1.194** |
| LDTF | 1.732 | 1.617 | 1.585 | 1.556 | 1.540 | 1.524 |
| LTTF | 1.749 | 1.596 | 1.564 | 1.537 | 1.521 | 1.505 |
| SDRF | 1.723 | 1.593 | 1.567 | 1.537 | 1.522 | 1.505 |
| STRF | 1.737 | 1.621 | 1.593 | 1.564 | 1.551 | 1.536 |
| LDRF | 1.437 | 1.277 | 1.254 | 1.227 | 1.213 | 1.200 |
| LTRF | 1.335 | 1.347 | 1.321 | 1.292 | 1.278 | 1.265 |
| SODF | 1.734 | 1.622 | 1.593 | 1.562 | 1.546 | 1.528 |
| SOTF | 1.746 | 1.628 | 1.602 | 1.575 | 1.561 | 1.546 |
| LODF | 1.761 | 1.623 | 1.593 | 1.562 | 1.545 | 1.530 |
| LOTF | 1.706 | 1.615 | 1.584 | 1.552 | 1.536 | 1.517 |

*6.5. Discussion*

While the presented research confirmed that the metaheuristic methods are effective with regards to makespan optimization for the considered MMRS, they suffer from vastly longer execution time compared to simple constructive algorithms. This is especially true for the TS metaheuristic, which is more effective but slower than ABC. Sometimes the running time is not an issue, but in some contexts (e.g., robot paths are unknown in advance) such a drawback might not be acceptable. Thus, here we discuss a few options to reduce this running times of both TS and ABC methods:

1. The solution evaluation procedure is by far the most time-consuming part of both methods. Thus, any reduction in its running time would benefit the overall algorithm. One possibility would be to abandon the "state safety" check. This would allow for a type I deadlock, but such solution would simply be treated as infeasible and rejected. It remains an open question how such change would affect the quality of both methods.

2. As mentioned in Section 4, it might be possible to reduce the solution size by forcing some of its element to always be equal to each other. This can be done optimally ("match" only elements that are sure to be matched in the optimal solution) or approximately (guess which elements could be matched, risking reduced solution quality).

3. Parallel computing: running time can be significantly reduced by dividing work among many processors. Both methods can be parallelized fairly well as neighbors in TS can be handled independent from each other. This is made easier by solution evaluation function being complex (this is unlike problems with $\mathcal{O}(1)$ neighbor evaluation which are harder to parallelize). This is similar for food sources in ABC.

4. Lastly, we can reduce running time significantly by decreasing the population size for ABC or by checking only a small (e.g., 10%) number of neighbors in TS. Moreover, we can simply reduce the iteration limit for either method. Both actions will generally affect the quality of obtained solutions, though it might be possible to reduce time significantly while still obtaining comparable results.

Nonetheless, a situation might occur in which metaheuristics will not be applicable due to time restriction or will not be able to provide satisfactory results in that time. In such cases, one needs to rely on constructive algorithms instead. For this, the best option appears to be the FIFO algorithm—for $\sigma = 0.2$ it is only 0.3% worse than TS. For the deterministic case, FIFO is considerably (30.1%) worse than TS, but still outclasses other algorithms. STTF and LDRF were also fairly effective, though to a lesser extent. The remaining constructive algorithms consistently proved to be considerably worse. Surprisingly, the STTF, LTTF,

STRF, LTRF, STF, and LOTF algorithms not only were not much better than their distance-based counterparts, but half the time they proved to be actually worse. This could be a coincidence, but it seems unlikely, as 50 random instances were used and robot speeds ranged considerably between each other.

## 7. Conclusions

In the paper we have considered the process of navigation for multiple mobile robots system with fixed paths operating in a real-valued two-dimensional space. Unlike existing works, we have tackled two problems at once: (1) optimization of robot moving schedules to minimize the makespan and (2) collision and deadlock avoidance. Based on the concept of sectors, we proposed a reformulation of the initial problem. This made it possible to disregard the specific shapes of robot paths, simplifying the problem.

We then introduced a binary solution representation based on resources derived from sectors. This allowed us to turn the original continuous optimization problem into discrete optimization formulation with vastly reduced solution space. However, we have shown that the resulting solution space is still large, even for relatively simple problem instances with only two or three robots. We have identified two types of deadlocks: (1) a classic deadlock caused by resources availability and (2) a deadlock caused by the solution-enforced robot queuing order. We have proposed approaches to avoid type I deadlocks while also resolving type II deadlocks.

For solving method, we have implemented two metaheuristic algorithms: taboo search (TS) and artificial bee colony (ABC). We have also implemented 14 constructive algorithms based on simple dispatch rules. Next, we provided 50 random problem instances as a testing benchmark, with number of resources ranging from dozens to several thousands. The results proved that TS outperforms other algorithms for both deterministic and uncertain cases. A few constructive algorithms (including FIFO, STTF, and LDRF) also obtained good results, lessening their gap to TS as uncertainty grows. ABC provided the best results for small instances, but its performance quickly dropped below that of TS, FIFO, STTF, and LDRF. Nonetheless, ABC managed to outperform the remaining constructive algorithms.

**Supplementary Materials:** The following supporting information can be downloaded at: https://www.mdpi.com/article/10.3390/pr10102087/s1.

**Author Contributions:** Conceptualization, J.R. and E.R.; methodology, J.R. and E.R.; software, J.R., R.I., and K.K.; validation, J.R. and R.I.; formal analysis, J.R. and E.R.; investigation, J.R. and R.I.; data curation, J.R.; writing—original draft preparation, J.R.; writing—review and editing, J.R. and R.I.; visualization, R.I. All authors have read and agreed to the published version of the manuscript.

**Funding:** This research was supported by the National Science Centre, Poland, under project number 2016/23/B/ST7/01441.

**Institutional Review Board Statement:** Not applicable.

**Informed Consent Statement:** Not applicable.

**Data Availability Statement:** Generated benchmark instances data are contained within the supplementary materials.

**Conflicts of Interest:** The authors declare no conflict of interest.

## Abbreviations

The following abbreviations are used in this manuscript:

| | |
|---|---|
| ABC | Artificial Bee Colony |
| FIFO | First-In First-Out |
| LDRF | Longest Distance Remaining First |
| LDTF | Longest Distance Traveled First |
| LNF | Lower Number First |
| LODF | Longest Overall Distance First |

|      |                                         |
|------|-----------------------------------------|
| LOTF | Longest Overall Time First              |
| LTRF | Longest Time Remaining First            |
| LTTF | Longest Time Traveled First             |
| MMRS | Multiple Mobile Robots System           |
| PRD  | Percentage Relative Deviation           |
| RCPS | Resource-Constrained Project Scheduling |
| SDRF | Shortest Distance Remaining First       |
| SDTF | Shortest Distance Traveled First        |
| SODF | Shortest Overall Distance First         |
| SOTF | Shortest Overall Time First             |
| STRF | Shortest Time Remaining First           |
| STTF | Shortest Time Traveled First            |
| TS   | Taboo Search                            |

## References


1. Roszkowska, E.; Jakubiak, J. Control synthesis for multiple mobile robot systems. *Trans. Inst. Meas. Control* **2021**. [CrossRef]
2. Roszkowska, E. Provably Correct Closed-Loop Control for Multiple Mobile Robot Systems. In Proceedings of the 2005 IEEE International Conference on Robotics and Automation, Barcelona, Spain, 18–22 April 2005; pp. 2810–2815. [CrossRef]
3. Reveliotis, S.; Roszkowska, E. Conflict Resolution in Free-Ranging Multivehicle Systems: A Resource Allocation Paradigm. *IEEE Trans. Robot.* **2011**, *27*, 283–296. [CrossRef]
4. Roszkowska, E.; Reveliotis, S. A Distributed Protocol for Motion Coordination in Free-Range Vehicular Systems. *Automatica* **2013**, *49*, 1639–1653. [CrossRef]
5. Kloetzer, M.; Mahulea, C.; Colom, J.M. Petri net approach for deadlock prevention in robot planning. In Proceedings of the 2013 IEEE 18th Conference on Emerging Technologies & Factory Automation (ETFA), Cagliari, Italy, 10–13 September 2013; pp. 1–4. [CrossRef]
6. Gakuhari, H.; Jia, S.; Takase, K.; Hada, Y. Real-Time Deadlock-Free Navigation for Multiple Mobile Robots. In Proceedings of the 2007 International Conference on Mechatronics and Automation, Harbin, China, 5–8 August 2007; pp. 2773–2778. [CrossRef]
7. Čáp, M.; Novák, P.; Kleiner, A.; Selecký, M. Prioritized Planning Algorithms for Trajectory Coordination of Multiple Mobile Robots. *IEEE Trans. Autom. Sci. Eng.* **2015**, *12*, 835–849. [CrossRef]
8. Ferrera, E.; Castaño, A.R.; Capitán, J.; Ollero, A.; Marrón, P.J. Decentralized collision avoidance for large teams of robots. In Proceedings of the 2013 16th International Conference on Advanced Robotics (ICAR), Montevideo, Uruguay, 25–29 November 2013; pp. 1–6. [CrossRef]
9. Ferrera, E.; Capitán, J.; Castaño, A.R.; Marrón, P.J. Decentralized safe conflict resolution for multiple robots in dense scenarios. *Robot. Auton. Syst.* **2017**, *91*, 179–193. [CrossRef]
10. Bhattacharjee, P.; Rakshit, P.; Goswami, I.; Konar, A.; Nagar, A.K. Multi-robot path-planning using artificial bee colony optimization algorithm. In Proceedings of the 2011 Third World Congress on Nature and Biologically Inspired Computing, Salamanca, Spain, 19–21 October 2011; pp. 219–224. [CrossRef]
11. Liang, J.H.; Lee, C.H. Efficient collision-free path-planning of multiple mobile robots system using efficient artificial bee colony algorithm. *Adv. Eng. Softw.* **2015**, *79*, 47–56. [CrossRef]
12. Mansury, E.; Nikookar, A.; Salehi, M.E. Artificial Bee Colony optimization of ferguson splines for soccer robot path planning. In Proceedings of the 2013 First RSI/ISM International Conference on Robotics and Mechatronics (ICRoM), Tehran, Iran, 13–15 February 2013; pp. 85–89. [CrossRef]
13. Kumar, S.; Muni, M.K.; Pandey, K.K.; Chhotray, A.; Parhi, D.R. Path planning and control of mobile robots using modified Tabu search algorithm in complex environment. In Proceedings of the International Conference on Artificial Intelligence in Manufacturing & Renewable Energy (ICAIMRE), Bhubaneswar, India, 25–26 October 2019.
14. Balan, K.; Luo, C. Optimal Trajectory Planning for Multiple Waypoint Path Planning using Tabu Search. In Proceedings of the 2018 9th IEEE Annual Ubiquitous Computing, Electronics & Mobile Communication Conference (UEMCON), New York, NY, USA, 8–10 November 2018; pp. 497–501. [CrossRef]
15. Lygeros, J.; Godbole, D.; Sastry, S. Verified hybrid controllers for automated vehicles. *IEEE Trans. Autom. Control* **1998**, *43*, 522–539. [CrossRef]
16. Tomlin, C.; Pappas, G.; Sastry, S. Conflict resolution for air traffic management: A study in multiagent hybrid systems. *IEEE Trans. Autom. Control* **1998**, *43*, 509–521. [CrossRef]
17. Pecora, F.; Cirillo, M.; Dimitrov, D. On mission-dependent coordination of multiple vehicles under spatial and temporal constraints. In Proceedings of the 2012 IEEE/RSJ International Conference on Intelligent Robots and Systems, Vilamoura, Algarve, Portugal, 7–12 October 2012; pp. 5262–5269. [CrossRef]
18. Andreasson, H.; Bouguerra, A.; Cirillo, M.; Dimitrov, D.N.; Driankov, D.; Karlsson, L.; Lilienthal, A.J.; Pecora, F.; Saarinen, J.P.; Sherikov, A.; et al. Autonomous Transport Vehicles: Where We Are and What Is Missing. *IEEE Robot. Autom. Mag.* **2015**, *22*, 64–75. [CrossRef]

19. Grover, J.S.; Liu, C.; Sycara, K. Deadlock analysis and resolution for multi-robot systems. In Proceedings of the International Workshop on the Algorithmic Foundations of Robotics, Oulu, Finland, 21–23 June 2021; pp. 294–312.
20. Akella, S.; Hutchinson, S. Coordinating the motions of multiple robots with specified trajectories. In Proceedings of the 2002 IEEE International Conference on Robotics and Automation (Cat. No. 02CH37292), Washington, DC, USA, 11–15 May 2002; Volume 1, pp. 624–631. [CrossRef]
21. Wang, Z.; Gombolay, M. Learning Scheduling Policies for Multi-Robot Coordination With Graph Attention Networks. *IEEE Robot. Autom. Lett.* **2020**, *5*, 4509–4516. [CrossRef]
22. Lin, S.; Liu, A.; Wang, J.; Kong, X. A Review of Path-Planning Approaches for Multiple Mobile Robots. *Machines* **2022**, *10*, 773. [CrossRef]
23. De Frene, E.; Schatteman, D.; Herroelen, W.; Van de Vonder, S. A Heuristic Methodology for Solving Spatial Resource-Constrained Project Scheduling Problems. 2007. Available online: https://ssrn.com/abstract=1089355 (accessed on 30 September 2022)
24. Prashant Reddy, J.; Kumanan, S.; Krishnaiah Chetty, O. Application of Petri nets and a genetic algorithm to multi-mode multi-resource constrained project scheduling. *Int. J. Adv. Manuf. Technol.* **2001**, *17*, 305–314. [CrossRef]
25. Lawley, M.; Reveliotis, S.; Ferreira, P. A correct and scalable deadlock avoidance policy for flexible manufacturing systems. *IEEE Trans. Robot. Autom.* **1998**, *14*, 796–809. [CrossRef]
26. Golmakani, H.; Mills, J.; Benhabib, B. Deadlock-free scheduling and control of flexible manufacturing cells using automata theory. *IEEE Trans. Syst. Man, Cybern. - Part A Syst. Humans* **2006**, *36*, 327–337. [CrossRef]
27. Sun, Y.; Chung, S.H.; Wen, X.; Ma, H.L. Novel robotic job-shop scheduling models with deadlock and robot movement considerations. *Transp. Res. Part E Logist. Transp. Rev.* **2021**, *149*, 102273. [CrossRef]
28. Dang, Q.V.; Nguyen, C.T.; Rudová, H. Scheduling of mobile robots for transportation and manufacturing tasks. *J. Heuristics* **2019**, *25*, 175–213. [CrossRef]
29. Sun, H.; Elghazi, R.; Gainaru, A.; Aupy, G.; Raghavan, P. Scheduling Parallel Tasks under Multiple Resources: List Scheduling vs. Pack Scheduling. In Proceedings of the 2018 IEEE International Parallel and Distributed Processing Symposium (IPDPS), Vancouver, BC, Canada, 21–25 May 2018; pp. 194–203. [CrossRef]
30. Gopalan, K.; Kang, K.D. Coordinated allocation and scheduling of multiple resources in real-time operating systems. In Proceedings of the Workshop on Operating Systems Platforms for Embedded Real-Time Applications, Pisa, Italy, 4–6 July 2007; p. 48.
31. Reveliotis, S.; Roszkowska, E. On the Complexity of Maximally Permissive Deadlock Avoidance in Multi-Vehicle Traffic Systems. *IEEE Trans. Autom. Control* **2010**, *55*, 1646–1651. [CrossRef]
32. Rudy, J.; Zelazny, D. Memetic algorithm approach for multi-criteria network scheduling. In Proceedings of the International Conference On ICT Management for Global Competitiveness And Economic Growth In Emerging Economies, Wroclaw, Poland, 17–18 September 2012; pp. 247–261.
33. Wang, Z.; Li, M.; Dou, L.; Li, Y.; Zhao, Q.; Li, J. A novel multi-objective artificial bee colony algorithm for multi-robot path planning. In Proceedings of the 2015 IEEE International Conference on Information and Automation, Lijiang, China, 8–10 August 2015; pp. 481–486. [CrossRef]
34. Lipowski, A.; Lipowska, D. Roulette-wheel selection via stochastic acceptance. *Phys. A Stat. Mech. Its Appl.* **2012**, *391*, 2193–2196. [CrossRef]
35. Rudy, J.; Pempera, J.; Smutnicki, C. Improving the TSAB algorithm through parallel computing. *Arch. Control. Sci.* **2020**, *30*, 411–435.
36. Idzikowski, R.; Rudy, J.; Gnatowski, A. Solving Non-Permutation Flow Shop Scheduling Problem with Time Couplings. *Appl. Sci.* **2021**, *11*, 4425. [CrossRef]

*Article*

# Decision-Making Model of Mechanical Components in a Lean–Green Manufacturing System Based on Carbon Benefit and Its Application

Xiaoyong Zhu [1], Yongmao Xiao [2,3,4,*] and Gongwei Xiao [1]

1  School of Economics & Management, Shaoyang University, Shaoyang 422000, China
2  School of Computer and Information, Qiannan Normal University for Nationalities, Duyun 558000, China
3  Key Laboratory of Complex Systems and Intelligent Optimization of Guizhou Province, Duyun 558000, China
4  Key Laboratory of Complex Systems and Intelligent Optimization of Qiannan, Duyun 558000, China
*  Correspondence: xym198302@163.com

**Abstract:** The key to achieving low-carbon manufacturing is to effectively reduce the carbon emissions of production systems and improve carbon benefits. The use of lean and green tools aids in measuring the added value of products, and increases the efficiency and sustainability of production systems. To address this problem and verify that the synergetic relationship between lean and green innovation increases the efficiency and sustainability in production systems, a new low-carbon manufacturing evaluation indicator—carbon benefit—in lean manufacturing systems was discussed. A low-carbon decision-making model of multiple processes aiming at carbon benefit maximization, as well as the dynamic characteristics of carbon benefit and sustainable process improvements in a lean production system, was established. A case study of a certain satellite dish parts manufacturing line was introduced to analyze and verify the feasibility of the proposed model. After improvement, the processing time of unit parts was reduced from 63 s to 54 s. The workstations were optimized again according to the lean–green manufacturing concept, and the number was reduced by 37.5%. The process was recombined and reduced from 8 to 5 to achieve continuous-flow processing. This reduced the distance by 77 m, and at the same time, the number of operating personnel was reduced, and the after-improvement carbon efficiency increased from 12.98 s/kg $CO_2$e to 36.33 s/kg $CO_2$e in comparison with that before the improvement. The carbon benefit after improvement was 193.92% higher than that before the improvement.

**Keywords:** carbon benefit; carbon efficiency; carbon emission; process decision-making; lean–green

**Citation:** Zhu, X.; Xiao, Y.; Xiao, G. Decision-Making Model of Mechanical Components in a Lean–Green Manufacturing System Based on Carbon Benefit and Its Application. *Processes* **2022**, *10*, 2297. https://doi.org/10.3390/pr10112297

Academic Editor: Sergey Y. Yurish

Received: 19 October 2022
Accepted: 3 November 2022
Published: 4 November 2022

**Publisher's Note:** MDPI stays neutral with regard to jurisdictional claims in published maps and institutional affiliations.

## 1. Introduction

According to the 2014 International Energy Agency's report on the "Five Key Technologies for Low Carbon Energy Outlook", more than 80% of nearly two-thirds of the greenhouse gas emissions worldwide are comprised of carbon dioxide [1]. Enterprises' strategies and consumer preferences have been focusing increasingly on environmental protection. To date, an increasing number of transnational companies have adopted the lean–green philosophy to maintain competitive power in the global market. The automobile manufacturing industry is also a pioneer in implementing lean–green manufacturing technology. For instance, Toyota has realized sustainable management using lean–green. In 2015, Toyota Motor Corporation made contributions to the realization of sustainable development society and released the strategy of "Toyota Environmental Challenge 2050". Aiming at global environmental problems such as climate change, water resource shortage, resource depletion, and biodiversity reduction, the strategy launched six challenges in three major areas, namely, "building better cars", "better production activities", and "a better city and a better society", with the goal of "making the negative impact of cars infinitely close to zero" and "bringing positive energy to society". Most manufacturing companies understand and accept this philosophy. The concept of lean production involves increasing the efficiency and quality of the entire manufacturing process while lowering the cost by eliminating waste in the product/service manufacturing process [2]. Lean production has a series of practical methods and tools to reduce manufacturing waste, eliminate negative environmental impacts, save resources and energy sources, and increase efficiency and benefits [3]. Furthermore, lean production is beneficial for relieving environmental pollution, removing barriers for enterprises to import and use new pollution control technologies, building a conducive atmosphere for enterprises to thrive, and emphasizing the benefits of enterprises by relieving environmental pollution. In addition, 3R technologies (reduce, reuse, and recycle) in green manufacturing present similar attributes to those of lean production. Enterprises should adopt green environmental production design and recycling parts in the manufacturing process while eliminating waste, thereby continuously improving the environment through lean production.

In recent years, market dynamics have changed with the rise of operations, environmental, social, and quality improvement methods (e.g., Lean, Six Sigma, and Green), and increasing attention has been paid to environmental and social responsibility [4–6]. Traditionally, productivity and profitability, as well as recent quality, customer satisfaction, and flexibility, have been the major concerns for organizations [7–11]. However, organizations are forced to reconsider how to manage their processes and operations and seek innovative ways of doing business in response to governmental environmental regulations and customer demands for environmentally sustainable services and products [12–14]. Thus, the challenge for the organization is to achieve economic success through strategies that are compatible with and support environmental and social sustainability to meet all stakeholder requirements [15]. To this end, lean and green technologies have become an important part of sustainable solutions [5,16]. Lean refers to eliminating waste in all production areas, supplier networks, design, and plant management [17], thus potentially increasing resource efficiency and reducing the environmental impact [18,19]. Green manufacturing is an integrated approach designed to reduce negative environmental impacts and waste in all areas of the product and service life cycle [20]. This enables companies to achieve a range of long-term performances, especially in reducing costs through a more efficient use of resources [21].

There are usually seven sources of wastes at a production site: overproduction, inventory, transportation, over-processing, over-action, waiting, and defects. Although the **EPA (United States Environmental Protection Agency)** has proposed a relationship between the seven wastes and their environmental impacts, most current literature only focuses on the seven wastes from the production management level [22]. Currently, only few studies have estimated the carbon emissions generated by analyzing and quantifying the seven wastes. Therefore, this study proposes an improved model of the process optimization method for collecting carbon emissions from production equipment, materials, transportation, and storage in the form of time flow, energy flow, material flow, transport flow, and carbon emissions: visualization of carbon emissions, quantification of value-added carbon emissions and non-value-added carbon emissions in production, and establishment of mathematical models based on carbon efficiency. The carbon emissions generated by the seven wastes were analyzed and quantified. Application analysis of the metal stamping part production process was conducted to verify the feasibility of the model.

Therefore, the objectives of this research were formulated as follows:

(1) To analyze the functional characteristics of three types of carbon emissions in the typical mechanical manufacturing process, such as material carbon, energy carbon, and process carbon, establish the carbon emission characteristic function of manufacturing system, quickly estimate and quantify the carbon emissions of products, and illustrate the application of this function in the selection of processing methods and processing sequences in process planning through a case study.

(2) To research low-carbon manufacturing process decision-making problems combined with a green ideas lean transformation product manufacturing system, explore the abil-

ity of different process route selections for low-carbon manufacturing of components, put forward the key decision-making indicators and quantitative analysis method (a process decision-making model based on low carbon efficiency), and improve the product line layout and the overall carbon efficiency improvement.

(3)  To demonstrate that environmental innovation and the transformation of production systems to lean systems can improve revenues, reduce costs, fulfill better social responsibility, and achieve corporate sustainability.

The paper consists of five parts. The rest of the paper is structured as follows: Section 2 presents the study background, namely the concepts of lean manufacturing, green manufacturing, and the lean–green. The research methodology is discussed in Section 3, followed by a presentation of the case study in Section 4. In the last section, some conclusions are drawn.

## 2. Background

This section presents some concepts of lean manufacturing, green manufacturing, and the lean–green synergetic model.

### 2.1. Lean Manufacturing

Lean Manufacturing (LM) can be viewed as a convergent sociotechnical approach aimed at reducing waste by minimizing manufacturing changes. Large corporations and SMEs strive to achieve considerable improvements in operational performance [23–25]. The implementation of lean manufacturing practices enables a business to remain competitive and thrive in the long term as it focuses on the pursuit of business operations that deliver value to customers.

Lean manufacturing is recognized for its long-term business goals of eliminating waste on the factory floor and in service. It is an organizational model that originated in Toyota Motor Corporation's Toyota Production System and enabled the company to remain competitive and thrive in a highly competitive automotive market under highly restrictive circumstances. MIT studied the success of TPS, characterized it in many success-based cases and events, and popularized the concept, which was then adopted by many manufacturing companies to improve the shop-floor-related management operating efficiency [23–25]. Lean thinking aims to achieve the goals of high quality, low cost, and shorter lead times by reducing waste in operational and supply chain processes. The seven types of waste involved are stock on hand, transportation, overproduction, processing, body motion, defects, and waiting [23–28]. To achieve this, academia and industry have summarized and developed many practical tools and techniques, including supplier quality management, lean supply chain, just-in-time, automation with herringbone, employee engagement, quality circles, hierarchical scheduling, Kanban, process manufacturing, one flow production, visual control, employee engagement, SMED, poka-yoke, VSM, TPM, and 5S [2,7,20,22,26,29–32].

### 2.2. Green Manufacturing

While implementing lean production, manufacturing enterprises must also be regulated by environmental laws and regulations in the production process to reduce its environmental impact, prevent pollution, and control pollution by adopting green manufacturing throughout the product life cycle. Green manufacturing (GM) refers to a modern manufacturing model that comprehensively considers environmental impact and resource efficiency on the premise of ensuring product function, quality, and cost. This ensures the design, manufacture, and use of scrap in the entire product life cycle, thereby minimizing or producing no environmental pollution in line with environmental protection requirements. It is also harmless or minimally harmful to the ecological environment, saving resources and energy and minimizing energy consumption [7,14,33–46].

In the product manufacturing process of manufacturing enterprises, the green manufacturing model mainly eliminates the negative environmental output of products through cleaner production, product life cycle management, an environmental management system (ISO14000 standard, among others), remanufacturing, and other methods and tools [47]. The goal of green manufacturing is to minimize the impact on the environment (negative effect) and maximize the utilization of resources in the entire product life cycle from design, manufacturing, packaging, transportation, and use of scrapping, and to coordinate and optimize the economic and social benefits of the enterprise [7,14,33–46]. In summary, the implementation of green manufacturing will provide many benefits to manufacturing enterprises.

### 2.3. The Integrated Relationship between Lean–Green Paradigms

Two paradigms, lean and green manufacturing, have been independently studied in the literature. However, the manufacturing system is highly complex, and a single approach to improve performance is often insufficient. The complex internal coupling relationship, key factors, and important interactions within the manufacturing system cannot be ignored, because the manufacturing performance is affected by multiple factors. The integration of lean and green practices is jointly promoted by internal factors such as risk management, profitability, and cost; changes in corporate culture; a focus on continuous innovation and process improvement; external factors such as customer and environmental pressure, government policies and regulations, and the potential for further profit by increasing customer value [7,14,33–38,40–46].

The practices of lean and green integration have many similarities, including reducing waste; practical tools to eliminate waste; shortening delivery and lead time; cooperating with suppliers; jointly improving environmental efficiency and profitability; realizing the economic, social, and environmental benefits of business organizations. Many scholars have verified that the key synergy between the two paradigms is reducing waste in the manufacturing process and/or operations in the supply chain [34–49]. In addition, lean and green practice aims to minimize the waste of transportation and handling to consequently save costs (lean), reduce $CO_2$ emissions (green), and shorten waiting and delivery times. Therefore, the synergy between lean and eco-efficient methods, commonly known as lean–green, is an arduous task.

## 3. Methodology

### 3.1. Mathematical Models

#### 3.1.1. Definition of Carbon Benefit and Carbon Efficiency

In the actual manufacturing process, after optimizing and improving the production line using the lean–green manufacturing methods and reforming the product mechanical process, a variety of different process selections can be used. Therefore, there are substantial differences in the production time, cost benefit, and energy consumption of the manufacturing process of the same production. To evaluate the actual improvement effect of lean–green manufacturing in the product manufacturing process, we must consider the main evaluation indicators of lean–green manufacturing theory—production efficiency, production benefit, production cycle time, and carbon emissions; this study uses two evaluation index systems of carbon efficiency and carbon benefit.

Carbon benefit ($CB$) refers to the economic benefit created by consuming unit time and producing unit carbon emission. Its mathematical expression is as follows:

$$CB = \left(\frac{B}{C}\right)/T \tag{1}$$

where $B$ is the economic benefit generated by the processing unit product, $C$ is the carbon emissions produced by the processing unit product, and $T$ is the production cycle or beat time of the processing unit product.

In the present study, an evaluation indicator of carbon efficiency was used to create more value and produce less environmental impact in lean production, thereby achieving certain production objectives [50–53]. According to [54], the formula can be expressed as:

$$C - efficiency = \frac{k \times t_{va}}{C_{total}} \tag{2}$$

where $k$ is a factor assigned to calculate the data, $t_{va}$ is the value-added time, and $C_{total}$ is total carbon emissions.

### 3.1.2. Calculation of Carbon Emissions

The conversion process of product manufacturing is analyzed from the input of various production factors (raw materials, auxiliary materials, and energy) and the output of products and waste. Simultaneously, part of the input energy is converted into useful work, and part of it is output into heat energy. The IPO process used in production is shown in Figure 1. The carbon emissions generated during the manufacturing process are composed of direct and indirect carbon emissions, as shown in Figure 2. $C_{direct}$ refers to the carbon emissions generated by fuel combustion in the production process, which directly changes the shape, size, and performance of raw materials, and the carbon emissions generated by the mutual chemical interaction of auxiliary materials in the production process, such as $CO_2$, $NO_x$, and other gases generated in the casting or heat treatment process. $C_{indirect}$ refers to carbon emissions calculated by the conversion coefficient of carbon emissions of raw materials, energy consumption of production equipment, and auxiliary materials used in the production process of products.

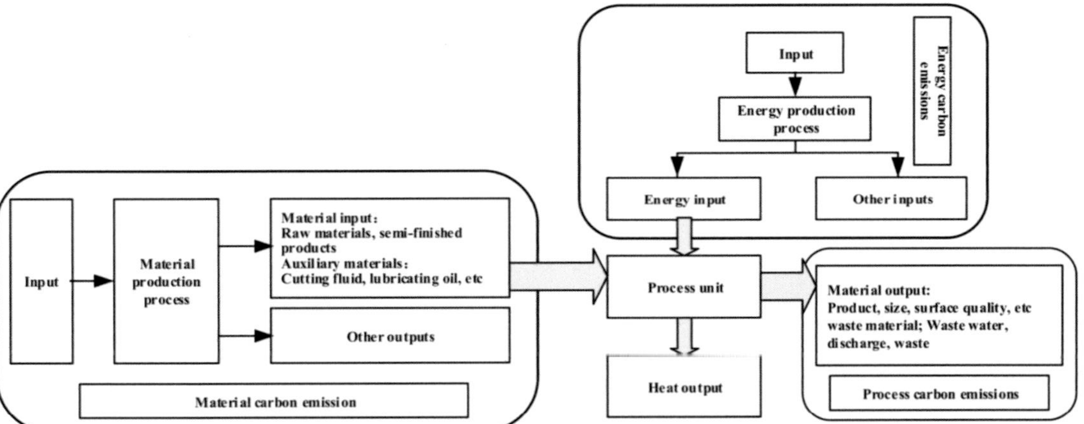


**Figure 1.** Schematic diagram of the IPO process of the product manufacturing conversion process.

The $C_{direct}$ of the product manufacturing process mainly comprises the direct carbon emission of each manufacturing process. The carbon emissions generated by the manufacturing process are primarily from three sources: from the plastic deformation process, the machining process such as cutting, and the assembly process. The formula for calculating $C_{direct}$ in the manufacturing process is as follows:

$$C_{direct} = \left( \sum_{i=1}^{D} \sum_{j=1}^{N1} W_{ij} \times E_{ij}^1 + \sum_{i=1}^{D} \sum_{j=1}^{N2} V_{ij} \times E_{ij}^2 + \frac{E_0 \times T_0}{\sum n_k \times T_k} \right) \times EF_{elec} \tag{3}$$

where $N_1$ is the type of plastic deformation and other processes; $W_{ij}$ is the weight of the $i$-th material using the $j$-th plastic deformation process; $E^1_{ij}$ is the embodied energy consumption of the $i$-th material for the $j$-th plastic deformation; $N_2$ is the type of cutting and other processes; $V_{ij}$ is the amount of removal or treatment of the $i$-th material for the $j$-th mechanical process; $E^2_{ij}$ is the specific energy consumption of the $i$-th material for the $j$-th process; $E_0$ is the total energy consumption of the assembly workshop; $T_0$ is the current calculated product assembly man-hour quota; $n_k$ is the quantity of $k$-type products in the workshop number; $T_k$ is the assembly man-hour quota of $k$-type *products*; $EF_{eleck}$ is the discharge coefficient of electric energy, whose value is 2.41 kgCO$_2$e/kWh.

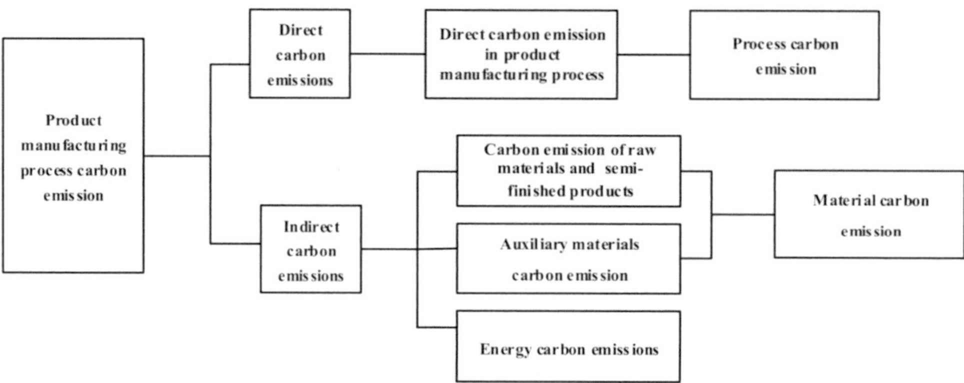


**Figure 2.** Classification of carbon emissions during product production.

$C_{indirect}$ represents the carbon emissions generated by the energy consumption of raw materials in the product manufacturing process, auxiliary materials, and production equipment. The formula for $C_{indirect}$ is as follows:

$$
\begin{aligned}
C_{indirect} &= C_m + C_{am} + C_E + C_{shop}/Q \\
&= q_m \times EF_{elec} + \sum_{i=1}^{n} \frac{C_{ami}}{N_i} + \sum_{j=1}^{k} \left( E_{idle,j} + E_j \times m_j \right) \times EF_{elec} \\
&+ \frac{E_{shop} \times EF_{elec}}{Q}
\end{aligned}
\tag{4}
$$

where $C$ is the carbon emissions of the raw materials, $C_{am}$ is the carbon emissions of the auxiliary materials, $C_E$ is the carbon emissions of the energy consumed by the equipment to maintain basic operation, $C_{shop}$ is the carbon emissions of workshop auxiliary equipment, $Q$ is the annual output of the workshop, $q_m$ is the weight of the initial input raw materials, $C_{ami}$ is the carbon emissions of the $i$-th type of auxiliary material, $N_i$ is the number of products in the $i$-th type of auxiliary material during its service life, $E_{idle,j}$ is the no-load energy consumption of the $j$-th type of equipment, $E_j$ is the specific energy consumption of the $j$-th processing method, and $m_j$ is the weight of the workpiece.

Carbon emissions emanate from multiple sources, such as indirect and direct carbon emissions, in mechanical manufacturing systems. In a lean production system, the production process is divided into value-creating and non-value-creating processes in the manufacturing system. In summary, this equation expresses the total carbon emissions (5).

$$
C_{total} = C_{direct} + C_{indirect} = C_{va} + C_{nva}
\tag{5}
$$

According to this study [23], the carbon emissions of the value-creating process calculation formula ($C_{va}$) are as follows:

$$
C_{va} = \sum_{i=1}^{P} \sum_{j=1}^{N} \left( Q_{i,j}^m \times EF_{i,j}^m \times M_{i,j} \right) + \sum_{i=1}^{P} \sum_{l=1}^{S} \left( E_{idle} + P_{i,l} \times t_{i,l}^{va} \right) \times EF_{elec}
\tag{6}
$$

where $Q_{i,j}^m$ is the weight of the $j$-th consumption of the raw material of the single piece product of the $i$-th process; $EF_{i,j}^m$ is the carbon emission coefficient of the $j$-th consumption of the raw material of the $i$-th process; $M_{i,j}$ is the material utilization ratio of the $j$-th consumption of the raw material of the $i$-th process; $E_{i,l}^{idle}$ is the no-load energy consumption of the $l$-th piece of equipment of the $i$-th process; $P_{i,l}$ is the rated power of the $l$th piece of equipment of the $i$-th process; $t_{i,l}^{va}$ is the effective working time of the $l$-th piece of equipment of the $i$-th process (value-added time).

### 3.1.3. Calculation of Carbon Emissions Generated by the Seven Wastes

According to the environmental impact shown in this study [54], carbon emissions generated ($C_{wastes}$) by the seven wastes can be calculated as follows:

$$C_{wastes}^v = \left[ \sum_{i=1}^{P} \sum_{j=1}^{N} \left( \Delta Q_{m_{i,j}}^v \times EF_{i,j}^m \right) \times Y_m^{v_k} + \sum_{i=1}^{P} \sum_{l=1}^{S} \left( E_{i,l}^{idle} + P_{i,l} \times \Delta t_{i,l}^v \right) \times EF_{elec} \times Y_E^{v_k} + \sum_{w=1}^{R} \Delta E_{Tw}^v \times EF_{elec} \times Y_T^{v_k} + \Delta E_I^v \times EF_{elec} \times Y_I^{v_k} \right] \times \eta^v \quad (7)$$

The meaning of each parameter in Formula (7) is explained in the papers [54–56]. During the manufacturing process, the raw materials or semi-finished products that need to be processed owing to the dispersion of equipment are moved between each process. Therefore, the transportation carbon emission $C_t$ is:

$$C_t = \sum_{t=1}^{t_o} D_t \times TE_t \quad (8)$$

where, $t_o$ is the type of transportation mode adopted in the process of transportation or movement, $D_t$ is the distance of the $t$-th mode of transportation, and $TE_t$ is the carbon emission coefficient of the $t$-th mode of transportation.

In conclusion, the manufacturing process of mechanical products includes parts machining and whole machine assembly. Manufacturing processes such as casting, extrusion, stamping, cold/hot rolling, turning, milling, grinding, surface hardening, annealing, and tempering may be involved in the processing of parts. According to statistics, in the process of mechanical manufacturing, carbon emissions mainly come from the consumption of electric energy [57]. Therefore, when calculating the carbon emissions of machine tool manufacturing processes, this paper focuses on the carbon emissions caused by the energy consumption of the process. For the carbon emissions of plastic deformation processes such as stamping, it can be calculated according to the embodied energy of the process [58]. The energy consumption of turning, milling, surface hardening, annealing, and tempering processes can be calculated according to the specific energy of the process [57]. Energy consumption in the assembly process shall be allocated according to the total energy consumption of the assembly workshop, the working hour quota, and the assembly quantity of various types of machine tools during a period of time after investigation.

## 4. Case Study

### 4.1. Problem Description and Data Collection

We selected the production line of a manufacturing unit that produced satellite dish kits as the case study. The daily production demand was 1000 pieces, of 0.7 mm thickness of 500 g weight, made of different materials such as copper, aluminum, brass, and galvanized steel. The production line adopted a double working shift, with an effective shift time of 7.5 h per shift. The production of the satellite dish kits included five main processes (Figure 3). Therefore, the takt time to manufacture each satellite dish kit was 54 s; the process cycle time of each satellite dish part is shown in Figure 4. The carbon emission coefficient of steel was 7.048 kg $CO_2e/kg$ [58], and $EF_{eleck}$ in the present study was 2.41 kg $CO_2e/kg/kWh$ (the discharge coefficient of electric energy was calculated based on the China energy conservation [59] and emission reduction development report and the 2050 China energy and $CO_2$ emission report [60]). All the data involved in the case were from the internal data of the factory. In this workshop, the staff were distributed into four white collars (for office tasks) and 23 blue collars (responsible for workshop work), and the building occupied a pavilion of 1000 m² in total area. Figure 5 depicts the layout of the underlying layers. The office is located on the upper floor such that one can directly observe the workshop below. The energy, water, and transportation distances consumed for each process are shown in Figure 6.

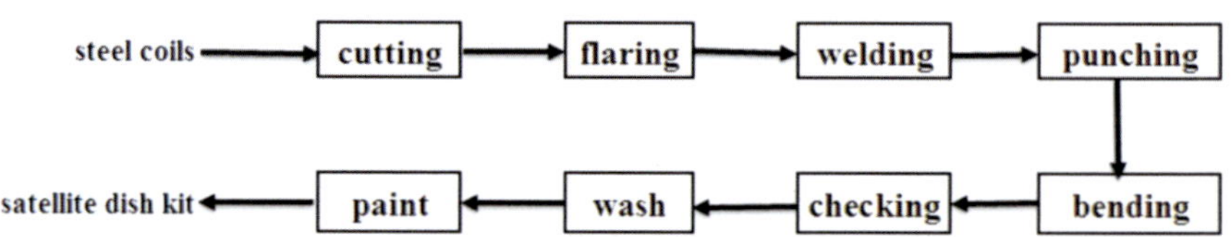


**Figure 3.** Process flow of satellite dish kit production line (before improvement).

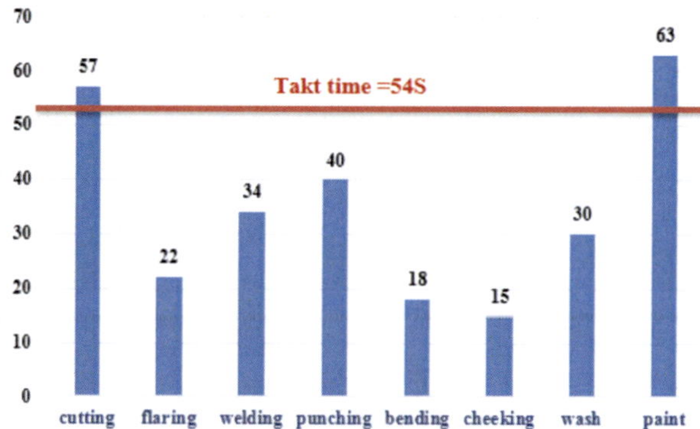


**Figure 4.** Cycle time of each satellite dish kit (before improvement; units: second).

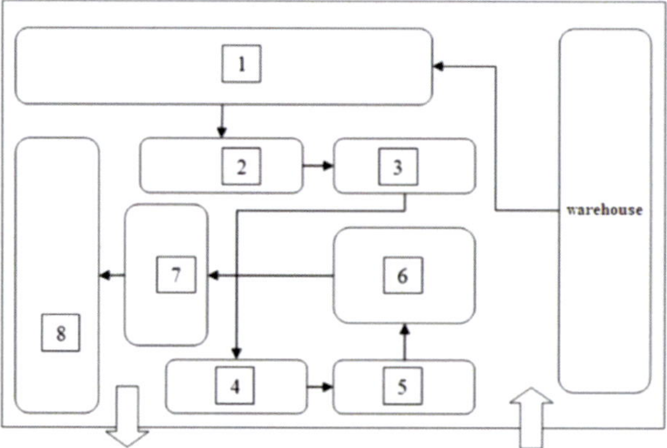


**Figure 5.** Initial layout of satellite dish kit production line (before improvement).

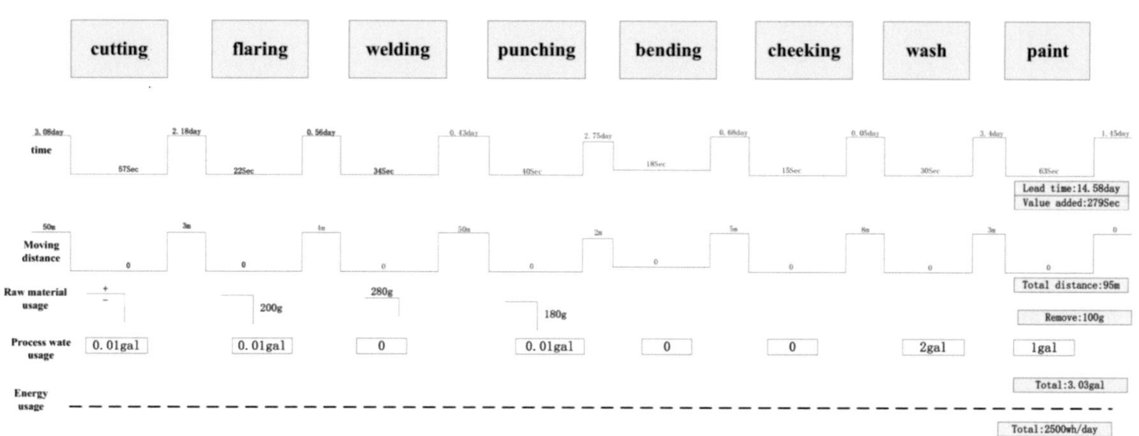


**Figure 6.** Resource consumption of satellite dish kit production line (before improvement).

In the satellite dish kit production line, using Formulas (3)–(8), the total carbon emissions were estimated to be 21.49 kg $CO_2$e, where 16.76 kg $CO_2$e was derived from the non-value-added carbon emissions and 4.73 kg $CO_2$e from value-added carbon emissions. Because the manufacturing process was considered as a defined system, in Formula (2), k was set to 1, and the current carbon efficiency was 12.98 s/kg $CO_2$e. From the lean production perspective, there were some wastes, including inventory, transportation, defects, and unbalanced production lines in the production process.

### 4.2. Future State Analysis and Improvement

Lean and green manufacturing methods should be used to solve problems in the production of satellite dish kits. Appropriate measures must be taken to optimize waste to improve carbon efficiency indicators. Lean improvement practices in the manufacturing process should be the new production operation mode. The process of implementing this new model shifts from the original batch push type to the one-piece flow pull method. In the new product manufacturing mode, the cycle production time of products in the production process must be based on the actual needs of customers determined by customer value, including its beat time or the cycle time of the manufacturing process, and must be produced according to the rhythm of the actual needs of the market. In accordance with the improvement strategy of the lean–green manufacturing perspective, the work improvement team should implement the following improvements:

(1) The production process was analyzed using the lean–green practice improvement tool, and its efficiency and sustainability were identified and quantified using the parameters of the sustainable value stream map (SVSM). Through SVSM combined with the five core values of lean manufacturing theory, the criteria of environment, society, economic development, and cost should be analyzed and considered separately in the improvement process of each production stage. For the analysis of each process, the number of workers, working space, process time, lead time, working time, shift, breaks, stop, product type, batch size, scrap, rework, first-pass rate, ideal cycle time, preparation time, scheduled time, and environmental impact involved in each processing process were analyzed first. Lean is an improvement from the original batch-push production type to the single-piece flow-pull method. In the new product manufacturing mode, the cycle production time of products in the production process was based on the actual needs of customers determined by customer value, including its beat time or the cycle time of the manufacturing process, and was produced according to the rhythm of the actual needs of the market. Combine adjacent operations

based on the takt time of the process, where the takt time for each satellite dish kit is 54 s.

(2) The improved production process implements a single-piece flow-pull methodology. Pull production is the technical carrier of "just in time", one of the two pillars of Toyota's production mode. Compared with the past push production, the former job "pushes" the parts to the latter job for processing. In pull production, the latter job requires the former job to manufacture the parts needed according to the number of products it needs to process. The production layout was improved according to the continuous flow processing and implementation of cellular manufacturing.

(3) Through direct observation of the processing site, the space for each process is scattered, and the scope of the operation space is large. Simultaneously, a large number of tools were replaced in the operation process, resulting in the loss of productivity and accumulation of inventory materials. Some processes were produced according to their capacity, resulting in excessive production. Through an analysis of the ECRS work improvement principle, we determined the improvement direction of the process flow and conceived a new working method to replace the current working method based on 5W1H analysis. Applying the four principles of ECRS work improvement, namely cancellation, merger, reorganization, and simplification, can help enterprises find better efficiency and better working methods. The production line was balanced based on the calculation of takt time to identify and improve bottlenecks by comparison with the actual cycle time.

(4) To reduce wait time waste and large inventories, improvements were achieved by building Kanban systems and using supermarket pull systems. The production process uses the pull method to implement the new production mode. The system has an advantage in that it cancels the intermediate warehouse. For the intermediate inventory existing between the original processes, excess inventory was accumulated between the workspaces of each process to ensure material flow and prevent material shortages in the process of mass production. Concurrently, in terms of transportation, there is much internal mobile transportation, and frequent transportation will damage the normal production of each process. Identify and remove or significantly reduce waste in a quantitative manner through lean–green approach tools supplemented by the environmental impact of innovation. Each process is currently being optimized to reduce overproduction, and the other processes are not considered. Each process should be performed in strict accordance with the order or shortage of goods in the previous process.

(5) 5S and TPM management systems were established in the workshop and production equipment maintenance, respectively, which have positive lean–green improvement environmental impacts.

By combining lean thinking and low-carbon manufacturing, the wastes of the current production line and the carbon efficiency indicator will be improved. The process was recombined and reduced from 8 to 5 to achieve continuous-flow processing. This reduced the distance by 77 m. The new layout is shown in Figure 7. The workstations were placed in the following manner: (1) cutting, (2) flaring and welding, (3) punching and bending, (4) checking and washing, and (5) painting. The improved process cycle time for each satellite dish is shown in Figure 8. The consumed energy, water, and transportation distance by each improved process are shown in Figure 9.

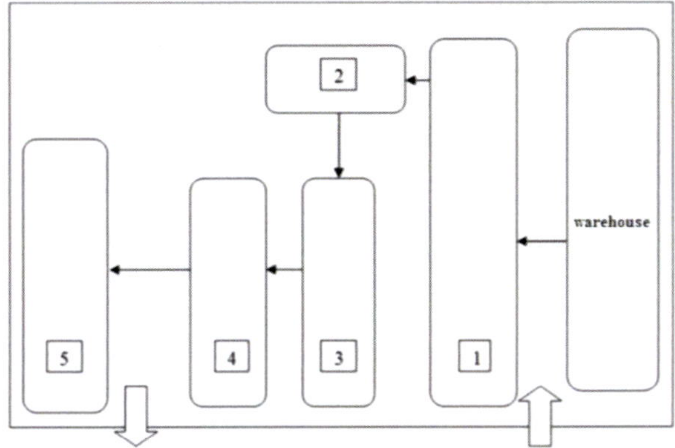

**Figure 7.** New layout of satellite dish kit production line (after improvement).

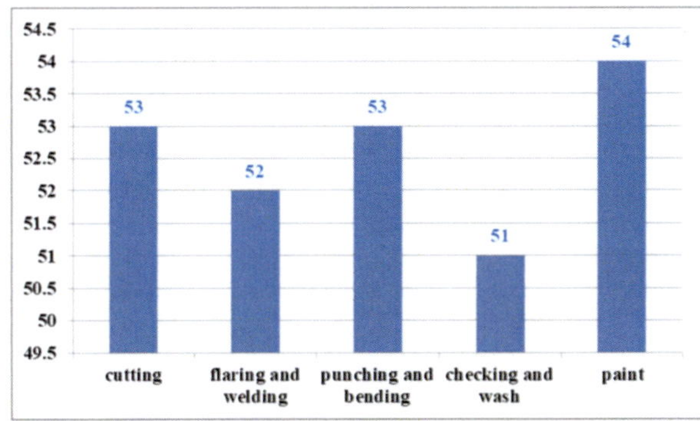

**Figure 8.** Improved cycle time of each satellite dish kit (after improvement, units: second).

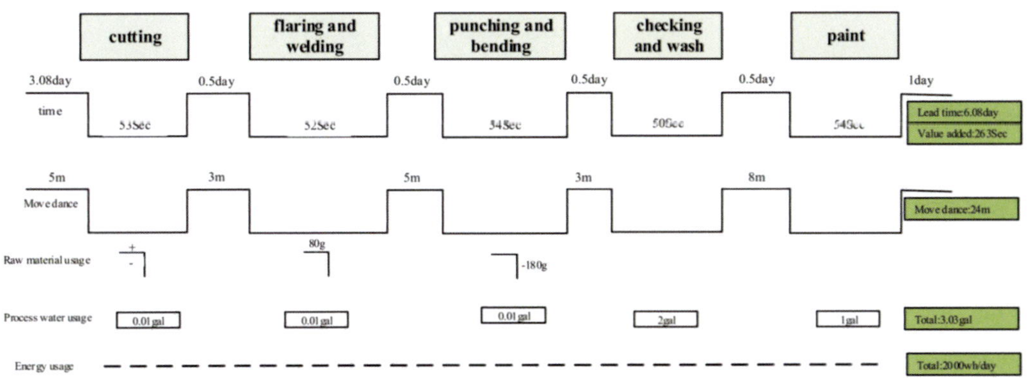

**Figure 9.** Resource consumption of satellite dish kit production line (after improvement).

The improvements in the production lines of the satellite dish kit are shown in Table 1. The comparisons of the effects are glaring after the improvement. The value-added time was approximately 263 s; the non-value-added time showed a very considerable drop from 14.58 days to 6.08 days. Because the implementation of the new pull production system simplifies current processes and reduces intermediate inventory, it also reduces inventory time at the same time. However, after improvement, using Formulas (3)–(8), the total carbon emissions were 7.24 kgCO$_2$e, and the value-added and non-value-added carbon emissions were 1.29 kgCO$_2$e and 5.65 kgCO$_2$e, respectively. Using Formulas (1) and (2), the after-improvement carbon efficiency increased from 12.98 s/kg CO$_2$e to 36.33 s/kg CO$_2$e in comparison with that before the improvement. The carbon benefit after improvement was 193.92% higher than that before the improvement.

**Table 1.** The result of satellite dish kit production line before and after improvement.

| Index | Before Improvement | After Improvement | Improvement (%) |
|---|---|---|---|
| Value-added time ($t_{va}$/s) | 279 | 263 | 5.73 |
| Moving distance (m) | 95 | 24 | 74.4 |
| Non-value-added time ($t_{nva}$/day) | 14.58 | 6.08 | 58.30 |
| Value-added carbon emission ($C_{va}$/kg CO$_2$e) | 4.73 | 1.29 | 72.73 |
| Non-value-added carbon emission ($C_{nva}$/kg CO$_2$e) | 16.76 | 7.24 | 56.80 |
| Total carbon emission ($C_{total}$/kg CO$_2$e) | 21.49 | 8.53 | 60.31 |
| Carbon benefit (CB) | 1 | 2.94 | 193.92 |
| Carbon efficiency ($C - efficiency$/s/kg CO$_2$e) | 12.98 | 36.33 | 179.89 |

## 5. Conclusions

The terms sustainability and efficiency reinforce each other. To achieve this goal, a decision model can be established to determine the synergy between sustainable production and efficient manufacturing systems. Through the analysis of the current value flow of satellite dish kits, the 5Why analysis method, the unitized production, the inventory supermarket pulling system, and the existing excess waste and inventory in production were eliminated. Waste of different types, including transportation, waiting, and unbalanced production lines, shortened the production cycle, effectively slowed down carbon emissions in the production process, and improved carbon efficiency indicators.

In this study, the energy flow, material flow, transport flow, and carbon emission flow were comprehensively considered, and the quantified carbon efficiency was used as the index to establish a mathematical model. Using the verified case of the carbon benefit, carbon efficiency, and sustainable value flow chart model established in this study, the company will be able to reduce costs by decreasing the material consumption and energy per unit output, and thereby gain a competitive advantage. The model can also reduce emissions and waste, optimize manufacturing processes, and improve the final inventory. In addition, the reduction in costs and improvement in efficiency will have a positive impact on the company's revenue. Finally, by reducing the consumption of raw materials and eliminating their negative impact on the environment, this approach shows better social responsibility and environmental sustainability. Therefore, this will be reflected in the increase in the added value of the product. The production process has a direct and significant environmental impact. If the appropriate tools and options are chosen, these methods may help solve environmental problems and lead to sustainable development. Therefore, green engineering and processes are an effective way to eliminate waste, improve carbon benefits and efficiency, and provide guidance and support for sustainable processes, products, and systems, while reducing risks to humans and the environment. Introducing a process that provides adequate environmental protection in a lean production system and realizing "greening" of the process is an essential part of sustainable development of the future society and improving environmental benefits. Therefore, in the new global economic framework of the 21st century, the company must integrate advanced management models to improve the organization's stand in the highly competitive market and achieve sustainability and efficiency, and introduce the integration of green manufacturing into the lean production system.

The quantitative model presented in this paper was designed for the specified scope of only one manufacturing company, and the framework must be tested in a variety of manufacturing processes to improve practicality. More relevant studies should be conducted to test its general application and limitations. For example, the proposed model can be used to improve not only production operations, but also other types of operations, such as healthcare, logistics and transportation, and services. In the future, augmented models and tools can be developed to visualize and evaluate process KPIs simultaneously from environmental, economic, and social perspectives. At the same time, dedicated expert systems can be developed to perform the analysis automatically. At the same time, the thorough integration of lean production and green manufacturing to achieve a thorough degree of lean–green involves the full cooperation of partner enterprises in the supply chain. Therefore, in future research, the relevant model constructed in this paper can be extended to the whole supply chain rather than limited to a single enterprise.

**Author Contributions:** Conceptualization, X.Z., Y.X. and G.X.; methodology, X.Z. and Y.X.; formal analysis, Y.X. and G.X.; investigation, X.Z.; writing—original draft preparation, X.Z., Y.X. and G.X.; writing—review and editing, X.Z. and Y.X. All authors have read and agreed to the published version of the manuscript.

**Funding:** This research was funded by the Natural Science Foundation of Hunan Province, China (Project number: 2022JJ50244); Education Department of Hunan Province (Project number: 21B0695; 21A0475); Project of Hunan social science achievement evaluation committee in 2022 (Project number: XSP22YBC081).

**Institutional Review Board Statement:** Not applicable.

**Informed Consent Statement:** Informed consent was obtained from all subjects involved in the study.

**Data Availability Statement:** Not applicable.

**Conflicts of Interest:** The authors declare no conflict of interest.

## References

1. IEA. The Way Forward: Five Key Action to Achieve a Low Carbon Energy Sector [EB/OL]. 2014. Available online: http://www.iea.org/publications/freepublications/publication/The_Way_forward.pdf (accessed on 10 September 2021).
2. Galeazzo, A.; Furlan, A.; Vinelli, A. Lean and green in action: Interdependencies and performance of pollution prevention projects. *J. Clean. Prod.* **2014**, *85*, 191–200. [CrossRef]
3. Fercoq, A.; Lamouri, S.; Carbone, V. Lean/Green integration focused on waste reduction techniques. *J. Clean. Prod.* **2016**, *137*, 567–578. [CrossRef]
4. Digalwar, A.K.; Tagalpallewar, A.R.; Sunnapwar, V.K. Green Manufacturing Performance Measures: An Empirical Investigation from Indian Manufacturing Industries. *Meas. Bus. Excell.* **2013**, *17*, 59–75. [CrossRef]
5. Garza-Reyes, J.A. Green lean and the need for six sigma. *Int. J. Lean Six Sigma* **2015**, *6*, 226–248. [CrossRef]
6. Cherrafi, A.; ElFezazi, S.; Govindan, K.; Garza-Reyes, J.A.; Benhida, K.; Mokhlis, A. A Framework for the Integration of Green and Lean Six Sigma for Superior Sustainability Performance. *Int. J. Prod. Res.* **2017**, *55*, 4481–4515. [CrossRef]
7. Garza-Reyes, J.A. Lean and green—A systematic review of the state-of-the-art literature. *J. Clean. Prod.* **2015**, *102*, 18–29. [CrossRef]
8. Green, K.W.; Zelbst, P.J.; Meacham, J.; Bhadauria, V.S. Green Supply Chain Management Practices: Impact on Performance. *Supply Chain Manag. Int. J.* **2012**, *17*, 290–305. [CrossRef]
9. Sartal, A.; Carou, D.; Dorado-Vicente, R.; Mandayo, L. Facing the challenges of the food industry: Might additive manufacturing be the answer? *Proc. Inst. Mech. Eng. Part B J. Eng. Manuf.* **2019**, *233*, 1902–1906. [CrossRef]
10. Baumers, M.; Wildman, R.; Wallace, M.; Yoo, J.; Blackwell, B.; Farr, P.; Roberts, C.J. Using total specific cost indices to compare the cost performance of additive manufacturing for the medical device's domain. *Proc. Inst. Mech. Eng. Part B J. Eng. Manuf.* **2019**, *233*, 1235–1249. [CrossRef]
11. Zhu, X.-Y.; Zhang, H. Construction of Lean-green coordinated development model from the perspective of personnel integration in manufacturing companies. *Proc. Inst. Mech. Eng. Part B J. Eng. Manuf.* **2020**, *234*, 1460–1470. [CrossRef]
12. Mccarty, T.; Jordan, M.; Probst, D. *Six Sigma for Sustainability—How Organizations Design and Deploy Winning Environmental Programs*; McGraw-Hill: New York, NY, USA, 2011.
13. Wong, W.P.; Wong, K.Y. Synergizing an Ecosphere of Lean for Sustainable Operations. *J. Clean. Prod.* **2014**, *85*, 51–66. [CrossRef]
14. Garza-Reyes, J.A.; Villarreal, B.; Kumar, V.; Molina Ruiz, P. Lean and green in the transport and logistics sector—A case study of simultaneous deployment. *Prod. Plan. Control* **2016**, *27*, 1221–1232. [CrossRef]
15. Shah, R.; Ward, P.T. Defining and Developing Measures of Lean Production. *J. Oper. Manag.* **2007**, *25*, 785–805. [CrossRef]
16. Zhu, X.-Y.; Zhang, H. A lean green implementation evaluation method based on fuzzy analytic net process and fuzzy complex proportional assessment. *Int. J. Circuits Syst. Signal Process.* **2020**, *14*, 646–655.
17. Chauhan, G.; Singh, T.P. Measuring Parameters of Lean Manufacturing Realization. *Meas. Bus. Excell.* **2012**, *16*, 57–71. [CrossRef]

18. King, A.; Lenox, M. Lean and Green: An Empirical Examination of the Relationship between Lean Production and Environmental Performance. *Prod. Oper. Manag.* **2001**, *10*, 244–256. [CrossRef]
19. Sergio, A.; Alvarez, R.; Domingo, R. Model of efficient and sustainable improvements in a lean production system through processes of environmental innovation. *J. Clean. Prod.* **2013**, *47*, 141–148.
20. Thanki, S.J.; Govindan, K.; Thakkar, J. An Investigation on Lean-Green Implementation Practices in Indian SMEs Using Analytical Hierarchy Process (AHP) Approach. *J. Clean. Prod.* **2016**, *135*, 284–298. [CrossRef]
21. Carvalho, H.; Govindan, K.; Azevedo, S.; Cruz-Machado, V. Modelling Green and Lean Supply Chain: An Eco-Efficiency Perspective. *Resour. Conserv. Recycl.* **2017**, *120*, 75–78. [CrossRef]
22. EPA. *The Lean and Environment Toolkit*; Environmental Protection Agency: Washington, DC, USA, 2007.
23. Womack, J.P.; Jones, D.T.; Roos, D. *The Machine that Changed the World*; Harper-Collins: New York, NY, USA, 1991.
24. Womack, J.P.; Jones, D.T. Beyond Toyota: How to root out waste and pursue perfection. *Harv. Bus. Rev.* **1996**, *74*, 140–151.
25. Womack, J.P.; Jones, D.T. *Lean Thinking: Banish Waste and Create Wealth in Your Corporation*; Free Press: New York, NY, USA, 2003.
26. Shah, R.; Ward, P.T. Lean manufacturing: Context, practice bundles, and performance. *J. Oper. Manag.* **2004**, *21*, 129–149. [CrossRef]
27. Abdulmalek, F.A.; Rojgopal, J. Analyzing the benefits of lean manufacturing and value stream mapping via simulation: A process sector case study. *Int. J. Prod. Econ.* **2007**, *107*, 223–236. [CrossRef]
28. Dora, M.; Kumar, M.; Gellynck, X. Determinants and barriers to lean implementation in food processing SMEs-a multiple case analysis. *Prod. Plan. Control.* **2016**, *27*, 1–23. [CrossRef]
29. Singh, B.J.; Khanduja, D. SMED: For quick changeovers in foundry SMEs. *Int. J. Product. Perform. Manag.* **2010**, *59*, 98–116. [CrossRef]
30. Wong, W.P.; Ignatius, J.; Soh, K.L. What is the leanness level of your organisation in lean transformation implementation? An integrated lean index using ANP approach. *Prod. Plan. Control.* **2014**, *25*, 273–287. [CrossRef]
31. Wu, L.; Subramanian, N.; Abdulrahman, M.D.; Liu, C.; Lai, K.-H.; Pawar, K.S. The Impact of Integrated Practices of Lean, Green, and Social Management Systems on Firm Sustainability Performance—Evidence from Chinese Fashion Auto-Parts Suppliers. *Sustainability* **2015**, *7*, 3838. [CrossRef]
32. Dües, C.M.; Tan, K.H.; Lim, M. Green as the new lean: How to use lean practices as a catalyst to greening your supply chain. *J. Clean. Prod.* **2013**, *40*, 93–100. [CrossRef]
33. Martinez-Jurado, P.J.; Moyano-Fuentes, J. Lean management, supply chain management and sustainability: A literature review. *J. Clean. Prod.* **2014**, *8*, 134–150. [CrossRef]
34. Piercy, N.; Rich, N. The relationship between lean operations and sustainable operations. *Int. J. Oper. Prod. Manag.* **2015**, *35*, 282–315. [CrossRef]
35. Colicchia, C.; Creazza, A.; Dallari, F. Lean and green supply chain management through intermodal transport: Insights from the fast-moving consumer goods industry. *Prod. Plan. Control* **2017**, *28*, 1221–1232. [CrossRef]
36. Negrao, L.L.L.; Filho, M.G.; Marodin, G. Lean practices and their effect on performance: A literature review. *Prod. Plan. Control* **2017**, *28*, 33–56. [CrossRef]
37. Cherrafi, A.; Elfezazi, S.; Chiarini, A.; Mokhlis, A.; Benhida, K. The Integration of Lean Manufacturing, Six Sigma and Sustainability: A Literature Review and Future Research Directions for Developing a Specific Model. *J. Clean. Prod.* **2016**, *139*, 828–846. [CrossRef]
38. Gandhi, N.S.; Thanki, S.J.; Thakkar, J.J. Ranking of drivers for integrated lean-green manufacturing for Indian manufacturing SMEs. *J. Clean. Prod.* **2018**, *171*, 675–689. [CrossRef]
39. Gandhi, N.S.; Thanki, S.J.; Thakkar, J.J. A model for Lean and Green integration and monitoring for the coffee sector. *Comput. Electron. Agric.* **2018**, *150*, 62–73.
40. Bhattacharya, A.; Nand, A.; Castka, P. Lean-green integration and its impact on sustainability performance: A critical review. *J. Clean. Prod.* **2019**, *236*, 117697.1–117697.16. [CrossRef]
41. Bhattacharya, A.; Nand, A.; Castka, P. Enhancing the Adaptability: Lean and Green Strategy towards the Industry Revolution 4.0. *J. Clean. Prod.* **2020**, *273*, 122870.
42. Leong, W.D.; Teng, S.Y.; How, B.S.; Ngan, S.L.; Rahman, A.A.; Tan, C.P.; Ponnambalam, S.; Lam, H.L. Simulation-based analysis of catalysers and trade-offs in Lean & Green manufacturing—Science Direct. *J. Clean. Prod.* **2020**, *242*, 118411–118436.
43. Kuo, S.Y.; Lin, C.P. Determinants of green performance in container terminal operations: A lean management. *J. Clean. Prod.* **2020**, *275*, 123105. [CrossRef]
44. Touriki, F.E.; Benkhati, I.; Kamble, S.S.; Belhadi, A.; Fezazi, S.E. An integrated smart, green, resilient, and lean manufacturing framework: A literature review and future research directions. *J. Clean. Prod.* **2021**, *319*, 128691. [CrossRef]
45. Cherrafi, A.; Garza-Reyes, J.A.; Belhadi, A.; Kamble, S.S.; Elbaz, J. A Readiness Self-Assessment Model for Implementing Green Lean Initiatives. *J. Clean. Prod.* **2021**, *309*, 127401. [CrossRef]
46. Miguel, L.; da Fonseca, C.M. ISO 14001: 2015: An improved tool for sustainability. *J. Ind. Eng. Manag.* **2015**, *8*, 37–50.
47. Mollenkopf, D.; Stolze, H.; Tate, W.; Ueltschy, M. Green, lean and global supply chains. *Int. J. Phys. Distrib. Logist. Manag.* **2010**, *40*, 14–41. [CrossRef]
48. Sheng, M.L.; Chien, I. Rethinking organizational learning orientation on radical and incremental innovation in high-tech firms. *J. Bus. Res.* **2016**, *69*, 2302–2308. [CrossRef]

49. Verrier, B.; Rose, B.; Caillaud, E. Lean and Green strategy: The Lean and Green House and maturity deployment model. *J. Clean. Prod.* **2016**, *116*, 150–156. [CrossRef]

50. Li, Y.; Cao, H.; Li, H.; Tao, G. The Carbon Emissions Dynamic Characteristic and Two Stage Optimization Scheduling Model for Job Shop. *Comput. Integr. Manuf. Syst.* **2015**, *21*, 2687–2693. (In Chinese)

51. Lv, J.; Peng, T.; Tang, R. Energy modeling and a method for reducing energy loss due to cutting load during machining operations. *Proc. Inst. Mech. Eng. Part B J. Eng. Manuf.* **2019**, *233*, 699–710. [CrossRef]

52. Ng, R.; Low, J.S.C.; Song, B. Integrating and implementing Lean and Green practices based on proposition of Carbon-Value Efficiency metric. *J. Clean. Prod.* **2015**, *95*, 242–255. [CrossRef]

53. Zhu, X.-Y.; Zhang, H.; Jiang, Z.-G. Application of green-modified value stream mapping to integrate and implement lean and green practices: A case study. *Int. J. Comput. Integr. Manuf.* **2020**, *33*, 716–731. [CrossRef]

54. Xiao, Y.; Jiang, Z.; Gu, Q.; Wei, Y.; Wang, R. A novel approach to CNC machining centre processing parameters optimization considering energy-saving and low-cost. *J. Manuf. Syst.* **2021**, *59*, 535–548. [CrossRef]

55. Xiao, Y.; Zhang, H.; Jiang, Z.; Gu, Q.; Wei, Y. Multiobjective optimization of machining center process route: Trade-offs between energy and cost. *J. Clean. Prod.* **2021**, *280*, 124171. [CrossRef]

56. Gutowski, T.; Dahmus, L.; Thiriez, A. Electrical Energy Requirements for Manufacturing Processes, [10 March 2010]. Available online: http://Web.Mit.Edu/ebm/www/publications/CIRP_2006.pdf (accessed on 11 September 2021).

57. Ashbly, M.F. *Materials and the Environment: Eco-Informed Material Choice*; Butterworth Heinemann: Brulington, VT, USA, 2009; pp. 112–291.

58. Li, X.; Xu, H. Life cycle evaluation of steel based on GaBi. *Environ. Prot. Circ. Econ.* **2009**, *6*, 15–18. (In Chinese)

59. China Energy Saving and Investment Company. *China Energy Conservation and Emission Reduction Development Report*; China Water Power Press: Beijing, China, 2009. (In Chinese)

60. 2050 China Energy and $CO_2$ Emission Research Group. *2050 China Energy and $CO_2$ Emission Report*; Science Press: Beijing, China, 2009. (In Chinese)

*Article*

# Low-Carbon and Low-Energy-Consumption Gear Processing Route Optimization Based on Gray Wolf Algorithm

**Yani Zhang** [1,2], **Haoshu Xu** [3,*], **Jun Huang** [1,2] **and Yongmao Xiao** [1,2,*]

1 School of Computer and Information, Qiannan Normal University for Nationalities, Duyun 558000, China
2 Key Laboratory of Complex Systems and Intelligent Optimization of Guizhou Province, Duyun 558000, China
3 Office of Academic Affairs, Qiannan Broadcast Television University, Duyun 558000, China
* Correspondence: zyn19800126@sina.com (H.X.); xym198302@163.com (Y.X.)

**Abstract:** The process of gear machining consumes a large amount of energy and causes serious pollution to the environment. Developing a proper process route of gear machining is the key to conserving energy and reducing emissions. Nowadays, the proper process route of gear machining is based on experience and is difficult to keep up with the development of modern times. In this article, a calculation model of low-carbon and low-energy consumption in gear machining processes was established based on an analysis of the machining process. With processing parameters as independent variables, the grey wolf algorithm was used to solve the problem. The effectiveness of the method was proven by an example of the machining process of an automobile transmission shaft.

**Keywords:** gear processing; process route optimization; gray wolf algorithm; low energy consumption; low carbon

**Citation:** Zhang, Y.; Xu, H.; Huang, J.; Xiao, Y. Low-Carbon and Low-Energy-Consumption Gear Processing Route Optimization Based on Gray Wolf Algorithm. *Processes* **2022**, *10*, 2585. https://doi.org/10.3390/pr10122585

Academic Editor: Sergey Y. Yurish

Received: 16 November 2022
Accepted: 2 December 2022
Published: 4 December 2022

**Publisher's Note:** MDPI stays neutral with regard to jurisdictional claims in published maps and institutional affiliations.

## 1. Introduction

The manufacturing process consumes a lot of energy and produces a great deal of pollution in the environment [1,2]. Environmental improvement is closely related to industrial development, and it is also inseparable from the development of energy saving and emission reduction technology and manufacturing technology innovation. Therefore, energy efficiency in manufacturing, as a global concept, has attracted increasing attention from academics, industry and government departments [3,4]. A process route is a means to guide the manufacturing workshop to complete the production task in accordance with the prescribed operation processes. The improvement of the product processing route can effectively achieve energy saving and lower carbon emissions. In gear machining, different process routes have a great impact on energy consumption, carbon emissions and the processing costs of gear machining [5–7]. In actual production, it is necessary to optimize the gear processing route with a reasonable optimization decision method.

In recent years, many scholars have studied process route decision making.. An et al. put forward a process route optimization method based on intuitionistic fuzzy number and CA-SPEA2. The method was verified by machining the transmission box [8]. Fan et al. constructed a process route decision space based on process constraints and solved it with the genetic algorithm [9]. Huang et al. proposed a process route generation method with dynamic updating of tabu manufacturing features, and combined it with an ant colony algorithm to optimize the problem of process route [10]. Cheng et al. proposed a bacterium-foraging ant colony optimization (BFACO) algorithm for process route planning, and compared it with the optimization results of other optimization algorithms. The results showed that the BFACO algorithm had high computational efficiency [11]. Li et al. introduced the concepts of feature element and processing element to process the features of parts, established an efficient and low-carbon optimization model of processing process route, and used the genetic algorithm to optimize the model. This method has been verified through the machining of an electric frame [12]. Zhai et al. adopted the minimum number of changes in manufacturing resources as the objective function to optimize the process route, and proposed a hybrid process route-sorting algorithm based on the ACO and SA algorithms [13]. Tang et al. established a process route optimization model with the objective of low energy consumption and high efficiency, and proposed a SA-QPSO algorithm to optimize the model [14]. Xiao et al. used a process element to express processing characteristics, and set up a low-energy consumption and low-cost process route optimization model. A combined algorithm of APSO and NSGA-II was proposed. The feasibility of this method was verified by comparing the process route of an emulsion pump box before and after optimization [15]. Milica et al. proposed a new algorithm combining PSO algorithm and chaos theory, and verified the flexibility and superiority of the algorithm for process route optimization through experiments and comparisons with other algorithms [16].

The above research on the process route decision optimization of mechanical manufacturing systems was mainly carried out by establishing the mapping between process route and optimization objective, and using intelligent algorithms and processing experiments.

These studies were used to optimize the process route from different angles and with different methods and then verify its efficiency and suitability. Gear machining is complex, and the research surrounding gear process route is limited; few people consider gear processing route optimization. Gear machining is a complicated process, the process route has an important influence on gear production. However, gear process route optimization is limited, and new optimization methods are emerging. In this paper, the process route of gear machining was studied, aiming to improve the energy consumption and carbon emissions of gear machining, and an optimization model of gear processing route based on low-carbon and low-energy consumption was established. An improved grey wolf algorithm was proposed to optimize the process route sequence, equipment allocation and tool allocation. The method was verified by machining the second gear of the intermediate shaft of automobile transmission.

## 2. Optimization Model of Gear Processing Route Based on Low-Carbon and Low-Energy-Consumption

Studies have shown that in discrete processing industries (turning, milling, etc.), 99% of the impact machine tools have on the environment is caused by power consumption, and machine tool energy consumption is one of the important indicators for evaluating machine tool environmental performance [17,18]. Therefore, reducing the processing energy consumption is one of the most important means to achieve low-carbon manufacturing. At the same time, with the development of green and low-carbon manufacturing, determining how to optimize the process route based on carbon emission and energy consumption is the focus of the research. Traditional gear machining only considers the function realization and machining quality, which is not suitable for the current green and low-carbon manufacturing. Therefore, a gear machining process route optimization model based on energy conservation and low-carbon is proposed [19–21]. In the process of dry cutting gear processing, a lot of carbon emissions will be produced in the processes of inputting and outputting various materials, energy conversion and consumption, waste discharge and treatment, etc.

### 2.1. Energy Consumption Optimization Model of Gear Processing Route

The production process of gear uses various machine tools, accompanied by the use of energy (electric energy, natural gas, etc.), cutting fluid, fixture and other materials under the assistance of refining, casting, rolling, cutting, i.e., the final formation of gear products. Depending on the energy used in the whole process, it can be divided into direct and indirect methods. The influencing factors of energy consumption in gear machining process are shown in Figure 1.

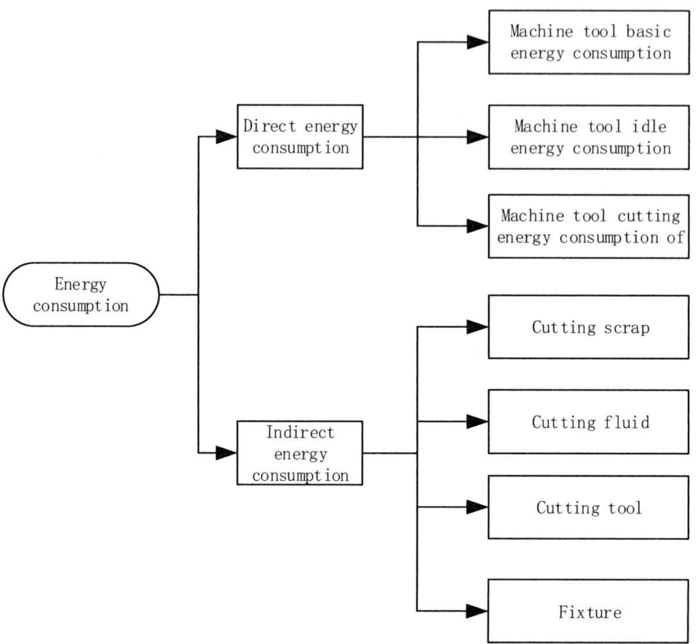



**Figure 1.** Influencing factors of energy consumption in gear machining process.

### 2.1.1. Direct Energy Consumption

The electrical energy consumed during machine tool processing is called direct energy consumption, which determines the size, shape and accuracy of the workpiece. In addition, machine tool lighting, transportation products, etc., can also be classified as direct energy consumption. The direct energy consumption is expressed as

$$E_{D,i} = E_B + E_I + E_C \tag{1}$$

where $E_{D,i}$ is the electric energy used by the I-step machine tool (kWh), $E_B$ is the basic energy consumption (kWh) of machine tool when clamping workpiece, $E_I$ is the energy consumed to keep the spindle running while adjusting the tool for machine tool (kWh), $E_C$ is energy consumption generated by cutting tool workpiece (kWh).

$$E_C = V * \delta * t_c \tag{2}$$

$V$ is the volume of material removed by cutting ($\text{mm}^3$), $\delta$ is specific energy consumption ($w{\cdot}s/\text{mm}^3$), and $t_c$ is working time of process.

Energy consumption for lighting and transportation is

$$E_A = P_L * t_L + \sum_{m=1}^{n} P_T * t_T \tag{3}$$

$E_A$ is lighting and transport energy consumption, $P_L$, $P_T$ are lighting and transport power, respectively, (kWh/s), $t_L$ is average lighting time of the workpiece (s), and $t_T$ is sum of average transit time and cutting time.

### 2.1.2. Indirect Energy Consumption

The indirect energy consumption in the production process of workpiece mainly comes from the consumption of auxiliary materials, such as cutting fluid, fixture, tool, etc. This type of energy consumption mainly comes from databases and literature, and can also be converted into electrical energy consumption by the intrinsic energy value of the workpiece, expressed as

$$E_{M,i} = \sum_{j=1}^{k} m_j * Eb_{m_j} * 1/\beta \tag{4}$$

$E_{M,i}$ is the work step $i$ indirect energy consumption (kWh); $m_a$ is the consumption of material j in this working step; $Eb_{m_j}$ is the intrinsic energy consumption of the j-th material (J/kg); the intrinsic energy consumption of a material refers to the total energy consumed to produce a certain material; $\beta$ is the work-electric energy conversion coefficient, and its value is 3,600,000 [22].

A part process route consists of $i$ steps, the average energy consumption of a workpiece is

$$E = \sum_{i=1}^{l} E_{D,i} + \sum_{i=1}^{l} E_{M,i} \tag{5}$$

*2.2. Carbon Emission Optimization Model Based on Gear Processing Route*

2.2.1. Material, Energy Consumption and Waste in Gear Processing

The gear blank will produce a lot of carbon emissions and consume a lot of energy in the process of processing. A carbon emission boundary is an effective means to calculate carbon emissions. The process of transferring gear blank to machine tool and finishing gear product is set as carbon emission boundary. The whole boundary contains carbon emissions from three aspects, namely material, energy and waste. The materials consumed in gear processing are mainly gear raw materials and various auxiliary materials, and the energy consumed is electricity, oil, natural gas, etc. [23,24]. A variety of materials $i$ ($i = 1, 2 \dots , I$) and energy $k$ ($k = 1, 2 \dots , K$) enter the workshop in turn according to the process route, carry out the gear machining process, and the finished gear products are obtained through machining. Each workshop shall discharge waste $l$ ($l = 1, 2 \dots , L$). The influencing factors of carbon emission during gear machining are shown in Figure 2.

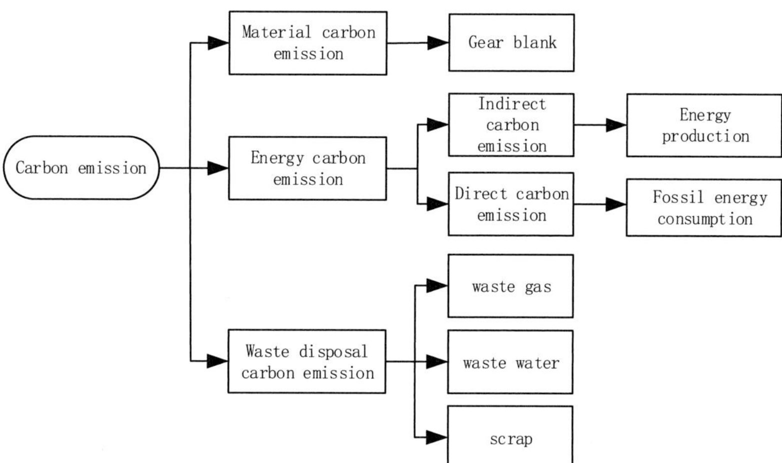


**Figure 2.** Influencing factors of carbon emission during gear processing.

$S(i, m)$, $E(k, m)$ and $W(l, m)$ represent workshop $m$ ($m = 1, 2 \dots , M$) materials, energy consumption and waste generated, respectively. The total carbon emission of materials, energy and waste within time $T$ can be expressed as:

$$M_T = \sum_{m=1}^{M} \sum_{i=1}^{I} S(i, m) \tag{6}$$

$$E_T = \sum_{m=1}^{M} \sum_{k=1}^{K} E(k,m) \tag{7}$$

$$W_T = \sum_{m=1}^{M} \sum_{l=1}^{L} W(l,m) \tag{8}$$

These three types of carbon emissions can be quantified

$$C_{em} = C_M + C_E + C_W \tag{9}$$

$C_{em}$ is the total carbon emission, $C_M$ $C_E$ and $C_W$ are the carbon emissions generated by materials, energy and waste, respectively. It can be calculated using the carbon emission factor method. $S_E$ is carbon emission factor ($S_E = e_C / t_E$, where $e_C$, $t_E$ are carbon emissions and standard coal volume, respectively).

### 2.2.2. Calculate Material Carbon Emissions

Material indirect carbon emission $C_M^{(i,m)}$ is generated by using material $i$ in workshop m, which can be expressed as

$$C_M^{(i,m)} = \sum_{c=1}^{N} S(i,m) J_c^i S_{E_c} \tag{10}$$

$J_c^i$ is the energy c amount required to produce one unit of material i (converted into standard coal amount), and $S_{E_c}$ is the energy c carbon emission factor.

The total indirect carbon emission generated by material $i$ is

$$C_M^i - \sum_{m=1}^{M} C_M^{(i,m)} - \sum_{m=1}^{M} \sum_{c=1}^{N} S(i,m) J_c^i S_{E_c} \tag{11}$$

The total indirect carbon emission of materials in gear processing is

$$C_M = \sum_{m=1}^{M} \sum_{i=1}^{I} C_M^{(i,m)} = \sum_{m=1}^{M} \sum_{i=1}^{I} \sum_{c=1}^{N} S(i,m) J_c^i S_{E_k} \tag{12}$$

### 2.2.3. Energy Carbon Emission Calculation

Energy consumption produces two kinds of carbon emissions, including indirect carbon emissions from preparation energy $C_{IE}$, and direct carbon emissions from machine tool processing energy $C_{DE}$, so $C_E = C_{IE} + C_{DE}$.

(1)  Indirect carbon emissions

Indirect carbon emissions $C_{IE}^{(k,m)}$ are generated by workshop $m$ using energy $k$, which can be expressed as

$$C_{IE}^{(k,m)} = \sum_{n=1}^{N} E(k,m) J_n^k S_{E_n} \tag{13}$$

$J_n^k$ is the energy $n$ amount required for preparation per unit of energy $k$ (converted into standard coal amount), and $S_{E_n}$ is the energy $n$ carbon emission factor.

The total indirect carbon emissions by using energy $k$ is

$$C_{IE}^k = \sum_{m=1}^{M} C_{IE}^{(k,m)} = \sum_{m=1}^{M} \sum_{n=1}^{N} E(k,m) J_n^k S_{E_n} \tag{14}$$

The total indirect carbon emissions by energy consumption in gear processing is

$$C_{IE} = \sum_{m=1}^{M} \sum_{k=1}^{K} C_{IE}^{(k,m)} = \sum_{m=1}^{M} \sum_{k=1}^{K} \sum_{n=1}^{N} E(k,m) J_n^k S_{E_n} \tag{15}$$

(2)  Direct carbon emissions

Direct carbon emission $C_{DE}^{(k,m)}$ is generated by workshop m using energy $k$, which can be expressed as

$$C_{DE}^{(k,m)} = E(k,m)P_k S_{E_k} \tag{16}$$

$P_k$, $S_{E_k}$ are the conversion coal coefficient and energy $k$ carbon emission factor respectively

The total direct carbon emission by using energy $k$ is

$$C_{DE}^k = \sum_{m=1}^{M} C_{DE}^{(k,m)} = \sum_{m=1}^{M} E(k,m)P_k S_{E_k} \tag{17}$$

The total direct carbon emissions by using energy in gear processing can be expressed as

$$C_{DE} = \sum_{m=1}^{M} \sum_{k=1}^{K} \sum_{n=1}^{N} E(k,m) J_n^k S_{E_k} \tag{18}$$

### 2.2.4. Waste Disposal Carbon Emission

Gear processing will produce some waste, such as waste gas, waste water and so on, which requires the consumption of energy to deal with the waste. The carbon emissions from waste $i$ discharged by workshop m is

$$C_W^{(l,m)} = \sum_{q=1}^{N} W(l,m) J_q^l S_{E_q} \tag{19}$$

where $J_q^l$ is the energy $q$ amount required for unit waste $i$ treatment (converted into standard coal amount), and $S_{E_q}$ is the carbon emission factor of energy $q$.

The total carbon emission generated by disposing waste l is

$$C_W^l = \sum_{m=1}^{M} C_W^{(l,m)} = \sum_{m=1}^{M} \sum_{q=1}^{N} W(l,m) J_q^l S_{E_q} \tag{20}$$

The total carbon emissions from gear processing waste can be expressed as

$$C_W = \sum_{m=1}^{M} \sum_{l=1}^{L} C_W^{(l,m)} = \sum_{m=1}^{M} \sum_{l=1}^{L} \sum_{q=1}^{N} W(l,m) J_q^l S_{E_q} \tag{21}$$

## 3. Optimization Model Solution Based on Hybrid Multi-Objective Gray Wolf Optimizer

A Grey Wolf algorithm (GWO) is a population intelligent optimization algorithm based on the study of grey wolf predation habits. Wolves have different social hierarchies, with low hierarchies subordinate to high hierarchies, so as to realize the whole process of finding, tracking, surrounding and even capturing prey [25–27]. Therefore, researchers proposed an optimization mechanism based on the predation process. Compared with other swarm intelligence algorithms, such as PSO and MODA, GWO has better global search capability. In this paper, a new update operator is designed, the cross and mutation operation is added, which can realize the optimization of energy consumption and carbon emission [28–30].

### 3.1. Description of Grey Wolf Algorithm

The grey wolf hierarchy has a strict system of management, similar to the form of a pyramid. In the social hierarchy of the grey wolf, the pack is divided into three tiers, $\alpha$ at the top, $\beta$ at the second, $\gamma$, at the third, and the rest $\delta$ at the bottom. During the hunt, the first three layers of wolves lead the pack, and the wolves $\delta$ obey the three of them, which leads to efficient hunting. The wolves first search for prey in this way, and surround it from all sides. As the encircling circle gradually shrinks, the wolf $\alpha$ leads the wolfs $\beta$ and $\gamma$ to attack the prey first, and the wolves $\delta$ guard around to catch the escaped prey [31,32]. This hunting mode can attack the prey in multiple directions, and finally capture the prey.

To form a circle, use the following formula to calculate the number of wolves between individual and prey

$$D = \left| C \cdot X_p(t) - X_w(t) \right| \tag{22}$$

$$X_w(t+1) = X_p(t) - A \tag{23}$$

$X_p(t)$, $X_w(t)$ are the location coordinates of the prey and the gray wolf, respectively, and t is iteration times. $A$ and $C$ are the convergence and oscillation factors, respectively.

$$A = 2a \cdot r_2 - a \tag{24}$$

$$C = 2 \cdot r_1 \tag{25}$$

$r_1$ and $r_2$ are two random vectors, with a value range of [0, 1]; a decreases from 2 to 0 as the number of iterations increases.

The best three wolves in each iteration are left as $(\alpha, \beta, \gamma)$ to guide the position update of other wolves. The formula for location update is as follows

$$D_\alpha = \left| C_1 \cdot X_\alpha(t) - X_w(t) \right| \tag{26}$$

$$D_\beta = \left| C_2 \cdot X_\beta(t) - X_w(t) \right| \tag{27}$$

$$D_\gamma = \left| C_3 \cdot X_\gamma(t) - X_w(t) \right| \tag{28}$$

$$X_1 = X_\alpha - A_1 \cdot D_\alpha \tag{29}$$

$$X_2 = X_\beta - A_2 \cdot D_\beta \tag{30}$$

$$X_3 = X_\gamma - A_3 \cdot D_\gamma \tag{31}$$

$$X_p(t+1) = (X_1 + X_2 + X_3)/3 \tag{32}$$

Prey search in GWO is divided into two aspects: prey location determination and gray wolf location update. First, the population is initialized to randomly generate the grey wolf population, and then excellent individuals $(\alpha, \beta, \gamma)$ are selected to guide the wolves. The value range of $A$ is $[-a, a]$, and the value is randomly taken within this interval, because the value of a gradually decreases with the increase of iteration, $A$ is ordered from large to small. When the value of $A > |1|$, the gray wolf encirclement of large wolves search range is larger, so the algorithm has better global searching ability; when the value of $A < |1|$, the gray wolf encirclement of smaller wolves to attack and capture prey, iterative output at the end of the optimal solution [33,34].

### 3.2. Algorithm Flow

The grey wolf algorithm process is shown in Figure 3.

### 3.3. Encoding and Decoding

Equipment selection, tool selection and process sequence have a great impact on process route optimization, and a reasonable coding mode should be selected in the algorithm, as shown in Figure 4. Each individual contains three substrings, namely processes, equipment and tools. The three substrings are the same length as the workpiece processing process. The sequential substring is used to represent the machining operation sequence of the workpiece. The sequential substring is kept in a continuous way, and the machining sequence of the workpiece is taken as the constraint. The equipment is numbered in sequence, and the corresponding equipment number is assigned to each process. The $j$-th aspect on the substring corresponds to the equipment number of the completed process $j$. Tools and equipment are coded in the same way [35].

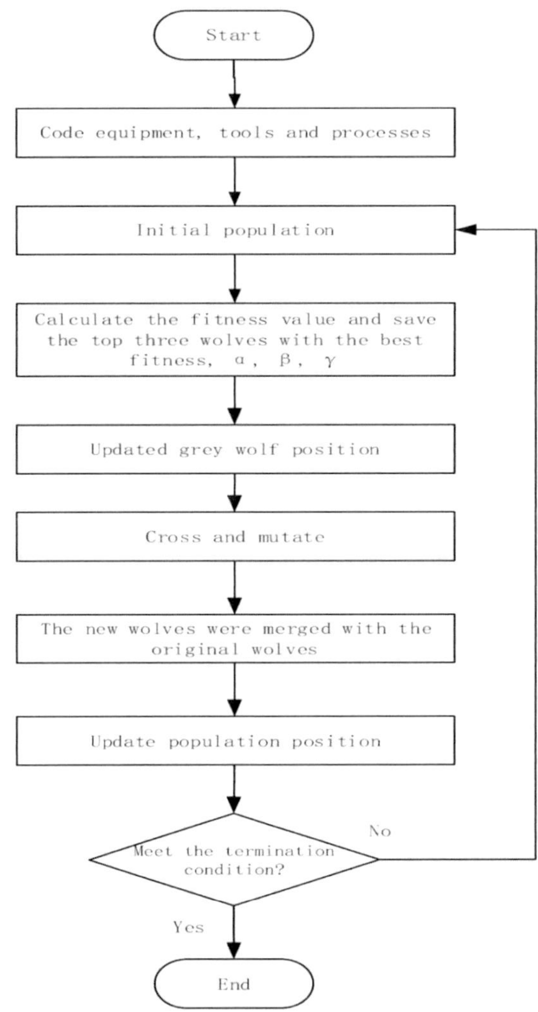

Figure 3. The flow of Grey Wolf Optimizer (GWO).

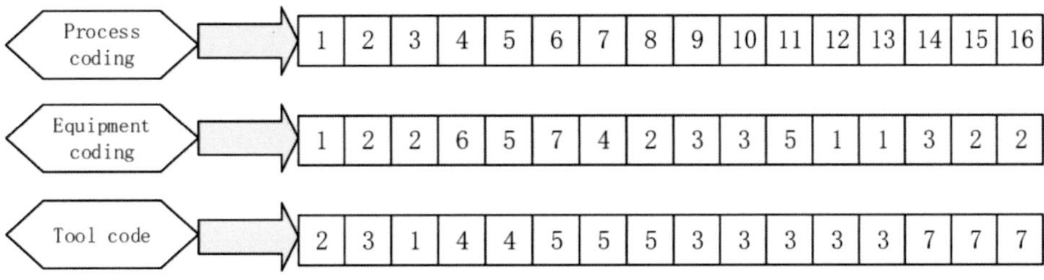

Figure 4. Coding method.

### 3.4. Fitness Function

Each solution represents a wolf, with the first initialization to obtain a random initial set of solutions. The fitness function of each solution was calculated to establish the rank of wolves. Wolves with higher fitness were retained as guides to lead wolves with lower fitness to hunt. There are two objective functions in this paper: $f_1$ (energy consumption), and $f_2$ (carbon emissions). The value of each objective function is calculated, respectively, and then the weight method is used to combine them into a function. The fitness function is,

$$min\ fitness_i = min(\omega_1 \frac{f_{1i} - f_{1min}}{f_{1max} - f_{1min}} + \omega_2 \frac{f_{2i} - f_{2min}}{f_{2max} - f_{2min}})$$ (33)

$$f_1 = min(\sum_{i=1}^{l} E_{D,i} + E_A + \sum_{i=1}^{l} E_{M,i})$$ (34)

$$f_2 = min(C_M + C_E + C_W)$$ (35)

$f_{1i}, f_{2i}$ represent the values of the ith wolf, $f_{1max}$ and $f_{1min}$ are, respectively, the energy consumption extremums when the energy consumption is independently optimized, $f_{2max}$ and $f_{2min}$ are, respectively, the carbon emission extremums when carbon emissions are individually optimized, $\omega_1$, $\omega_2$ are, respectively, the weight of energy consumption and carbon emissions, and satisfy $\omega_1 + \omega_2 = 1$. The values of $\omega_1$, $\omega_2$ can be evaluated by fuzzy evaluation method, analytic hierarchy process and other methods.

### 3.5. Constraints

Parameter selection of gear cutting process should follow the following constraints

(1) Machine tool speed constraints

The spindle speed has an important effect on the quality of the workpiece. The spindle speed should be within the allowable range of the machine tool

$$n_{min} \leq n \leq n_{max}$$ (36)

where $n_{min}$ and $n_{max}$ represent the machine tool limit speed, respectively.

(2) Feed limit constraint

The feed rate has an important effect on the machining accuracy.

$$f_{min} \leq f \leq f_{max}$$ (37)

where $f_{min}$ and $f_{max}$ are respectively the limit feed amount of machine tool.

(3) Cutting force constraint

Cutting force has an important effect on machining accuracy and tool wear. The total cutting force $F_i$ includes three parts, main cutting force $F_c$, backside force $F_f$ and feed force $F_p$. $F_i$ cannot exceed the maximum cutting force $F_{k_{max}}$.

$$F_i = \sqrt{F_{c_i}^2 + F_{f_i}^2 + F_{p_i}^2} \leq F_{k_{max}}$$ (38)

$$\begin{cases} F_c = C_{F_c} a_p^{x_{F_c}} f^{y_{F_c}} v_c^{n_{F_c}} k_{F_c} \\ F_f = C_{F_f} a_p^{x_{F_f}} f^{y_{F_f}} v_c^{n_{F_f}} k_{F_f} \\ F_p = C_{F_p} a_p^{x_{F_p}} f^{y_{F_p}} v_c^{n_{F_p}} k_{F_p} \end{cases}$$ (39)

in turning, for example, $C_{F_c}, x_{F_c}, y_{F_c}, n_{F_c}, k_{F_c}$ are the main cutting force correlation coefficient $C_{F_f}, x_{F_f}, y_{F_f}, n_{F_f}, k_{F_f}$ are the feed force correlation coefficient, $C_{F_p}, x_{F_p}, y_{F_p}, n_{F_p}, k_{F_p}$ are the backward force correlation coefficient, these coefficients are determined by material, tool and other processing conditions. The feeding force must meet the following conditions

$$F_f = C_{F_f} a_p^{x_{F_f}} f^{y_{F_f}} v_c^{n_{F_f}} k_{F_f} \leq F_{f_{max}}$$ (40)

(4) Machine tool power constraint

The machine power should be within certain conditions.

$$\frac{F_c v_c}{\tau} \leq P_{max} \tag{41}$$

$P_{max}$ is the maximum machine power, $F_c$, $v_c$, $\tau$ are cutting force, cutting speed and machine tool power coefficient, respectively.

(5)  Roughness constraint

Surface quality $R_a$ is an important evaluation index of parts, the surface roughness of parts should be less than the maximum allowed surface roughness $R_{a_{max}}$.

$$R_a = \frac{0.0312 f^2}{r} \leq R_{a_{max}} \tag{42}$$

$r$ is the radius of the tool tip.

### 3.6. Population Classification and Location Update

GWO leads the pack to search with three optimal solutions $\alpha, \beta, \gamma$. $\alpha, \beta, \gamma$ are producing randomly as the population non-dominated series is 1; as the non-dominated grade is 2, $\alpha$ is produced from grade 1, $\beta, \gamma$ are obtained from grade 2. As the non-dominant grade is 3 or more, $\alpha, \beta, \gamma$ are produced from the above three grades, respectively [36,37].

This paper improves the operator update of the algorithm based on the transformation law to solve the process route optimization problem, and selects one of them as the child according to a certain probability.

$$X(\pi)_i^{t+1} = \begin{cases} shift\left(X(\pi)_i^t, C \cdot \left(X(\pi)_\alpha^t - X(\pi)_i^t\right)\right) \; if \; 0 \leq rand \leq \frac{1}{3} \\ shift\left(X(\pi)_i^t, C \cdot \left(X(\pi)_\beta^t - X(\pi)_i^t\right)\right) \; if \; \frac{1}{3} \leq rand \leq \frac{2}{3} \\ shift\left(X(\pi)_i^t, C \cdot \left(X(\pi)_\gamma^t - X(\pi)_i^t\right)\right) \; if \; \frac{2}{3} \leq rand \leq 1 \end{cases} \tag{43}$$

where the shift function helps the wolves to update their position. $X(\pi)_i^t$ is the individual wolf pack. $shift\left(\vec{x}, \vec{d}\right)$ means individual wolves can move from side to side. $\vec{d}$ represents the distance the element has traveled. $rand$ randomly generated in [0, 1] in [0, 1], and $C = 1$.

### 3.7. Genetic Operations

When genetic information is inherited, there are usually two kinds of operation: crossover and mutation. Different substrings can have different genetic manipulations. In this paper, two points are selected to cross the equipment and tool substring [38,39], as shown in Figure 5. Duplication and omission can be avoided by using an improved two-point crossover method based on priority order. Two points are randomly selected in the substring as the intersection points. In parent P1, the genes before point 1 and after point 2 are retained to the same position as offspring O1. The existing genes in O1 were removed from the parent P2, and the remaining genes were copied to the remaining site of O1 in the order of P2. The offspring O2 performed the same operation. The mutation operation is shown in Figure 6, which randomly selects a process to replace a certain site that can be replaced while ensuring the process constraints [40–42].

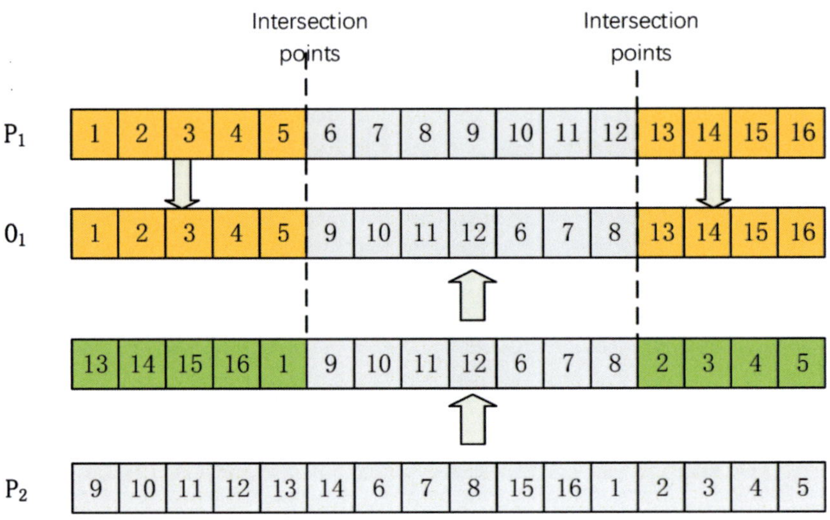

Figure 5. Sequence substring crossing.

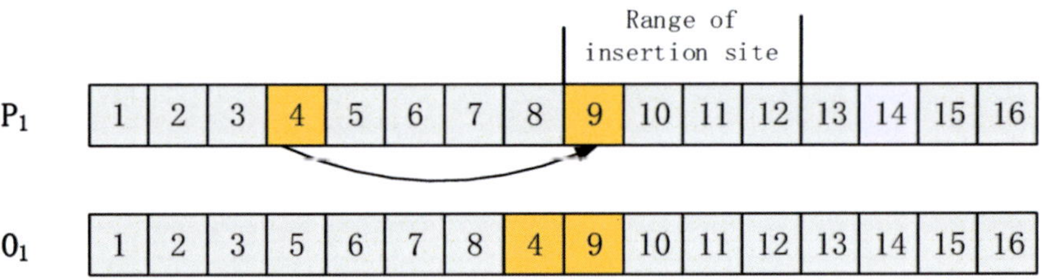

Figure 6. Sequential substring variation.

## 4. Method

### 4.1. Instance Parameters

There are 2 CNC lathes (M1, M2) involving turning processing in the example workshop. The processing material is the second gear of the intermediate shaft of the automobile transmission, the drawing of gear to be machined is shown in Figure 7, gear parameters are in Table 1.

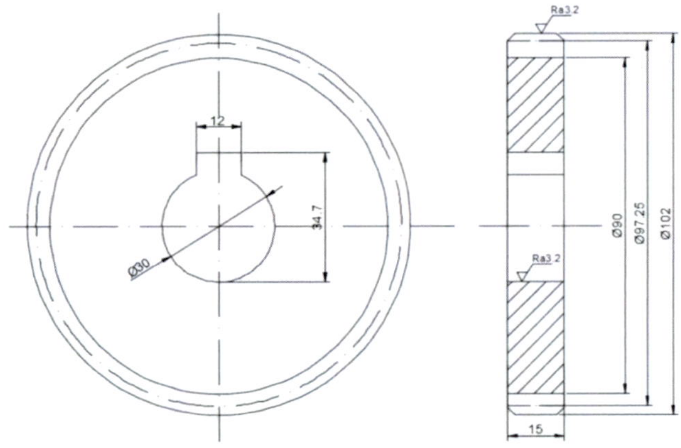

Figure 7. Gear dimension drawing.

**Table 1.** Gear parameters.

| Material | Outer Diameter/mm | Tooth Thickness/mm | Modulus/mm | Number of Teeth | Weight/kg |
|---|---|---|---|---|---|
| 20CrMnTiH | 97.25 | 15 | 1.75 | 46 | 0.665 |

The machine parameters are shown in Table 2.

**Table 2.** Machine parameters.

| Type | Serial Number | $n$ ($r/min$) | $f$ ($mm/r$) | $\tau$ | $F_{max}$ (N) | $P_{max}$ (kW) |
|---|---|---|---|---|---|---|
| lathe | M1 | 100–1400 | 0.1–0.25 | 0.85 | 1700 | 8.0 |
|  | M2 | 120–1600 | 0.1–0.35 | 0.8 | 1700 | 10 |

Tool: the tool material K1 is high-speed steel, the tool main deflection angle is 45°, hook angle is 20°, tool edge inclination is 5°, and corner radius $r_\theta = 0.8$ mm. The tool material K2 is cemented carbide. The main deflection angle of the tool is 45°, hook angle is 20°, tool edge inclination is 5°, and corner radius $r_\theta = 0.8$ mm.

The cutting force coefficients are in Table 3.

**Table 3.** Cutting force coefficient.

| | Main Cutting Force Coefficient | | | | | Feed Force Coefficient | | | | | Backward Force Coefficient | | | | |
|---|---|---|---|---|---|---|---|---|---|---|---|---|---|---|---|
| | $C_{F_c}$ | $x_{F_c}$ | $y_{F_c}$ | $n_{F_c}$ | $k_{F_c}$ | $C_{F_f}$ | $x_{F_f}$ | $y_{F_f}$ | $n_{F_f}$ | $k_{F_f}$ | $C_{F_p}$ | $x_{F_p}$ | $y_{F_p}$ | $n_{F_p}$ | $k_{F_p}$ |
| M1 | 1750 | 0.9 | 0.75 | 0 | 1 | 580 | 1.1 | 0.65 | 0 | 1 | 1100 | 0.9 | 0.65 | 0 | 1 |
| M2 | 2855 | 1 | 0.75 | −0.1 | 1 | 2920 | 1 | 0.5 | −0.35 | 1 | 1930 | 0.9 | 0.6 | −0.35 | 1 |

The gear processing carbon emission factors are shown in Tables 4–7.

**Table 4.** Material preparation process carbon emission factor.

| Carbon Emission Category | Material $i$ Consumption | Production Process Consumes Energy $c$ | Energy $c$ Carbon Emission Factor $S_{Ec}$ |
|---|---|---|---|
| Material preparation process carbon emissions $C_M$ | Steel | Raw coal | 2.653 |

**Table 5.** Indirect carbon emission factors in energy preparation process.

| Carbon Emission Category | The $n$th Energy Type Consumed by Energy $k$ | Production Process Consumes Energy | Energy $n$ Carbon Emission Factor $S_{En}$ |
|---|---|---|---|
| Indirect carbon emissions in the energy production process $C_{IE}$ | Electricity | Raw coal | 2.565 |
| | | Crude | 2.221 |
| | | Natural gas | 1.642 |
| | Coal | Crude | 2.221 |
| | | Natural gas | 1.642 |
| | | Electricity | 8.220 |
| | Natural gas | Raw coal | 2.565 |
| | | Crude | 2.221 |
| | | Natural gas | 1.642 |
| | Fuel/Circulating oil/Lubricant | Raw coal | 2.565 |
| | | Crude | 2.221 |
| | | Natural gas | 1.642 |

### 4.2. Grey Wolf Algorithm Settings

All programs are written by Matlab R2019b and run on a Windows 10 host configured with 16.0G RAM, AMD Ryzen 3700X 3.6Ghz, and a 64-bit operating system. Grey wolf algorithm parameters are: the total population is 100, the iteration times is 500, the crossover rate is 0.8, and the mutation rate is 0.1, the number of leading wolves is 3, the coefficient of affecting the search times of the neighborhood is 2, the coefficient of choosing the global search operator is 0.5.

**Table 6.** Direct carbon emission factor from fossil energy.

| Carbon Emission Category | Consumption Type of Material $k$ | Energy Carbon Emission Factor $S_{E_k}$ |
|---|---|---|
| Processing direct carbon emissions $C_{DE}$ | Coal | 0.6764 |
| | Natural gas | 0.4593 |
| | Fuel/Circulating oil/Lubricant | 0.6878 |

**Table 7.** Waste disposal carbon emission factor.

| Carbon Emission Category | Waste $l$ Discharge Type | Energy Consumed Type in the Waste Treatment Process | Energy Carbon Emission Factor $S_{E_q}$ |
|---|---|---|---|
| Waste treatment carbon emissions $C_W$ | Waste water/waste oil | Electricity | 8.221 |
| | Scraps | Electricity | 8.221 |

*4.3. Optimization Results and Analysis*

In this paper, the energy consumption and carbon emission of gear machining process are taken as the optimization objectives, and a concrete calculation model can be obtained according to the proposed process route optimization model and the parameters in 4.1. The grey wolf algorithm was used to solve the calculation model, and the selection of equipment, tool and process sequencing and other process routes were taken as variables, and they were reasonably coded in the program. Figure 8 shows the iterative convergence curve of carbon emissions, and Figure 9 shows the energy consumption convergence iteration. With the increase in the number of iterations, energy consumption and carbon emissions gradually decrease and become stable.

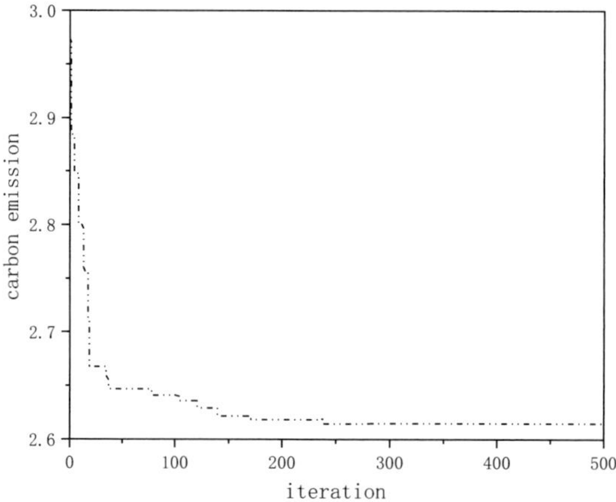

**Figure 8.** Carbon emission convergence algebraic diagram.

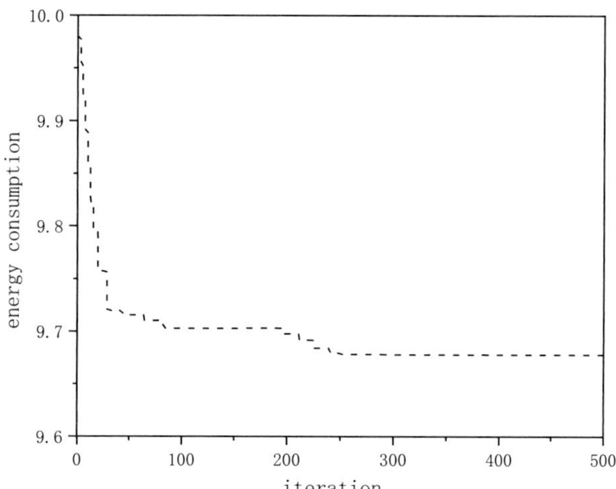

**Figure 9.** Energy consumption convergence algebraic diagram.

### 5. Discussion

*5.1. Results Analysis*

The comparison data between the results and the optimization results of low carbon and low energy consumption alone are shown in Table 8.

**Table 8.** Optimization results.

| Optimization Results | Low Carbon | Low Energy | Low Carbon and Low Energy |
|---|---|---|---|
| Carbon emission/kg | 2.612 | 2.813 | 2.975 |
| Energy consumption/kW· h | 10.064 | 9.689 | 9.989 |

The comparison results show that the carbon emission and energy consumption are higher. As the process route is optimized with low energy consumption, there will be higher carbon emissions. As the process route is optimized with low carbon, both energy consumption and carbon emissions have been reduced to some extent, but there are also obvious shortcomings. As the process route is optimized with two objectives simultaneously, a process route for balancing carbon emission and energy consumption is available.

*5.2. Comparison with Previous Works*

In the introduction, references [8–16] put forward many methods of process route optimization, effectively achieving their objectives. They mainly established the relationship between process parameters and the objective to be optimized, and used intelligent algorithms to optimize the solutions. However, gear processing is complicated, and few people study the gear process route. Different process routes have great influence on machining results. The core of reference [17] is through the blank production and uses the process parameters to design an energy-saving and low-carbon gear blank dimension optimization method. The theme in this paper is to optimize the process route of gear which is occur after blank choose, and make reasonable arrangements for the process route of gear blank cutting equipment, tools and processes. In this paper, the carbon emissions and energy consumption of gear machining were analyzed systematically, and the optimization model of gear machining process route was established. An improved grey wolf algorithm was proposed to optimize the gear process route for equipment selection, tool selection and process sequencing. The algorithm improves the updating operator, and the solution accuracy is higher. A reasonable processing route can reduce carbon emissions and energy consumption.

*5.3. Research Significance and Future Steps*

In this paper, a low carbon and low energy consumption optimization method of gear process route was proposed, and verified by the machining process of automobile transmission gear, which can help designers choose the best process route. This study is helpful to improve the cognition level of energy consumption and carbon emission in gear processing, which can enable enterprises to choose reasonable processing process routes, help manufacturing industries to save energy and reduce emissions, and provide ideas for the green development of the manufacturing industries. The tool wear and precision state of machine tools are also factors that affect the energy efficiency of processing route. Determining how to comprehensively consider the tool life and precision state of machine tools will be the focus of the next research.

## 6. Conclusions

The energy saving and emission reduction in gear machining is a complicated problem, which not only affects the production of enterprises, but also has important significance for the green development of society. This paper made the following research on gear processing:

1.  The carbon emission and energy consumption of gear processing were systematically analyzed, and the optimization model of gear processing route with the low-carbon and low-energy-consumption was established.
2.  An improved grey wolf algorithm was proposed to solve the multi-objective optimization model, optimize the equipment selection, tool selection and process sequencing.
3.  Taking the second gear of the intermediate shaft of an automobile transmission as an example, the results of optimizing the process routes of energy consumption and carbon emission with comprehensive consideration of these three objectives were compared, and the validity of the method was proven.

The results show that this method can comprehensively consider the carbon emission and energy consumption of gear processing, and provide the process route guidance for enterprises to process gear, and make contributions to social energy saving and emission reduction. The influencing factors of process route are very large; this paper only studies energy consumption and carbon emissions. Tool wear and the state of machine tool accuracy have a great influence on workpiece quality. The influence of tool wear and machine tool accuracy on process route will be studied in the future.

**Author Contributions:** Conceptualization, Y.Z., H.X. and Y.X.; methodology, Y.Z.; Software, Y.Z.; validation, H.X.; formal analysis, H.X.; data curation, J.H.; writing—original draft, J.H. and Y.X.; writing—review & editing, Y.X.; funding acquisition, J.H. All authors have read and agreed to the published version of the manuscript.

**Funding:** This study was supported by the development project of young scientific and technological talents in colleges and universities of Guizhou Province ([2020]195); the program of Qiannan Normal University for Nationalities under Grant (Nos. QNSY 2018025).

**Data Availability Statement:** The study did not report any data.

**Conflicts of Interest:** The authors declare no conflict of interest.

## References

1.  Liu, P.; Tuo, J.; Liu, F.; Li, C.; Zhang, X. A novel method for energy efficiency evaluation to support efficient machine tool selection. *J. Clean. Prod.* **2018**, *191*, 57–66. [CrossRef]
2.  Sihag, N.; Sangwan, K.S. A systematic literature review on machine tool energy consumption. *J. Clean. Prod.* **2020**, *275*, 123125.
3.  Li, X. Design of energy-conservation and emission-reduction plans of China's industry: Evidence from three typical industries. *Energy* **2020**, *209*, 118358.

4.  McKenna, P.; Prashant, S.; Heejin, C. The current state of the industrial energy assessment and its impacts on the manufacturing industry. *Energy Rep.* **2022**, *8*, 7297–7311.
5.  Mahapatra, B.; Irfan, M. Asymmetric impacts of energy efficiency on carbon emissions: A comparative analysis between developed and developing economies. *Energy* **2021**, *227*, 120485.
6.  Liu, Z.; Lu, H.; Xia, L. CNC high-speed dry hobbing optimization design process parameters of planetary gears. *Mach. Des. Manuf.* **2020**, *8*, 239–242.
7.  Álvarez, Á.; Calleja, A.; Ortega, N.; De Lacalle, L.N.L. Five-Axis Milling of Large Spiral Bevel Gears: Toolpath Definition, Finishing, and Shape Errors. *Metals* **2018**, *8*, 353. [CrossRef]
8.  An, X.H.; Chen, T. Optimization of process route based on intuitionistic fuzzy number and multi-objective optimization algorithm. *Comput. Integr. Manuf. Syst.* **2019**, *25*, 1180–1191.
9.  Fan, S.; Wang, J.; Shijie, L.I. Decision and optimization of process routing based on genetic algorithm. *Manuf. Technol. Mach. Tool* **2012**, *3*, 95–99.
10. Huang, F.; Gu, J.; Zhang, L.; Xu, C.; Wang, H. Ant colony optimization of STEP-NC process route based on dynamic update of taboo manufacturing features. *China Mech. Eng.* **2016**, *27*, 596–602.
11. Cheng, B.; Jing, B.X. Process route optimization based on bacteria foraging and ant colony algorithm. *Chin. J. Eng. Des.* **2020**, *27*, 9.
12. Li, C.B.; Li, P.Y.; Liu, F.; Cui, L.G.; Shui, H. Multi-objective machining process route optimization model for high efficiency and low carbon. *J. Mech. Eng.* **2014**, *50*, 9.
13. Zhai, D.; Zhang, F.; Gao, B.; Han, W.; Zhang, T.; Zhang, J. Ant Colony Algorithm and Simulated Annealing Algorithm Based Process Route Optimization. In Proceedings of the 2014 Enterprise Systems Conference, Shanghai, China, 2–3 August 2014; pp. 102–107. [CrossRef]
14. Tang, Y.; Yang, Q.; Li, C.; Xiao, Q.; Chen, X. Process Route Optimization for Generalized Energy Efficiency and Production Time in Machining System. In Proceedings of the 2019 IEEE International Conference on Service Operations and Logistics, and Informatics (SOLI), Zhengzhou, China, 6–8 November 2019; pp. 110–115. [CrossRef]
15. Xiao, Y.; Zhang, H.; Jiang, Z.; Gu, Q.; Yan, W. Multiobjective optimization of machining center process route: Tradeoffs between energy and cost. *J. Clean. Prod.* **2021**, *280*, 124171.
16. Petrović, M.; Mitić, M.; Vuković, N.; Miljković, Z. Chaotic particle swarm optimization algorithm for flexible process planning. *Int. J. Adv. Manuf. Technol.* **2016**, *85*, 2535–2555. [CrossRef]
17. Xiao, Y.; Zhou, J.; Wang, R.; Zhu, X.; Zhang, H. Energy-Saving and Low-Carbon Gear Blank Dimension Design Based on Business Compass. *Processes* **2022**, *10*, 1859.
18. Wang, Y.; Liu, C.L.; Ji, Z.C. Energy Consumption Model of the Discrete Manufacturing System. In *Quantitative Analysis and Optimal Control of Energy Efficiency in Discrete Manufacturing System*; Springer: Singapore, 2020.
19. Li, Y.; Liu, Q. Research on Service-Oriented Green Efficient Milling Optimization Method. *J. Mech. Eng.* **2015**, *51*, 89–98.
20. Xiao, Y.; Jiang, Z.; Gu, Q.; Yan, W.; Wang, R. A novel approach to CNC machining center processing parameters optimization considering energy-saving and low-cost. *J. Manuf. Syst.* **2021**, *59*, 535–548.
21. LI, C.B. Multi-objective NC Machining Parameters Optimization Model for High Efficiency and Low Carbon. *J. Mech. Eng.* **2014**, *49*, 87. [CrossRef]
22. Hammond, G.P.; Jones, C.I. Embodied Energy and Carbon in Construction Materials. *Constr. Mater.* **2009**, *162*, 87–98.
23. Li, L.; Deng, X.; Zhao, J.; Zhao, F.; Sutherland, J.W. Sutherland, Multi-objective optimization of tool path considering efficiency, energy-saving and carbon-emission for free-form surface milling. *J. Clean. Prod.* **2018**, *172*, 3311–3322.
24. Yi, Q.; Li, C.; Tang, Y.; Chen, X. Multi-objective parameter optimization of CNC machining for low carbon manufacturing. *J. Clean. Prod.* **2015**, *95*, 256–261.
25. Fu, Q.; Wang, P. Improved Grey Wolf Algorithm Based on Parallel Search Strategies. *Appl. Res. Comput.* **2016**, *33*, 1662–1665.
26. Lu, C.; Gao, L.; Li, X.; Xiao, S. A hybrid multi-objective grey wolf optimizer for dynamic scheduling in a real-world welding industry. *Eng. Appl. Artif. Intell.* **2017**, *57*, 61–79.
27. Du, H.B.; Ge, Z.Z. Inverse kinematics solution algorithm of electric climbing robot based on improved beetle antennae search algorithm. *Control. Decis.* **2022**, *37*, 9.
28. Lv, Y.; Mo, Y. Improved beetle antennae search algorithm with mutation crossover in TSP and engineering application. *Appl. Res. Comput.* **2021**, *38*, 3662–3666.
29. Han, S. Modified Grey-Wolf Algorithm Optimized Fractional-Order Sliding Mode Control for Unknown Manipulators with a Fractional-Order Disturbance Observer. *IEEE Access* **2020**, *8*, 18337–18349.
30. Jarndal, A. On modeling of substrate loading in GaN HEMT using grey wolf algorithm. *J. Comput. Electron.* **2020**, *19*, 576–590.
31. Nadimi-Shahraki, M.H.; Taghian, S.; Mirjalili, S.; Zamani, H.; Bahreininejad, A. GGWO: Gaze cues learning-based grey wolf optimizer and its applications for solving engineering problems. *J. Comput. Sci.* **2022**, *61*, 101636.
32. Rajammal, R.R.; Mirjalili, S.; Ekambaram, G.; Palanisamy, N. Binary Grey Wolf Optimizer with Mutation and Adaptive K-nearest Neighbour for Feature Selection in Parkinson's Disease Diagnosis. *Knowl. Based Syst.* **2022**, *246*, 108701.
33. Gujarathi, P.K.; Shah, V.A.; Lokhande, M.M. Combined Rule Based-Grey Wolf Optimization Energy Management Algorithm for Emission Reduction of Converted Plug-In Hybrid Electric Vehicle. *SAE Int. J. Passeng. Cars Electron. Electr. Syst.* **2019**, *12*, 101–116.
34. Chen, Y.; Luca, G.D. Technologies Supporting Artificial Intelligence and Robotics Application Development. *J. Artif. Intell. Technol.* **2021**, *1*, 1–8.

35. Tu, Q.; Chen, X.; Liu, X. Hierarchy Strengthened Grey Wolf Optimizer for Numerical Optimization and Feature Selection. *IEEE Access* **2019**, *7*, 78012–78028. [CrossRef]
36. Zhang, G.; Li, H.; Xiao, C.; Sobhani, B. Multi-aspect analysis and multi-objective optimization of a novel biomass-driven heat and power cogeneration system; utilization of grey wolf optimizer. *J. Clean. Prod.* **2022**, *355*, 131442. [CrossRef]
37. Luo, S.; Zhang, L.; Fan, Y. Energy-efficient scheduling for multi-objective flexible job shops with variable processing speeds by grey wolf optimization. *J. Clean. Prod.* **2019**, *234*, 1365–1384.
38. Feng, X.; Huang, S. Research and Application of Beetle Antennae Genetic Hybrid Algorithm. *Comput. Eng. Appl.* **2021**, *57*, 90–100.
39. Rezaeipanah, A.; Mojarad, M. Modeling the Scheduling Problem in Cellular Manufacturing Systems Using Genetic Algorithm as an Efficient Meta-Heuristic Approach. *J. Artif. Intell. Technol.* **2021**, *1*, 228–234.
40. Wolff, S.; Seidenfus, M.; Brönner, M.; Lienkamp, M. Multi-disciplinary design optimization of life cycle eco-efficiency for heavy-duty vehicles using a genetic algorithm. *J. Clean. Prod.* **2021**, *318*, 128505. [CrossRef]
41. Peng, D.; Tan, G.; Fang, K.; Chen, L.; Agyeman, P.K.; Zhang, Y. Multiobjective Optimization of an Off-Road Vehicle Suspension Parameter through a Genetic Algorithm Based on the Particle Swarm Optimization. *Math. Probl. Eng.* **2021**, *2021*, 9640928.
42. Martowibowo, S.Y.; Damanik, B.K. Optimization of Material Removal Rate and Surface Roughness of AISI 316L under Dry Turning Process using Genetic Algorithm. *Manuf. Technol.* **2021**, *21*, 373–380. [CrossRef]

*Article*

# Research on Driving Factors of Collaborative Integration Implementation of Lean-Green Manufacturing System with Industry 4.0 Based on Fuzzy AHP-DEMATEL-ISM: From the Perspective of Enterprise Stakeholders

Xiaoyong Zhu [1], Yongmao Xiao [2,3,4,*], Gongwei Xiao [1] and Xiaojuan Deng [1]

[1]  School of Economics & Management, Shaoyang University, Shaoyang 422000, China
[2]  School of Computer and Information, Qiannan Normal University for Nationalities, Duyun 558000, China
[3]  Key Laboratory of Complex Systems and Intelligent Optimization of Guizhou Province, Duyun 558000, China
[4]  Key Laboratory of Complex Systems and Intelligent Optimization of Qiannan, Duyun 558000, China
*   Correspondence: xym198302@163.com



**Citation:** Zhu, X.; Xiao, Y.; Xiao, G.; Deng, X. Research on Driving Factors of Collaborative Integration Implementation of Lean-Green Manufacturing System with Industry 4.0 Based on Fuzzy AHP-DEMATEL-ISM: From the Perspective of Enterprise Stakeholders. *Processes* **2022**, *10*, 2714. https://doi.org/10.3390/pr10122714

Academic Editor: Sergey Y. Yurish

Received: 29 November 2022
Accepted: 11 December 2022
Published: 15 December 2022





**Abstract:** The existing research and practices have shown that the coordinated implementation of lean-green manufacturing can have a positive impact on the economic and environmental benefits, which is an effective means to ensure the environmental protection of the production process of manufacturing without damaging their profitability. Within the field of lean-green research, there is still a lack of research to analyze the driving factors for the collaborative implementation of integrated lean and green integration. Although, some scholars and researchers have studied lean and green integration paradigms, their research has mostly focused on lean-green integration practices and their impact on environmental performance and their respective operations. In the context of Industry 4.0, this article investigates the driving forces behind the collaborative integration implementation of a lean-green manufacturing system from the viewpoint of stakeholders. Specifically addressing the issues of correlation and ambiguity in the identification of driving factors, this manuscript proposes an Interpretation Structure Model (ISM) of fuzzy comprehensive Analytic Hierarchy Process (AHP), based on Decision-Making Trial and Evaluation Laboratory (DEMATEL), to determine the importance of the driving factors. Combined with the complex network theory, the evaluation index system is divided into four levels from eight factor categories, including endogenous lean-green driving factors and exogenous driving factors. The fuzzy AHP-DEMATEL-ISM is used to analyze the relationship between indicators and the structure of the indicator system. The complex network which is composed of the indicator system is divided into different levels. The importance of indicators is analyzed from the perspective of the global network, and key factors affecting the driving of lean-green system is analyzed. The integration of the lean-green manufacturing system and organizational synergy are promoted to jointly lead the enterprise toward sustainable development by paying particular attention to the primary impact indicators and aggressively cultivating the key impact indicators.

**Keywords:** driving factors; lean-green manufacturing; Industry 4.0; fuzzy AHP-DEMATEL-ISM; enterprise stakeholders


## 1. Introduction

Smart manufacturing/Industry 4.0 is going to be the future development trend of manufacturing enterprises [1]. In the various stages of the implementation of Industry 4.0, each sector of the manufacturing industry in various nations will have its own major directions and development aspects [2]. The core feature of the industrial internet in the United States, Germany's Industry 4.0 strategy, and made in China 2025 is interconnection, the essence of which is to shift from economies of scale to economies of scope through the automated flow of data, and to build out heterogeneous and customized industries at a homogeneous and scaled cost, thus promoting the reform of industrial structure [3–8]. According to the data compiled by some of the scholars, Chinese manufacturing companies have great enthusiasm and expectation for Industry 4.0. A total of 76% of Chinese manufacturing companies believe that the use of Industry 4.0 strategy will greatly enhance the competitiveness of manufacturing industry, compared to 54% in the United States, 51% in Japan, and a low of 47% in Germany [9]. Chinese manufacturers are indeed very enthusiastic about Industry 4.0 and have high expectations for it, yet there are significant concerns and challenges with its implementation. While interviewing Chinese manufacturing companies, only 53% of them said that they are fully prepared for Industry 4.0 strategy, compared to 71% and 68%, in the US and Germany, respectively. Among them, state-owned enterprises are the most conservative, only 44% of the surveyed state-owned enterprises said that they are ready for Industry 4.0 strategy; the proportion of private manufacturing enterprises is as high as 68% [10]. One of the reasons is that the digital foundation for the development of intelligent manufacturing is relatively weak. The development of manufacturing industry as a whole is still in the transition stage from mechanical automation to digital automation. As far as the Industry 4.0 is considered in Germany as a reference system, the overall is still in the 2.0 era, but some enterprises are moving towards the 3.0 era. Therefore, most of China's manufacturing industry still needs to use lean management to reduce costs and increase efficiency, as well as use Industry 4.0 technology to upgrade and transform. Lean management is the soft aspect of overall optimization of manufacturing and management processes; while intelligent manufacturing is the hard aspect of intelligent upgrading of production factors and information systems. Chinese manufacturing companies must combine "hard and soft" and employ lean management to build a strong foundation for manufacturing companies to execute smart manufacturing in this wave of global manufacturing transformation and upgrading.

The rapid development of industry and the excessive use and waste of resources by human beings have led to the depletion of resources. In addition, a large amount of industrialized production also causes environmental pollution [11]. Existing manufacturing methods are causing climate warming and resource depletion, and are unsustainable. In order to solve this global challenge, Industry 4.0 is developed. Therefore, the first connotation of Industry 4.0 is to be smart, green, lean, and humanized. The pursuit of personalized items is gradually becoming more and more challenging for traditional production methods to generate personalized or customized products in large quantities due to the changing needs of consumers. The smart factory in Industry 4.0 environment aims to produce precise, high-quality, and personalized smart products, so that the efficiency and cost of single-piece small batch production can reach the same level of mass production. It can be customized for enterprise customers in large scale and small batch, and also for individual users in small batch and single product. The logistics and transportation system from raw materials to final products is completed by intelligent logistics. The lean production method is a great way to produce high quality and low consumption, under mixed production conditions of multiple varieties and small batches. Green first term refers to the use of alternative, non-traditional clean energy. It can reduce traditional energy consumption, effectively alleviate resource depletion, but also produces fewer pollutants, meaning effective protection of the environment. No matter how a product is made, its use and disposal have little effect on the environment across its entire life cycle, and they could be recovered and repurposed to promote sustainable development. Therefore, lean management and green manufacturing are the cornerstones for digital transformation to bring benefits. At the same time, John et al. (2021) showed that the goal of implementing lean-green manufacturing in manufacturing companies is at the same time the goal to be achieved by Industry 4.0, where lean-green manufacturing promotes a more time-efficient and resource-efficient of Industry 4.0 factory, which in turn further enhances lean-green manufacturing [12]. It can help substantiate this theory: the integration of lean-green manufacturing into the same framework in an Industry 4.0 context.

In summary, it is crucial for manufacturing companies to integrate and collaborate on the implementation of lean-green manufacturing systems in the process of Industry 4.0 transformation and upgrading. However, the lack of confidence and preparedness of Chinese manufacturing companies regarding the Industry 4.0 strategy is largely due to two major challenges: high-quality development, and improving total factor productivity. Therefore, it is necessary for the vast majority of manufacturing companies in China to determine how to integrate the collaborative implementation of lean-green manufacturing systems in a smart manufacturing/Industry 4.0 scenarios, and what are their drivers and relationships. In order for effectively collaborative integration implementation of lean-green manufacturing system under Industry 4.0, this research framework addresses the following research questions:

**Question 1.** *What are the drivers for integrated and collaborative implementation of lean-green manufacturing systems?*

**Question 2.** *What are the most critical drivers?*

**Question 3.** *What is the cause-and-effect relationship between them?*

**Question 4.** *What are the steps that need to be taken to better motivate these drivers to work? How about to achieve the goal?*

In the context of Industry 4.0, this manuscript analyzes the driving factors for collaboratively integrating the implementation of a lean-green manufacturing system. Aiming at the uncertainty and correlation problems in the process of driving factors identification, this study proposes an Interpretation Structure Model (ISM) of fuzzy comprehensive Analytic Hierarchy Process (AHP), based on decision making trial and evaluation laboratory (DEMATEL) to determine the importance of the driving factors. Combined with the complex network theory, the evaluation index system is divided into four levels from eight categories of factors, including endogenous lean-green driving factors and exogenous driving factors. The fuzzy AHP-DEMATEL-ISM is used to analyze the relationship between indicators and the structure of the indicator system, and the complex network composed of the indicator system is divided into different levels. The importance of indicators is analyzed from the perspective of the global network, and the important factors and key factors affecting the driving of lean-green system are analyzed. By paying special attention to the main impact indicators and actively cultivating the key impact indicators, the lean-green manufacturing system integration and synergy of the organization are promoted to jointly drive the enterprise to achieve sustainable development.

The paper consists of five parts. The rest of the paper is structured as follows: Section 2 presents the analysis of the driving factors of collaborative integration implementation of lean-green manufacturing system. The research methodology is discussed in Section 3, followed by a presentation of the case study in Section 4. In the last section, some conclusions are drawn.

## 2. Analysis of the Driving Factors of Collaborative Integration Implementation of Lean-Green Manufacturing System

There have been related research conducted by foreign scholars on the integrated implementation of lean-green manufacturing, for example, Mittal et al. (2017) proposed a new manufacturing strategy for manufacturing companies to increasing customer choice, address environmental issues in the manufacturing process and enhance their own competition among global manufacturers, thus adopting a lean-green-agile manufacturing system to coordinate trade-offs to meet the economic, environmental and social demands of modern manufacturing systems [13]. A recent study by Teresa et al. (2022) used a questionnaire to verify the operational, environmental, and financial performance of manufacturing companies implementing lean and green practices in Portugal, ultimately confirming that the widespread adoption of lean and green practices produced better overall operational, environmental and financial performance, and that the integrated implementation of lean and green management practices in companies resulted in superior return on investment [14]. A new paradigm of lean-green manufacturing is made possible by the Industry 4.0 era in the manufacturing sector. Industry 4.0 can be understood as digital lean with an emphasis on green. That is to say, there will always be pressure to raise standards of productivity, quality, agility, environmental friendliness, and customer service in order to stay profitable and competitive in today's corporate environment. Lean-green manufacturing is the foundation for achieving smart factories. A comprehensive review of the lean-green literature by the author team and related scholars in this paper indicates that there is a lack of research to analyze the drivers of integrated lean-green integrated synergistic implementation [15–20]. Several scholars and researchers have studied the integration paradigm of lean and green, and their research has mainly focused on the practical approach of the integration of lean and green and its impact on operational and environmental performance. Some scholars have also studied the internal barriers to the implementation of lean-green manufacturing, and have taken countermeasures to eliminate the existing barriers [18–20]. The corresponding initiative is to study and activate the drivers of lean-green manufacturing system implementation in order to better collaborate and integrate the development of these systems.

The existing literature has confirmed the conclusion that the core concept of lean manufacturing is the creation of value and the elimination of activities that do not add value to the product in the manufacturing process. The lean model is seen to reduce waste, reduce costs, improve quality and productivity, make better use of resources, and create value for customers. The green manufacturing is designed to reduce negative environmental risks and impacts throughout the product life cycle process, while increasing resource productivity and eliminating environmental waste in the organization. The overlap (synergy) between lean-green manufacturing models consists of the following common attributes: waste and manufacturing process waste reduction or elimination techniques, people and organization, lead time reduction and thus production cycle time reduction, supply chain relationships, KPI: service levels and other specific practices. The main commonality between the two can be found in the target attributes of waste elimination and waste reduction in manufacturing processes. The waste (waste) reduction techniques of the two advanced manufacturing models, lean-green, are often similar, with a focus on operational and production practice processes. Both lean and green manufacturing models look at how to integrate product and process redesign to extend product life, make products easily recyclable and make processes more efficient (i.e., reduce waste). The development and success of improvement projects during the lean-green manufacturing model practices require a high level of employee involvement, encouragement of employee participation and empowerment of responsibility to streamline the realization of lean and green practices. When it comes to supply chain relationships, both models rely on close collaboration with supply chain partners, with collaboration supporting the sharing of information and best practices across the chain to serve the goal of an integrated supply chain.

While the two have much in common, there are also incompatible differences between lean-green. Green manufacturing is now no longer optional for manufacturing companies and by introducing green practices into a lean operating environment will have to make certain trade-offs between multiple objectives that are not entirely compatible. The differences between lean and green practices are: their focus, what is considered waste, customers, product design and manufacturing strategy, end of product lifecycle, KPIs, costs, key tools used, and certain specific practice approaches, for example, replenishment frequency, where lean emphasizes multiple batches and multiple frequencies, whereas there is an inconsistency in green's emphasis on reducing carbon emissions.

Lean-green manufacturing systems are an emerging area of research in the introduction and implementation of modern advanced manufacturing models in manufacturing enterprises. Therefore, the integration of lean and green manufacturing models is an important part of the sustainable development of enterprises, especially in the context of China's manufacturing industry, which has become a supporting industry, and there is a need to explore and study the integration of lean and green manufacturing systems in the context of China's economic development and the current situation of enterprise operations. Lean-green manufacturing's integrated and collaborative implementation is driven by both endogenous and exogenous factors, which together drive the company to achieve sustainable development. Endogenous lean-green drivers include managers and internal employees. However, to study the synergistic effect generated by the integrated implementation of lean-green manufacturing also requires exogenous driving elements including shareholders, upstream and downstream companies, consumers, competitors, the public, and government agencies, which can be used as four levels for dividing evaluation indicators (or evaluation indicators can be divided into four categories).

The drive intensity evaluation indicators were divided into categories according to the different types of elements in the endogenous and exogenous green drives. It should be especially noted that since these indicators are derived from published research results, their scientific validity and effectiveness have been widely accepted by experts and scholars. These identified evaluation indicators are both scientifically valid and systematically comprehensive. The above factor indicators can be illustrated in Figure 1.

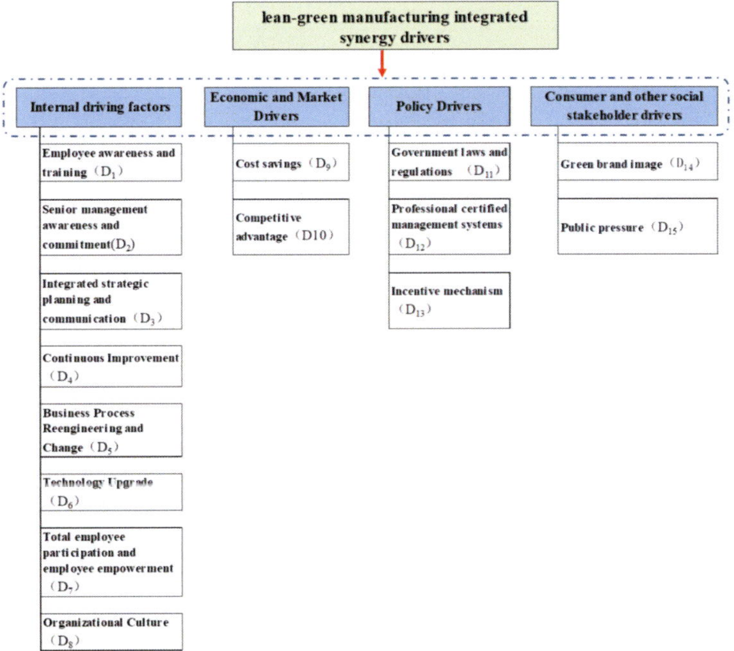


**Figure 1.** Classification of lean-green manufacturing integrated synergy drivers.

In today's highly competitive global environment, business organizations need to improve not only their operational performance but also their eco-efficiency. This has led many scholars to explore the possible merging and convergence of the lean and green paradigms, which in the traditional approach to business organization management practices are often implemented separately and deployed to achieve different corporate strategic goals. Dozens of corresponding papers were found through the corresponding databases (Elsevier, Springer, Emerald, Google Scholar, T&F, Wiley, IEEE, etc.) that delve into the compatibility, synergies, and success key indicator systems between the two. In terms of the search strings, they are specified based on the main topics of the phenomena under investigation. Thus, we searched strings with titles, abstracts, keywords, and articles from January 1997 to 30 September 2022, including (lean green), (lean Industry 4.0), (green Industry 4.0), (lean-green Industry 4.0), (lean environment), (lean environmental), and (lean sustainability). The driver indicators in this thesis are sourced from existing research papers. In the schematic diagram of the classification of driver evaluation indicators shown in Figure 1, the specific meaning of each evaluation factor indicator is as follows.

### 2.1. Internal Driving Factors

Managers and enterprises are directly driven by external pressure, or by their own awareness of the management benefits of carrying out lean and green manufacturing, or they can also be driven by the spontaneous actions of internal employees based on their awareness of environmental efficiency and cost of their own economic and environmental interests.

**Employee awareness and training ($D_1$):** Awareness and acceptance of the Lean-green concept by employees within the company and the company's training to raise awareness of Lean-green among internal employees [15,18,20–23].

**Senior management awareness and commitment ($D_2$):** Awareness and public support from senior management and their commitment to the long-term competitive advantage of lean and green management in the company [18,22–28].

**Integrated strategic planning and communication ($D_3$):** The extent to which top management incorporates lean manufacturing, green production, and environmental protection into its planning; the degree to which middle and lower-level managers are aware of and support the application of lean-green; and the level of collaboration between corporate departments to advance lean-green management [15,19,23–30].

**Continuous Improvement ($D_4$):** Using lean-green tools for continuous improvement to solve internal problems and continuous improvement to eliminate waste, improve efficiency, reduce costs, etc., and lead to business sustainability and reduce the seven major wastes in the lean-green domain [15,18,20–28].

**Business Process Reengineering and Change ($D_5$):** Lean and green thinking is a useful strategy for streamlining business operations. Lean and green tools help companies to follow and use globally accepted methods and standards, to promote the streamlining of business processes, and fit the requirements of organizational lean and green change [15,18,19,26–33].

**Technology Upgrade ($D_6$):** Using energy and resource efficient advanced technologies to achieve production technology, equipment performance, and product performance that meet environmental requirements and industry leading levels [25,28,34–38].

**Total employee participation and employee empowerment ($D_7$):** Manufacturing companies attempting to build lean and green enterprises need all employees to share responsibility for all business functions, all employees to review and suggest ideas to solve problems that arise in business operations, and companies to motivate and empower employees to carry out lean and green project improvement activities in their daily business activities and provide institutional and financial support [15–48].

**Organizational Culture ($D_8$):** Establish a dynamic company culture that is open to new concepts and management models, and create a good work environment and atmosphere that promotes lean-green initiatives [31,35,45,48].

### 2.2. Economic and Market Drivers

Through differentiated "lean-green" competitive strategies, companies can drive cost savings and quickly deliver green products to customers to meet market demand.

**Cost savings ($D_9$):** Reduce energy and resource consumption, eliminate all work methods that do not add value, and even if the work adds value, the time and resource effort spent does not exceed the minimum threshold [15–51].

**Competitive advantage ($D_{10}$):** Maintain a competitive advantage in the market through the concept of greening products, while giving the company a competitive advantage by saving costs, optimizing product quality, and responding quickly to customer needs [15–51].

### 2.3. Policy Drivers

The lean and green role of companies is driven by government agencies and several other organizations, mostly through required and industry-specific rules and industry barriers, such as, industry certifications, and through internal incentive systems that promote the active promotion and implementation of lean and green-related strategies, systems and plans by all departments and employees.

**Government laws and regulations ($D_{11}$):** Government laws and regulations related to environmental protection, pollution control, landfill tax, emission standards, and other measures required by enterprises [15,18,21,35,48].

**Professional certified management systems ($D_{12}$):** The adoption, implementation, improvement, and certification of ISO9000 quality management system, ISO14000 environmental management system, OHSAS80000 occupational health and safety management system, and ISO50000 energy management system promote the continuous lean and green direction of the enterprise [49–55].

**Incentive mechanism ($D_{13}$):** Government agencies reward and punish managers and companies in the industry in which they operate for carrying out green management [21,23,43–48].

### 2.4. Consumer and Other Social Stakeholder Drivers

Consumers use their "monetary votes" to drive the lean-green of enterprises through price, responsiveness, and green consumption. By increasing public knowledge, keeping an eye on the government's environmental management responsibilities, and supporting company greening, the public indirectly drives the greening of companies.

**Green brand image ($D_{14}$):** The lean-greening of enterprises is driven by the awareness of building a positive brand image through green products, as well as the positive degree of eco-trademark or brand image of products and the degree of customer recognition of the company's green trademark [15,21,29,46,48,51,53].

**Public pressure ($D_{15}$):** It includes demands and monitoring from the local community where the company is located, such as partners in the supply chain, shareholders of the company, NGOs, and media [15,18,21,35,48,56,57].

The effect factor indicators of lean-green drivers on the collaborative implementation of lean-green manufacturing systems in manufacturing enterprises have varying degrees of influence. Therefore, the magnitude of influence, i.e., weight, of each influence factor indicator on the lean-green system will directly affect the overall evaluation results, so it is important to choose the weight calculation method reasonably. AHP, which is highly preferred for its simplicity and flexibility, uses a combination of qualitative and quantitative analysis to construct a two-by-two judgment matrix between each influence factor and calculate the initial weight value of each influence factor.

There may be correlative influence relationships among the impact evaluation indicators of lean-green drivers. Assuming that the 15 indicators in the four categories are kept independent of each other, and do not overlap with each other, then it does not mean that there is no interrelated influence relationship among these indicators. For example, at the internal driver level, if top management has sufficient awareness of the company's green competitive advantage, and has made commitments to government agencies and the public, then the degree to which green production combined with lean production and environmental protection are included in the planning will increase, and the company's green management activities will receive sufficient attention and support. Therefore, the awareness and commitment of top management ($D_2$) will have an impact on other factors, such as lean-green integrated strategic planning and communication ($D_3$), continuous improvement ($D_4$), and employee awareness and training ($D_1$). Other indicators' adoption and the outcomes attained will, in turn, have an impact on top management's knowledge and commitment. The implementation of other indicators and the results achieved will in turn influence the awareness and commitment of top management ($D_2$), and strengthen the belief and will of top management, forming a virtuous circle that will facilitate the further progress of the company on the road to lean and green. At the same time, there may be a correlative influence relationship between the driving evaluation indicators belonging to different lean-green driving factors. For example, if there is a substantial demand from end users for quick delivery, high quality, and ecologically friendly products, then it may promote a strong willingness of manufacturing companies to establish a good market image of the brand based on green products through the advantages that can be created by lean production. In addition, on the other hand, it may cause sufficient attention and support for lean-green business management activities within the company, so that the formation of a green brand image ($D_{14}$) at the level of consumer and other social stakeholder drivers can affect the establishment and maintenance of competitive advantage ($D_{10}$) at the level of economic and market drivers, etc.

In light of the foregoing, the four categories and 15 evaluation indicators used to reflect the relationship between supply-lean-green drive can be thought of as a complex network with intricate correlated influence relationships between elements. The significance effect that an indicator or factor has by connecting other indicators or factors determines the importance of the indicator in a complex network [58]. The decision laboratory analysis (DEMATEL) method in complex network theory is able to reveal important influencing factors as well as internal constructions by analyzing the logical relationships between factors in the system with the direct influence matrix to calculate the degree of influence, the degree of being influenced, the degree of cause, and centrality of the factor [59]. Therefore, quantitative evaluation based on the DEMATEL method was used to analyze the lean-green drive intensity evaluation index system.

## 3. Research Methodology

### 3.1. Fuzzy AHP-DEMATEL Model

AHP calculates the weights based on the relative relevance of the various indicators, which is highly accurate and valid, but it ignores the interdependence of the various indicators. It assumes that each indicator is independent of each other, while the driver indicators of collaborative integrated implementation of lean-green manufacturing system consider more factors and cannot completely guarantee the absolute independence among indicators. The DEMATEL method can solve this problem by determining the weights using the degree of mutual influence between indicators. Therefore, the initial weights of indicators are determined by using the AHP method first, and then the DEMATEL method is adopted to correct the initial weights, reduce the subjective one-sidedness in the process of weight assignment, improve the accuracy and rationality, and make the evaluation results more scientific. Basílio et al. (2022) searched and reviewed papers on the use of multi-criteria decision-making methods between January 1977 and 29 April 2022, including titles, abstracts, keywords, and articles, and found that methods such as AHP and DEMATEL are the most popular multi-criteria decision-making methods [60].

### 3.1.1. Calculation of Initial Weights Based on Fuzzy AHP

Analytic Hierarchy Process is an effective tool for dealing with complex decision-making problems. It is a strategy for making decisions that breaks down the components that are usually involved in making decisions into levels, such as objectives, criteria, and options; on the basis of which qualitative and quantitative analysis is performed. This method is a hierarchical weighted decision analysis method proposed by Professor Satie of University of Pittsburgh, an American operations researcher, in the early 1970s when he was working for the U.S. Department of Defense on the topic of "power allocation based on the contribution of each industrial sector to national welfare", applying network system theory and multi-objective comprehensive evaluation methods. Analytic hierarchy process is a simple decision-making method for complex decision-making problems with multiple objectives, multiple criteria, or unstructured characteristics, by mathematizing the thinking process of decision-making with less quantitative information, based on an in-depth analysis of the nature of complex decision-making problems, influencing factors and their intrinsic relationships.

The decision problem is divided into several hierarchical structures according to the total objective, each level of sub-objectives, evaluation criteria, etc., using the analytical hierarchy process approach. Then we can use the method of solving the eigenvectors of the judgment matrix to find the priority weight of each element at each level to a certain element of the previous level, and finally the method of weighted sum to the final weight of the total objective. In establishing a hierarchical structure model and constructing a pairwise comparison array by introducing fuzzy set theory to address the limitations of cognitive uncertainty expression under limited information, this study combines the traditional nine-level scale method and triangular fuzzy number for the ratio of the two-two importance degree of each element of the same level regarding a criterion in the previous level in AHP to establish a nine-level scale table, as shown in Table 1.

**Table 1.** Nine-level fuzzy scale of expert judgment terms.

| Scale | Definition | Triangular Fuzzy Number $(l_{ij}, m_{ij}, u_{ij})$ | Countdown $(1/u_{ij}, 1/m_{ij}, 1/l_{ij})$ |
|---|---|---|---|
| 1 | Equally important | (1,1,1) | (1,1,1) |
| 2 | Between equally important and slightly more important | (1,2,3) | (1/3,1/2,1) |
| 3 | Slightly more important | (2,3,4) | (1/4,1/3,1/2) |
| 4 | Between slightly more important and obviously important | (3,4,5) | (1/5,1/4,1/3) |
| 5 | Obviously important | (4,5,6) | (1/6,1/5,1/4) |
| 6 | Between obviously important and strongly important | (5,6,7) | (1/7,1/6,1/5) |
| 7 | Strongly Important | (6,7,8) | (1/8,1/7,1/6) |
| 8 | Between strongly important and extremely important | (7,8,9) | (1/9,1/8,1/7) |
| 9 | Extremely important | (8,9,9) | (1/9,1/9,1/8) |

According to the characteristics of the problem and the overall aim that is supposed to be attained, the first phase of the analytical hierarchy process model is to list all the contributing aspects of the decision problem. First, to establish an orderly hierarchy according to the interrelationship, mutual influence, and affiliation between the influencing factors, so as to obtain a structural analysis model of multiple levels. Secondly, to reflect the subjective judgment about the importance of each factor in a quantitative way, and on this basis, to compare the factors on the same level with each other and to establish a judgment matrix. In this manuscript, Table 1 is used to collect experts' opinions on the two relative importance of each factor. Assuming that there are '$n$' factors in an index layer, and the relative importance of the $i$-th factor to the $j$-th factor as determined by the $K$-th expert, the fuzzy judgment matrix of this index layer is shown in Equation (1).

$$\widetilde{A}^{(K)} = \widetilde{a}_{ij}^{(k)} = \begin{bmatrix} (1,1,1) & (l_{12}^{(k)}, m_{12}^{(k)}, u_{12}^{(k)}) & \cdots & (l_{1n}^{(k)}, m_{1n}^{(k)}, u_{1n}^{(k)}) \\ (1/u_{12}^{(k)}, 1/m_{12}^{(k)}, 1/l_{12}^{(k)}) & (1,1,1) & \cdots & (l_{2n}^{(k)}, m_{2n}^{(k)}, u_{2n}^{(k)}) \\ \vdots & \vdots & \ddots & \vdots \\ (1/u_{1n}^{(k)}, 1/m_{1n}^{(k)}, 1/l_{1n}^{(k)}) & (1/u_{2n}^{(k)}, 1/m_{2n}^{(k)}, 1/l_{2n}^{(k)}) & \cdots & (1,1,1) \end{bmatrix} \tag{1}$$

$$k=1,2,\cdots,K \ i,j=1,2,\cdots,n$$

For the evaluation of the $K$-th expert, the triangular fuzzy number of each factor weight is calculated using the modified formula proposed by W.Y et al. [61] as shown in Equation (2).

$$\dot{X}_i^{(k)} = \left( \frac{\sum_{j=1}^{n} l_{ij}^{(k)}}{\sum_{j=1}^{n} l_{ij}^{(k)} + \sum_{z=1,z\neq i}^{n} \sum_{j=1}^{n} u_{zj}^{(k)}}, \frac{\sum_{j=1}^{n} m_{ij}^{(k)}}{\sum_{z=1}^{n} \sum_{j=1}^{n} m_{zj}^{(k)}}, \frac{\sum_{j-1}^{n} l_{ij}^{(k)}}{\sum_{j=1}^{n} u_{ij}^{(k)} + \sum_{z=1,z\neq i}^{n} \sum_{j=1}^{n} l_{zj}^{(k)}} \right) \tag{2}$$

Equation (2) denotes the triangular fuzzy number of the single ranking weight of the $i$-th factor, determined by the $k$-th expert. By iteratively computing the single ranking weights of each component, it is possible to determine the weight of a compared factor in relation to its upper target layer, and finally the total weight of each factor relative to the target is shown in Equation (3).

$$\widetilde{W}_i^{(k)} = \prod_{m=1}^{n-1} S_i^{(k)(m)}, k = 1, 2, \cdots n \tag{3}$$

In Equation (3), $S_i^{(k)(m)}$ is the weight of the $m$-th layer of indicators, judged by the $k$-th expert, and finally, the weight of the attributes is DE-fuzzified according to the fuzzy mean method, assuming that the triangular fuzzy number of their final weights is $\widetilde{w}_{A_i} = \left(\widetilde{w}_{Ai}^L, \widetilde{w}_{Ai}^M, \widetilde{w}_{Ai}^U\right)$, and the explicit weight value is $w_{A_i}$ obtained by DE-fuzzifying $\widetilde{W}_i^{(k)}$ according to the fuzzy mean method, and the calculation formula is shown in Equation (4).

$$w_{A_i} = \frac{w_{A_i}^L + 2w_{A_i}^M + w_{A_i}^U}{4} \tag{4}$$

In the construction of the judgment matrix $\widetilde{A}^{(K)}$, in order to prevent the phenomenon that "factor A is more important than factor B, factor B is more important than factor C, and factor C is more important than factor A", which does not conform to normal logic, a consistency test is conducted on W. Take the elements of the criterion $F$ as an example.

Calculate the maximum characteristic root $\lambda_{\max}$ of the judgment matrix $\widetilde{A}^{(K)}$, as shown in Equation (5), where $\left(\widetilde{A}^{(K)}W\right)_i$ denotes the $i$-th component of the vector, as shown in Equation (6).

$$\lambda_{\max} = \frac{1}{n} \sum_{i=1}^{n} \frac{\left(\widetilde{A}^{(K)}W\right)_i}{W_i} \tag{5}$$

$$\widetilde{A}^{(K)}w = \begin{bmatrix} (1,1,1) & (l_{12}^{(k)}, m_{12}^{(k)}, u_{12}^{(k)}) & \cdots & (l_{1n}^{(k)}, m_{1n}^{(k)}, u_{1n}^{(k)}) \\ (1/u_{12}^{(k)}, 1/m_{12}^{(k)}, 1/l_{12}^{(k)}) & (1,1,1) & \cdots & (l_{2n}^{(k)}, m_{2n}^{(k)}, u_{2n}^{(k)}) \\ \vdots & \vdots & \ddots & \vdots \\ (1/u_{1n}^{(k)}, 1/m_{1n}^{(k)}, 1/l_{1n}^{(k)}) & (1/u_{2n}^{(k)}, 1/m_{2n}^{(k)}, 1/l_{2n}^{(k)}) & \cdots & (1,1,1) \end{bmatrix} \begin{bmatrix} w_1 \\ w_2 \\ \vdots \\ w_n \end{bmatrix} = \begin{bmatrix} \sum_{i=j=1}^{n} a_{1j}w_i \\ \sum_{i=j=1}^{n} a_{2j}w_i \\ \vdots \\ \sum_{i=j=1}^{n} a_{ij}w_i \\ \vdots \\ \sum_{i=j=1}^{n} a_{nj}w_i \end{bmatrix} \tag{6}$$

The fuzzy value of attribute $\left(\widetilde{A}^{(K)}W\right)_i$ according to the fuzzy mean method can be DE-fuzzified according to Formula (4), and the Consistency Index $C.I.$ of the judgment matrix is calculated, where n indicates the order of the judgment matrix. The calculation formula is determined by Equation (7).

$$C.I. = \frac{\lambda_{\max} - n}{n - 1} \tag{7}$$

The judgement matrix's consistency indicator is then determined using the stochastic consistency ratio C.R. The calculation formula is given in Equation (8).

$$C.R. = \frac{C.I.}{R.I.} \tag{8}$$

The average randomness index of the judgement matrix, or *R.I.*, is proposed to assess if the judgement matrix of various orders exhibits adequate consistency. For the random consistency index of the judgment matrix, the *R.I.* values are shown in Table 2.

**Table 2.** Random consistency index.

| n | 4 | 5 | 6 | 7 | 8 | 9 | 10 | 11 | 12 | 13 | 14 | 15 |
|---|---|---|---|---|---|---|----|----|----|----|----|----|
| *R.I.* | 0.90 | 1.12 | 1.24 | 1.32 | 1.41 | 1.45 | 1.49 | 1.52 | 1.54 | 1.56 | 1.58 | 1.59 |

If C.R. < 0.1, the judgment matrix is considered to have satisfactory consistency, and the relative importance, calculated based on this judgment matrix, is acceptable. If this condition is not satisfied, the judgment matrix needs to be revised again until satisfactory agreement is obtained.

The consistency index of the total ranking of the hierarchy relative to the recursive hierarchy is calculated in Equation (9):

$$CI_G = \sum_{k=1}^{h} \sum_{i=1}^{nk} w_{ik} \times CI_{ik+1} \tag{9}$$

The average random consistency index of the total hierarchical ranking, relative to the total recursive hierarchy, is calculated in Equation (10):

$$RI_G = \sum_{k=1}^{h} \sum_{i=1}^{nk} w_{ik} \times RI_{ik+1} \tag{10}$$

Finally, the total relative consistency index of the recursive hierarchy is calculated, as given in Equation (11):

$$CR_G = \frac{CI_G}{RI_G} \tag{11}$$

Similarly, when $CR_G < 0.1$, it indicates that the consistency of the overall judgment matrix is acceptable.

### 3.1.2. Calculation of Centrality Based on Fuzzy DEMATEL

Since the DEMATEL method is based on expert experience for scoring, the results are influenced by individual differences and expert subjectivity. So, combining fuzzy theory and the DEMATEL method can eliminate the problems such as semanticization and fuzzification of expert evaluation information, by converting the expert scoring into the corresponding Triangular Fuzzy Number (TFN) as well to obtain the direct influence matrix, and then fuzzifying it, using the defuzzification method of Opricovic et al. [62] to convert the triangular fuzzy number into an accurate value, by the following steps.

(1)　Firstly, the mapping relationship between linguistic variables and fuzzy is established, as shown in Table 3.

**Table 3.** Semantic transformation table.

| Language Variables | Triangular Fuzzy Number ($l_{ij}, m_{ij}, u_{ij}$) |
|---|---|
| No effect (NO) | (0,0,1) |
| Very low impact (VL) | (0,1,2) |
| Low impact (L) | (1,2,3) |
| High impact (H) | (2,3,4) |
| Very high impact (VH) | (3,4,4) |

(2)　Conduct research for the relevant personnel and experts of lean-green manufacturing projects in manufacturing enterprises, and $\tilde{x}_{ij}^{(k)}$ is the result of the determination of the $k$-th evaluator, thus establishing the direct impact matrix $\tilde{M}^{(k)} = \left[\tilde{x}_{ij}^{(k)}\right]_{n \times n}$. The initial direct impact matrix is as follows:

$$\tilde{M}^{(k)} = \left[\tilde{x}_{ij}^{(k)}\right]_{n \times n} = \begin{bmatrix} (0,0,1) & (l_{12}^{(k)}, m_{12}^{(k)}, u_{12}^{(k)}) & \cdots & (l_{1n}^{(k)}, m_{1n}^{(k)}, u_{1n}^{(k)}) \\ (l_{21}^{(k)}, m_{21}^{(k)}, u_{21}^{(k)}) & (0,0,1) & \cdots & (l_{2n}^{(k)}, m_{2n}^{(k)}, u_{2n}^{(k)}) \\ \vdots & \vdots & \ddots & \vdots \\ (l_{n1}^{(k)}, m_{n1}^{(k)}, u_{n1}^{(k)}) & (l_{n2}^{(k)}, m_{n2}^{(k)}, u_{n2}^{(k)}) & \cdots & (0,0,1) \end{bmatrix} \tag{12}$$
$$i, j = 1, 2, \cdots n$$

(3)   Calculate the left and right standard values $xls_{ij}^k$ and $xus_{ij}^k$.

$$xls_{ij}^k = \frac{xm_{ij}^k}{1 + xm_{ij}^k - xl_{ij}^k}, \quad xus_{ij}^k = \frac{xu_{ij}^k}{1 + xu_{ij}^k - xm_{ij}^k} \tag{13}$$

(4)   Calculate the integrated standardized value $x_{ij}^k$.

$$x_{ij}^k = \frac{xls_{ij}^k \times (1 - xls_{ij}^k) + xus_{ij}^k \times xus_{ij}^k}{1 - xls_{ij}^k + xus_{ij}^k} \tag{14}$$

(5)   Calculate the value of the influence of the $i$-th factor on the $j$-th factor.

$$x_{ij}^{(k)} = \min_{1 \leq k \leq n} l_{ij}^k + x_{ij}^k \Delta_{\min}^{\max} \tag{15}$$

(6)   The average direct impact matrix is averaged over the processed direct impact matrix according to $M^{(k)} = \left[x_{ij}^{(k)}\right]_{n \times n}$ to obtain the average direct impact matrix $M = \left[x_{ij}\right]_{n \times n}$. Then the average direct impact matrix is normalized to calculate the combined impact matrix $M = \left[x_{ij}\right]_{m \times n}$.

(7)   In order to analyze the indirect influence relationship between the factors, it is necessary to solve the integrated influence matrix $M''$, $I$ is the unit matrix. This can be found using Equations (16) and (17).

$$M' = \frac{x_{ij}}{\max(\sum_{j=1}^{n} x_{ij})} \tag{16}$$

$$M'' = M' + M'^2 + \cdots + M'^n = \frac{M'(I - M'^n)}{(I - M')} = M'(I - M')^{-1} \tag{17}$$

(8)   Calculate the cause degree ($R_i - C_i$) and the center degree ($R_i + C_i$) of the driving strength between the lean-green drivers.

The degree of influence and the degree of being influenced of each factor indicator can be determined from the integrated influence matrix as $Ri$ and $Ci$, respectively (see Equation (18), and then the centrality $m_i = R + C$, which is used to indicate the magnitude (importance) of the role of each factor in all evaluation indicators, and the cause $r_i = R - C$, which is used to indicate the internal construct can be deduced.

$$R_i = \sum_{j=1}^{n} t_{ij} \quad C_j = \sum_{i=1}^{n} t_{ij} \tag{18}$$

### 3.1.3. Comprehensive Weight Calculation Based on Fuzzy AHP-DEMATEL

The degree of mutual influence between components and the relationship between criteria or elements can both be successfully determined by DEMATE. However, the DEMATEL method does not take into account the high or low weights among the evaluation indicators. The AHP method is different from the DEMATEL method in the aspect that the AHP method determines the weights by two-by-two comparison among the influencing factors. However, the premise of determining the weights is that each influencing factor is independent of the others. While the lean-green manufacturing system involves a wide range of aspects, it is difficult to ensure the independence of each influencing factor, and if only the AHP may cause bias in the results. When it comes to the DEMATEL method, it obtains the comprehensive weights by calculating the initial weights, influence degree, and influenced degree, and integrating the influenced degree into the influence weights. Therefore, the advantages of both these methods are combined resulting in AHP-DEMATEL, and the combined weights of factor '$i$' are constructed based on the results of AHP and DEMATEL analysis. It is calculated in Equation (19).

$$z_i = \frac{w_i \times m_i}{\sum_{i=1}^{n} w_i \times m_i}, \quad i = 1, 2, \cdots n \tag{19}$$

where, $w_i$ is the weight of each influencing factor calculated by the AHP method, and $m_i$ is the centrality of each influencing factor calculated by the AHP method.

### 3.2. DEMATEL-ISM Model

ISM model is based on graph theory, and it is an effective network analysis tool to analyze and deal with the structure of complex systems with the help of logical matrix operations. ISM develops multi-layer recursive structural models by breaking down complex systems with the aid of computer assistance and human empirical expertise. Based on this, DEMATEL and ISM can be used together to create the hierarchical structure of the indicator system in addition to identifying its major components and level of influence. The specific steps of the DEMATEL-ISM model are as follows:

(1) Based on the final integrated impact matrix $M''$ obtained in the DEMATEL method in Section 3.1, the unit matrix is introduced into the overall impact matrix $D$, as shown in Equation (20).

$$D = I + M'' = \left[ d_{ij} \right]_{n \times n} \tag{20}$$

(2) Set the threshold value $\lambda$, and calculate the reachability matrix $H$. Based on the experience as well as the extrapolation of the threshold value, it can be calculated by Equation (21).

$$\lambda = \mu + \delta \tag{21}$$

where, $\mu$ and $\delta$ are the mean and standard deviation of all elements in the overall impact matrix $M''$, respectively. Equation (22) determines the elements in the final reachable matrix.

$$h_{ij} = \begin{cases} 0, d_{ij} \prec \lambda \\ 1, d_{ij} \succ \lambda \end{cases} \quad (i = 1, 2, \cdots, n; j = 1, 2, \cdots, n) \tag{22}$$

(3) Calculate the reachable set, current set, and common set of the overall impact matrix $H$. The elements in the final reachable matrix are hierarchically divided to build the ISM model.

## 4. Case Study

### 4.1. Calculation of Initial Weights Based on Fuzzy AHP

In order to analyze the relationship between the drivers of lean-green manufacturing system in a manufacturing company S, the S company invited senior scholars, experts, senior management of lean-green manufacturing system, and senior management consultants to form an evaluation expert group according to the evaluation criteria in Table 1. The evaluation team was composed of many experts and scholars from manufacturing enterprises, including senior management of lean-green manufacturing system and senior management consultants, who used the questionnaire survey method to score each influence factor according to its relative importance on a scale of 1–9, and established a judgment matrix. A total of 70 questionnaires were distributed in this study, and 65 questionnaires were received. The information about the questionnaire participant population is shown in Table 4. Taking the judgment results of $D_{11}$–$D_{13}$ indicators under policy-driven conditions as an example for calculation, the judgment results of expert 1 is shown in Table 5.

**Table 4.** Distribution of experts participating in the questionnaire.

| Category | | Distribution of Questionnaires | | Valid Questionnaires Collected | |
|---|---|---|---|---|---|
| | | **Number** | **Percentage** | **Number** | **Percentage** |
| Gender | Man | 65 | 92.86% | 60 | 92.31% |
| | Woman | 5 | 7.14% | 5 | 7.69% |
| Company size | Small enterprises | 30 | 42.86% | 28 | 43.08% |
| | Medium-sized enterprises | 25 | 35.71% | 24 | 36.92% |
| | Large enterprises | 15 | 21.43% | 13 | 20.00% |
| Seniority | 5 years and below | 10 | 14.29% | 10 | 15.38% |
| | 6–10 years | 10 | 14.29% | 10 | 15.38% |
| | 11–15 years | 30 | 42.86% | 27 | 41.54% |
| | 16 years and above | 20 | 28.57% | 18 | 27.69% |
| Degree | undergraduate | 35 | 50.00% | 32 | 49.23% |
| | Master | 30 | 42.86% | 28 | 43.08% |
| | Doctor | 5 | 7.14% | 5 | 7.69% |
| Lean-Green Management Levels | Senior | 20 | 28.57% | 19 | 29.23% |
| | Medium | 30 | 42.86% | 29 | 44.62% |
| | Junior | 20 | 28.57% | 17 | 26.15% |
| Level of digitization | Senior | 10 | 14.29% | 10 | 15.38% |
| | Medium | 40 | 57.14% | 36 | 55.38% |
| | Junior | 20 | 28.57% | 19 | 29.23% |

**Table 5.** Opinions on the importance of each indicator factor under policy-driven conditions two-by-two judgment.

| Experts | $C_1$ | $D_{11}$ | $D_{12}$ | $D_{13}$ |
|---|---|---|---|---|
| Expert 1 | $D_{11}$ | (1,1,1) | (7,8,9) | (1,2,3) |
| | $D_{12}$ | (1/9,1/8,1/7) | (1,1,1) | (1/4,1/3,1/2) |
| | $D_{13}$ | (1/3,1/2,1) | (2,3,4) | (1,1,1) |

Based on Table 4, construct the fuzzy judgment matrix for this hierarchy.

$$\widetilde{A}_{C_3}^{(1)} = \begin{bmatrix} (1,1,1) & (7,8,9) & (1,2,3) \\ (1/9,1/8,1/7) & (1,1,1) & (1/4,1/3,1/2) \\ (1/3,1/2,1) & (2,3,4) & (1,1,1) \end{bmatrix}$$

The fuzzy judgment matrix $\widetilde{A}_{C_3}^{(1)}$ is calculated using Equation (2) to obtain the triangular fuzzy number of the weight vector of each indicator factor under the policy-driven condition.

$$\widetilde{S}_{C_3}^{(1)} = \begin{Bmatrix} \widetilde{S}_{D_{11}}^{(1)} \\ \widetilde{S}_{D_{12}}^{(1)} \\ \widetilde{S}_{D_{13}}^{(1)} \end{Bmatrix} = \begin{bmatrix} (0.5408, 0.6486, 0.7347) \\ (0.0668, 0.0860, 0.1175) \\ (0.1854, 0.2654, 0.3667) \end{bmatrix}$$

The matrix $\widetilde{A}_{C_3}^{(1)}$ is also tested for consistency at the same time. In addition, after the Formula (4) and defuzzification, we obtain from the Formulas (7) and (8): $\lambda_{max}$ = 3.0092, Ri = O.58, C.I. = 0.0046, C.R. = 0.0079 < 0.10, and the judgment matrix satisfies the consistency. Similarly, according to Equations (9) and (10), the consistency index C.R. = 0.0921 < 0.10 of the total ranking of the hierarchy relative to the recursive hierarchy, the comprehensive ranking satisfies the consistency test, and the weights of each index are reasonably assigned.

Formula (3) was used to determine the triangle fuzzy values of each index's weights, and the mean values of each expert's determination findings were obtained. Formula (4) then was used to carry out the fuzzification procedure, and the results are shown in Table 6.

**Table 6.** Defuzzified values of the initial weights of the drivers.

| Driving Factors | Initial Weights | Driving Factors | Initial Weights | Driving Factors | Initial Weights |
|---|---|---|---|---|---|
| $D_1$ | 0.0415 | $D_6$ | 0.0685 | $D_{11}$ | 0.0536 |
| $D_2$ | 0.0962 | $D_7$ | 0.0538 | $D_{12}$ | 0.0282 |
| $D_3$ | 0.0223 | $D_8$ | 0.0361 | $D_{13}$ | 0.0415 |
| $D_4$ | 0.0586 | $D_9$ | 0.1564 | $D_{14}$ | 0.0596 |
| $D_5$ | 0.0826 | $D_{10}$ | 0.1696 | $D_{15}$ | 0.0315 |

*4.2. Calculation of Index Weights Based on Fuzzy DEMATEL*

According to Table 2, the experts in Section 4.1 are invited to give their judgment on the degree of mutual influence of each index. The judgment result of expert 1, according to Equation (5) direct influence matrix $\widetilde{M}^{(1)} = \left[ \widetilde{\chi}_{ij}^{(1)} \right]_{15 \times 15}$, is shown below:

$$\widetilde{M}^{(1)} = \begin{bmatrix} (0,0,1) & (0,0,1) & \cdots & (0,0,1) \\ (2,3,4) & (0,0,1) & \cdots & (0,0,1) \\ \vdots & \vdots & \ddots & \vdots \\ (0,0,1) & (2,3,4) & \cdots & (0,0,1) \end{bmatrix}_{15 \times 15}$$

Combining the opinions of other experts, the direct influence matrix after defuzzification was obtained using Equations (13)–(15), and the mean values were taken for all experts, as shown in Table 7.

**Table 7.** Lean Green Driver Indicator Composite Impact Matrix.

| | $D_1$ | $D_2$ | $D_3$ | $D_4$ | $D_5$ | $D_6$ | $D_7$ | $D_8$ | $D_9$ | $D_{10}$ | $D_{11}$ | $D_{12}$ | $D_{13}$ | $D_{14}$ | $D_{15}$ |
|---|---|---|---|---|---|---|---|---|---|---|---|---|---|---|---|
| $D_1$ | 0.015 | 0.003 | 0.003 | 0.154 | 0.018 | 0.009 | 0.141 | 0.113 | 0.165 | 0.146 | 0.001 | 0.019 | 0.001 | 0.062 | 0.007 |
| $D_2$ | 0.171 | 0.013 | 0.135 | 0.206 | 0.170 | 0.152 | 0.200 | 0.171 | 0.152 | 0.177 | 0.002 | 0.131 | 0.002 | 0.169 | 0.020 |
| $D_3$ | 0.059 | 0.007 | 0.007 | 0.131 | 0.101 | 0.055 | 0.077 | 0.081 | 0.089 | 0.105 | 0.001 | 0.063 | 0.001 | 0.087 | 0.010 |
| $D_4$ | 0.017 | 0.007 | 0.007 | 0.036 | 0.055 | 0.050 | 0.061 | 0.071 | 0.148 | 0.170 | 0.001 | 0.055 | 0.001 | 0.133 | 0.016 |
| $D_5$ | 0.028 | 0.009 | 0.010 | 0.173 | 0.031 | 0.021 | 0.154 | 0.125 | 0.096 | 0.153 | 0.001 | 0.103 | 0.001 | 0.065 | 0.008 |
| $D_6$ | 0.090 | 0.005 | 0.006 | 0.147 | 0.015 | 0.013 | 0.026 | 0.027 | 0.157 | 0.176 | 0.001 | 0.051 | 0.001 | 0.059 | 0.007 |
| $D_7$ | 0.093 | 0.007 | 0.007 | 0.162 | 0.027 | 0.015 | 0.039 | 0.156 | 0.167 | 0.188 | 0.001 | 0.059 | 0.001 | 0.110 | 0.013 |
| $D_8$ | 0.018 | 0.006 | 0.007 | 0.120 | 0.097 | 0.014 | 0.108 | 0.040 | 0.080 | 0.095 | 0.001 | 0.061 | 0.001 | 0.083 | 0.010 |
| $D_9$ | 0.001 | 0.003 | 0.003 | 0.004 | 0.002 | 0.002 | 0.002 | 0.013 | 0.002 | 0.124 | 0.001 | 0.003 | 0.001 | 0.138 | 0.017 |
| $D_{10}$ | 0.001 | 0.002 | 0.002 | 0.003 | 0.002 | 0.002 | 0.002 | 0.012 | 0.002 | 0.003 | 0.001 | 0.002 | 0.001 | 0.123 | 0.015 |
| $D_{11}$ | 0.080 | 0.123 | 0.138 | 0.106 | 0.037 | 0.149 | 0.044 | 0.041 | 0.060 | 0.068 | 0.000 | 0.032 | 0.000 | 0.046 | 0.005 |
| $D_{12}$ | 0.166 | 0.087 | 0.098 | 0.233 | 0.167 | 0.151 | 0.161 | 0.165 | 0.148 | 0.210 | 0.002 | 0.054 | 0.002 | 0.172 | 0.021 |
| $D_{13}$ | 0.050 | 0.125 | 0.100 | 0.202 | 0.044 | 0.153 | 0.170 | 0.100 | 0.165 | 0.107 | 0.001 | 0.043 | 0.001 | 0.078 | 0.009 |
| $D_{14}$ | 0.008 | 0.019 | 0.021 | 0.026 | 0.015 | 0.014 | 0.017 | 0.096 | 0.015 | 0.027 | 0.010 | 0.019 | 0.010 | 0.026 | 0.123 |
| $D_{15}$ | 0.057 | 0.152 | 0.167 | 0.138 | 0.060 | 0.105 | 0.073 | 0.109 | 0.076 | 0.165 | 0.082 | 0.122 | 0.082 | 0.157 | 0.019 |

Finally, ranking analysis was carried out based on the centrality pairs of each factor index, and the influence degree, influenced degree, cause degree, and centrality of each factor were determined using the conventional DEMATEL approach, and the results are shown in Table 8.

**Table 8.** Influence, Influenced, Cause and Centrality of lean-green driving factors.

| Driving Factors | Influence | Influenced | Cause | Centrality | Sort |
|---|---|---|---|---|---|
| Continuous Improvement ($D_4$) | 0.830 | 1.840 | −1.010 | 2.669 | 1 |
| Professional certified management systems ($D_{12}$) | 1.838 | 0.819 | 1.019 | 2.657 | 2 |
| Senior management awareness and commitment ($D_2$) | 1.872 | 0.567 | 1.305 | 2.438 | 3 |
| Total employee participation and employee empowerment ($D_7$) | 1.047 | 1.277 | −0.230 | 2.324 | 4 |
| Competitive advantage ($D_{10}$) | 0.174 | 1.915 | −1.741 | 2.088 | 5 |
| Organizational Culture ($D_8$) | 0.739 | 1.318 | -0.579 | 2.058 | 6 |
| Green brand image ($D_{14}$) | 0.447 | 1.507 | −1.060 | 1.953 | 7 |
| Public pressure ($D_{15}$) | 1.562 | 0.301 | 1.261 | 1.863 | 8 |
| Cost savings ($D_9$) | 0.314 | 1.524 | −1.209 | 1.838 | 9 |
| Business Process Reengineering and Change ($D_5$) | 0.978 | 0.842 | 0.136 | 1.820 | 10 |
| Employee awareness and training ($D_1$) | 0.856 | 0.853 | 0.003 | 1.709 | 11 |
| Technology Upgrade ($D_6$) | 0.780 | 0.905 | −0.125 | 1.685 | 12 |
| Integrated strategic planning and communication ($D_3$) | 0.872 | 0.710 | 0.162 | 1.583 | 13 |
| Incentive mechanism ($D_{13}$) | 1.348 | 0.104 | 1.244 | 1.452 | 14 |
| Government laws and regulations ($D_{11}$) | 0.930 | 0.104 | 0.826 | 1.034 | 15 |

The difference between the row and column sums is because of the degree of the factor, as shown in Figure 2, which represents the cause degree of each factor indicator. The sum of row and column is called the centrality of the factor, and Figure 3 shows the centrality of each factor.

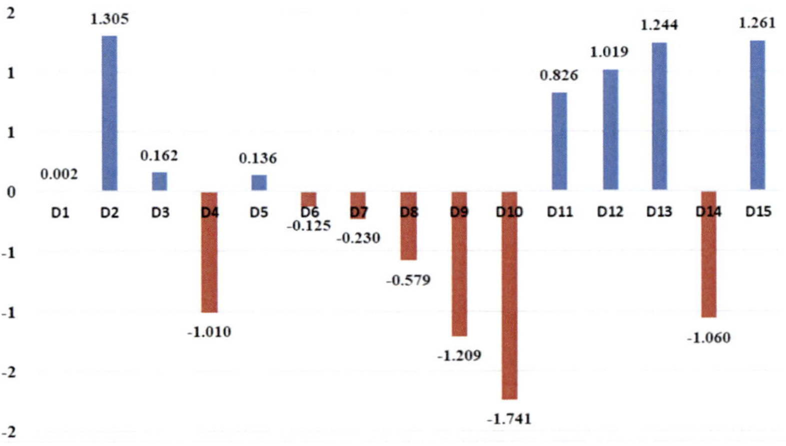


**Figure 2.** Distribution of cause degree of driving factors.

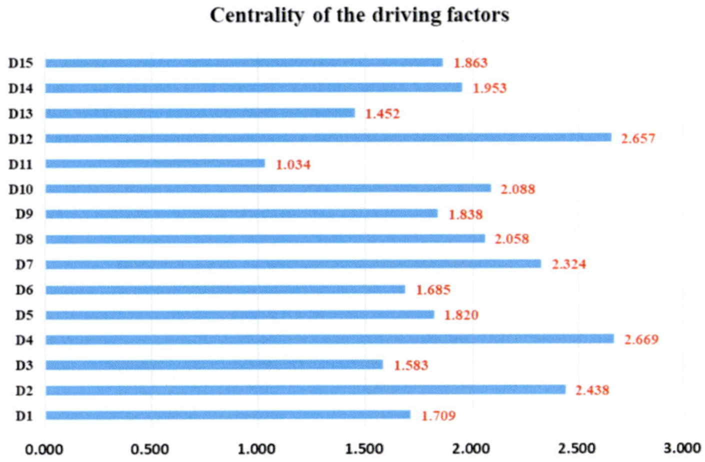

**Centrality of the driving factors**



**Figure 3.** Centrality distribution of driving factors.

### 4.3. Comprehensive Weight Calculation Based on Fuzzy AHP-DEMATEL

The initial weights and centrality of the risk factors are integrated according to Equation (15), such that the integrated weight values are obtained and ranked. The results are shown in Table 9. A comparison of the initial weights and combined weights of the lean green integration driver factors for manufacturing companies as can be seen below in Figure 4.

**Table 9.** Combined weights of driving factors.

| Factors | Combined Weights | Sort | Factors | Combined Weights | Sort | Factors | Combined Weights | Sort |
|---|---|---|---|---|---|---|---|---|
| $D_1$ | 0.0360 | 11 | $D_6$ | 0.0586 | 8 | $D_{11}$ | 0.0281 | 14 |
| $D_2$ | 0.1191 | 3 | $D_7$ | 0.0635 | 6 | $D_{12}$ | 0.0380 | 9 |
| $D_3$ | 0.0179 | 15 | $D_8$ | 0.0377 | 10 | $D_{13}$ | 0.0306 | 12 |
| $D_4$ | 0.0794 | 4 | $D_9$ | 0.1460 | 2 | $D_{14}$ | 0.0591 | 7 |
| $D_5$ | 0.0763 | 5 | $D_{10}$ | 0.1798 | 1 | $D_{15}$ | 0.0298 | 12 |

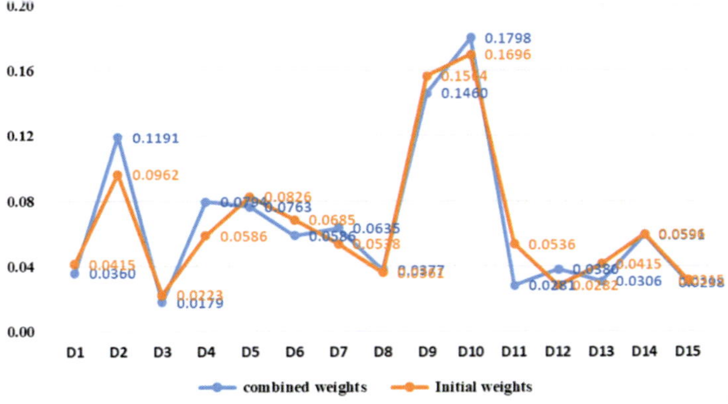


**Figure 4.** Initial, combined weights of driving factors.

### 4.4. Analysis Process Based on the DEMATEL-ISM Method

Using MATLAB software, the overall impact matrix D is first calculated according to Equation (20), and then the mean $\mu$, and standard deviation $\delta$ of all elements in the matrix $M''$ are found with the following threshold values.

$$\mu = 0.0648 \ \delta = 0.0531 \ \lambda = \mu+\delta=0.1179$$

Execute Equation (22) to calculate the reachable matrix $H$, as shown in Table 10.

**Table 10.** Reachable matrix of driving factors.

| | $D_1$ | $D_2$ | $D_3$ | $D_4$ | $D_5$ | $D_6$ | $D_7$ | $D_8$ | $D_9$ | $D_{10}$ | $D_{11}$ | $D_{12}$ | $D_{13}$ | $D_{14}$ | $D_{15}$ |
|---|---|---|---|---|---|---|---|---|---|---|---|---|---|---|---|
| $D_1$ | 1 | 0 | 0 | 1 | 0 | 0 | 1 | 0 | 1 | 1 | 0 | 0 | 0 | 0 | 0 |
| $D_2$ | 1 | 0 | 1 | 1 | 1 | 1 | 1 | 1 | 1 | 1 | 0 | 1 | 0 | 1 | 0 |
| $D_3$ | 0 | 0 | 0 | 1 | 0 | 0 | 0 | 0 | 0 | 0 | 0 | 0 | 0 | 0 | 0 |
| $D_4$ | 0 | 0 | 0 | 0 | 0 | 0 | 0 | 0 | 1 | 1 | 0 | 0 | 0 | 1 | 0 |
| $D_5$ | 0 | 0 | 0 | 1 | 0 | 0 | 1 | 1 | 0 | 1 | 0 | 0 | 0 | 0 | 0 |
| $D_6$ | 0 | 0 | 0 | 1 | 0 | 0 | 0 | 0 | 1 | 1 | 0 | 0 | 0 | 0 | 0 |
| $D_7$ | 0 | 0 | 0 | 1 | 0 | 0 | 0 | 1 | 1 | 1 | 0 | 0 | 0 | 0 | 0 |
| $D_8$ | 0 | 0 | 0 | 1 | 0 | 0 | 0 | 0 | 0 | 0 | 0 | 0 | 0 | 0 | 0 |
| $D_9$ | 0 | 0 | 0 | 0 | 0 | 0 | 0 | 0 | 0 | 1 | 0 | 0 | 0 | 1 | 0 |
| $D_{10}$ | 0 | 0 | 0 | 0 | 0 | 0 | 0 | 0 | 0 | 0 | 0 | 0 | 0 | 1 | 0 |
| $D_{11}$ | 0 | 1 | 1 | 0 | 0 | 1 | 0 | 0 | 0 | 0 | 0 | 0 | 0 | 0 | 0 |
| $D_{12}$ | 1 | 0 | 0 | 1 | 1 | 1 | 1 | 1 | 1 | 1 | 0 | 0 | 0 | 1 | 0 |
| $D_{13}$ | 0 | 1 | 0 | 1 | 0 | 1 | 1 | 0 | 1 | 0 | 0 | 0 | 0 | 0 | 0 |
| $D_{14}$ | 0 | 0 | 0 | 0 | 0 | 0 | 0 | 0 | 0 | 0 | 0 | 0 | 0 | 0 | 1 |
| $D_{15}$ | 0 | 1 | 1 | 1 | 0 | 0 | 0 | 0 | 0 | 1 | 0 | 1 | 0 | 1 | 0 |

All rows and columns of the reachable matrix are summed up separately, and then sorted from largest to smallest according to rows, and then sorted from largest to smallest according to columns. This sorting is performed to obtain the reconstructed reachable matrix. The influencing factors with the same serial numbers are divided into uniform orders, and the hierarchical structure diagram of each factor index is obtained, and the hierarchical structure diagram shown in Figure 5 is constructed.

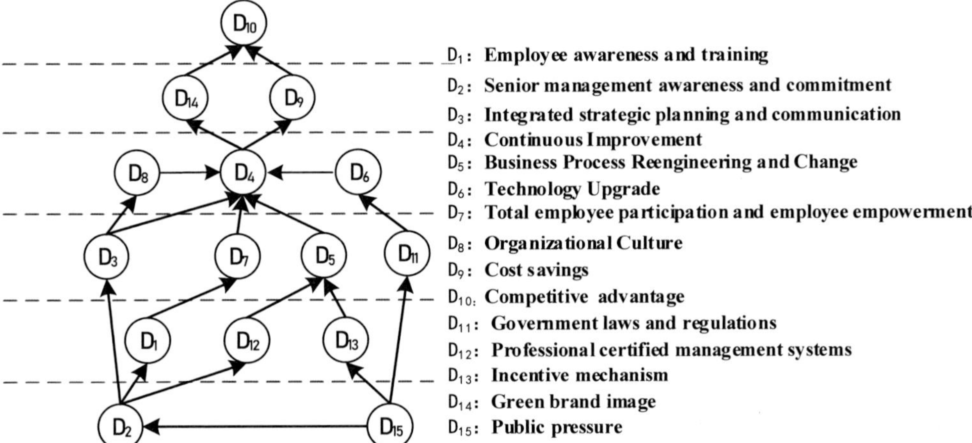


**Figure 5.** Hierarchy of indicators of driving factors.

## 5. Results, Discussion, and Implications

*5.1. Degree of Influence and Degree of Influenced Analysis*

Through the comprehensive influence matrix analysis, it can be concluded from Table 8 that the causal factors (factors with causality degree greater than zero) affecting Lean-green implementation are, in descending order of importance, top management awareness and commitment ($D_2$), public pressure ($D_{15}$), incentives ($D_{13}$), professional certification management system ($D_{12}$), laws and regulations set by the government ($D_{11}$), integrated strategic planning and communication ($D_3$), Business Process Reengineering and Change ($D_5$), and Employee Awareness and Training ($D_1$).

As can be seen from Figure 2, the awareness and commitment of top management ($D_2$) has the greatest degree of cause. Therefore, the top management of manufacturing companies must be aware of and committed to making decisions regarding the introduction of lean-green advanced manufacturing models in their organizations if the analysis of the drivers of lean-green implementation in manufacturing companies is to be successful. It can be used to provide continuous attention and resources to support them during the implementation promotion process, and to establish a process change and organizational corporate culture for companies to adapt to Lean-green. This also indicates whether the implementation of a lean-green manufacturing system is valued by the company. The company can be guaranteed to support the lean-green manufacturing system if top management values and supports it in this area and provides adequate financial, human, and policy assistance. Next, public pressure ($D_{15}$), incentives ($D_{13}$), professional certification management system ($D_{12}$), government laws and regulations ($D_{11}$), and integrated strategic planning and communication ($D_3$) are also key drivers. The key factors of public pressure ($D_{15}$), incentive mechanism ($D_{13}$), and laws and regulations set by the government ($D_{11}$) are also the external facilitators of external pressure on companies to respond to external voices of government, society, and informal organizations in order to better expand market share and build green brands and competitive advantages. The professional certification management system ($D_{12}$), and integrated strategic planning and communication ($D_3$) are the internal pressure and action guide and direction for the implementation of lean-green manufacturing system within the company. To adopt internal lean-green measures to accomplish the integration of economic, environmental, and social benefits, the top management set a lean-green strategy as well as certification requirements for each individual management system.

The outcome factors (factors with less than zero cause) are, in descending order of importance: competitive advantage ($D_{10}$), cost savings ($D_9$), green brand image ($D_{14}$), continuous improvement ($D_4$), organizational culture ($D_8$), full participation and employee empowerment ($D_7$), and technology upgrades ($D_6$). The outcome factor is the influence of other factors on the implementation of lean-green. So, the root cause can be traced to identify the most primitive influencing factors, so as to realize the smooth, convenient, and profitable implementation of lean-green in the enterprise from the root.

*5.2. Centrality and Comprehensive Weighting Analysis*

From Table 8, we can conclude that the importance of centrality in descending order is continuous improvement ($D_4$), professional certification management system ($D_{12}$), top management awareness and commitment ($D_2$), full participation and employee empowerment ($D_7$), competitive advantage ($D_{10}$), organizational culture ($D_8$), green brand image ($D_{14}$), public pressure ($D_{15}$), cost saving ($D_9$), business process Re-engineering and Change ($D_5$), Employee Awareness and Training ($D_1$), Technology Renewal and Upgrading ($D_6$), Integrated Strategic Planning and Communication ($D_3$), Incentives ($D_{13}$), and Laws and Regulations Established by the Government ($D_{11}$).

Table 9 and Figure 4 show the initial weights and centrality of each factor indicator considering the interaction between the internal factors to obtain the combined weight and ranking. It can be seen that competitive advantage ($D_{10}$) has the largest weight value of 0.1798, followed by cost savings ($D_9$) and top management awareness and commitment

($D_2$) with 0.146 and 0.1191, respectively. Continuous improvement ($D_4$) also has a large influence, with a weight of 0.08 or more. After considering the correlation, the weight of business process reengineering and change ($D_5$) increases, and the weight of professional certification management system ($D_{12}$) and full participation and employee empowerment ($D_7$) decreases. Manufacturing businesses operate in a competitive environment, and the market- and economy's external drivers serve as the initial impetus for implementing various management operation models. Following this, the businesses make internal adjustments to strengthen their competitive position and satisfy external customers' needs and desires while achieving sustainable business practices.

*5.3. Analysis of ISM Results*

As can be seen in Figure 5, the 15 factor indicators are divided into 6 levels, with a clear hierarchy of each influencing factor indicator. The uppermost factor indicators are the direct causes that result in the integration of collaborative lean-green manufacturing systems, and the middle factor indicators are the indirect causes that drive the integrated collaborative lean-green system implementation. The closer to the bottom level indicates that the driving factors are more fundamental. The drivers are more fundamental the closer we get to the bottom. When motivating the contributing drivers, it is important to improve from the fundamental factors to the direct factors, and to change and motivate the intermediate factors to facilitate the positive cycle of lean-green manufacturing system integration and collaboration.

Therefore, the following points are primarily focused on in order to determine whether the implementation of a lean-green manufacturing system in manufacturing enterprises can be successful and obtain the necessary economic, environmental, and social benefits: (1) Continuous improvement in internal processes, the elimination of all consumption, the conservation of raw materials and energy, among other things, are necessary for the adoption of lean-green in manufacturing enterprises to be successful. This is also used to reduce costs, improve quality, and increase efficiency while achieving more green products. (2) Manufacturing enterprises to implement lean-green measures within the enterprise, should introduce ISO9000 quality management system, ISO14000 environmental management system, OHSAS80000 occupational health and safety management system, and ISO50000 energy management system as far as possible. The implementation and use of these management systems may unintentionally or intentionally give the impression that the business is going green. (3) The support of top management plays a very important role in the introduction and implementation of lean and green management models. The establishment of a lean and green organizational culture can be a champion and a model, thus promoting sustainable development. (4) For government agencies, to give full play to the incentive-driven role of government agencies, the focus of policy incentives should be on managers.

*5.4. Conclusions*

From the perspective of stakeholders, this paper studies the driving factors of collaborative Integration implementation of lean-green manufacturing system in the context of Industry 4.0. Through literature analysis, the lean-green manufacturing integrated and collaborative implementation drivers are identified as being composed of endogenous and exogenous factors that come together to drive a manufacturing company towards sustainability. The 15 factor indicators identified are derived from published research papers and other results, and their scientific validity and effectiveness have been widely accepted by experts and scholars as the drivers for the integrated and collaborative implementation of lean-green manufacturing systems in Chinese manufacturing companies. The 15 drivers include endogenous lean green drivers including managers and internal staff, and exogenous drivers including shareholders, upstream and downstream companies, consumers, competitors, the public and government agencies, which are divided into four levels of evaluation indicators.

In order to identify the most critical drivers, the relationships between indicators and the structure of the indicator system are analyzed using fuzzy AHP-DEMATEL-ISM, the hierarchy of the complex network constituted by the indicator system is divided, the importance of the indicators is analyzed from a global perspective of the network, and the important factors and major factors influencing the driving lean green system are analyzed. The factors at the bottom of the hierarchy are the awareness and commitment of top management ($D_2$) and public pressure ($D_{15}$). The factors at the bottom of the hierarchy influence the other hierarchies through direct and indirect transmission, and are the essential causes of the integrated implementation of lean-green manufacturing systems in Chinese manufacturing enterprises. This is influenced by employee awareness and training ($D_1$), integrated strategic planning and communication ($D_3$), government regulations ($D_{11}$), professional certification management systems ($D_{12}$) and incentives ($D_{13}$). The two middle layers are the deep causal layers, which are based on continuous improvement ($D_4$) and have a complex relationship between other factors and this base, so that if there is a change, it can be passed on quickly, triggering a series of changes and linking to the transitional causal factors at the higher levels through the base factors, thus achieving a coupling of endogenous and exogenous factors. Green brand image ($D_{14}$) and cost savings ($D_9$) are transitional drivers, influenced by numerous other indirect factors, necessary for deeper drivers to act on top-level factors, potential influencers of system drivers that cannot be ignored, and the basis for top-level direct drivers. The top-level factor indicator Competitive Advantage ($D_{10}$) is the direct inducement driver for manufacturing companies to implement a lean-green manufacturing system, the proximate causal factor, and the factor with the largest combined weighting, and the key to achieving sustainable business.

The causal relationships of the 15 driver indicators for the integrated implementation of lean-green manufacturing systems in Chinese manufacturing companies can be explained and illustrated by the explanatory structural progressive order model diagram in Figure 5. The driver indicator system can be divided into six tiers, the structure of which is related to the characteristics of the factors. The drivers are located in the lower two tiers of the hierarchy and are the deeper causal factors of the driver indicator system, driving the other factors. The indicators with high correlation to other factors are located in the middle level of the skeleton diagram, they come out to be closely related to each other, and also carry on the top and bottom to link the factors indicators to form an interlinked system. Top-level factors largely influence the key to the implementation of lean-green manufacturing systems in manufacturing companies, directly affecting the drive, firmness, and sustainability of the driving system from economic, social and environmental aspects.

### 5.5. Managerial Implications

In conclusion, the cornerstone and secret to sustainable business in a worldwide economy is for manufacturing enterprises to generate a green brand image and their own distinct competitive advantage. However, the integrated and collaborative implementation of a lean-green manufacturing system within a manufacturing company is the result of a combination of external stakeholders, such as internal and external consumers and government, especially external public pressure (consumers) that further drives the determination and will of managers within the company. The fuzzy AHP-DEMATEL-ISM analysis method is constructed to explore the internal linkage of the drivers of integrated and collaborative implementation of lean-green system and to clearly obtain the weights, hierarchy, and influence paths where the key factors are located. This allows for a thorough and scientific analysis of the drivers of integrated and collaborative implementation of lean-green manufacturing systems. This method provides a new analysis idea for the causal analysis of similar responsible system events. Although the method still has some subjectivity, but it is still an effective method for analyzing the drivers of integrated collaborative implementation of lean-green manufacturing systems. For example, in order to describe the relationship between the factors as non-independent and correlated at a later stage in the research process, the ANP method can be used instead of the AHP method. Moreover, in order to classify the system of driving indicators, the Cluster Analysis method can be used to manage the classification. There are input class indicators and output class indicators in the driving indicators, which can be analyzed using Data Envelopment Analysis in order to the relative validity of a particular unit. Moreover, in order to rank the indicator system, TOPSIS or VIKOR methods can be used. These multi-criteria decision tools can be used individually or in combination. The combined use of these methods allows for more scientific, verifiable, and robust research results.

However, this research project can also identify the root causes that drive the synergistic implementation of lean-green manufacturing system integration from other methods, which will lead to better solutions. In addition, more research is required to complete and further validate the elements that drive the synergistic deployment of lean-green manufacturing systems as this project is still at the preliminary level of quantitative research. Further validation is also needed for different industrial environments, such as process industries, and especially across industries, such as service industries. Moreover, the analytical approach will be extended to other areas, such as improving energy efficiency and reducing water consumption. More research is therefore needed to show that the metrics system driving the integrated and collaborative implementation of lean-green manufacturing systems can be applied across the entire product lifecycle and across all economic activities. In the future, similar studies can be conducted in other developing economies, such as Southeast Asia and India. The structural equation modeling approach can be used to validate the results. In addition, a nonlinear approach with artificial neural networks can also be used. In order to effectively adopt lean-green systems under smart manufacturing in manufacturing companies in China, research on people and other corresponding role factors is needed.

**Author Contributions:** Conceptualization, X.Z., Y.X. and G.X.; methodology, X.Z. and Y.X.; formal analysis, Y.X., X.D. and G.X.; investigation, X.Z.; writing—original draft preparation, X.Z., Y.X., X.D. and G.X.; writing—review and editing, X.Z. and Y.X. All authors have read and agreed to the published version of the manuscript.

**Funding:** This research was funded by the Natural Science Foundation of Hunan Province, China (Project number: 2022JJ50244); Education Department of Hunan Province (Project number: 21B0695;21A0475); Project of Hunan social science achievement evaluation committee in 2022 (Project number: XSP22YBC081); Project of shaoyang social science achievement evaluation committee in 2022 (Project number: 22YBB10).

**Data Availability Statement:** Not applicable.

**Conflicts of Interest:** The authors declare no conflict of interest.

## References

1. Narwane, V.S.; Raut, R.D.; Gardas, B.B.; Narkhede, B.E.; Awasthi, A. Examining smart manufacturing challenges in the context of micro, small and medium enterprises. *Int. J. Comput. Integr. Manuf.* **2022**, *35*, 1395–1412. [CrossRef]
2. Huang, Q.; He, J. The "Third Industrial Revolution" and the Strategic Adjustment of China's Economic Development. *China Ind. Econ.* **2013**, *1*, 5–18. (In Chinese)
3. Pei, C.; Yu, Y. Germany's "Industry 4.0" and the new development of Sino-German manufacturing cooperation. *Res. Financ. Econ.* **2014**, *10*, 27–33. (In Chinese)
4. Rui, M.J. "Industry 4.0": A new generation of intelligent production methods. *World Sci.* **2014**, *5*, 19–20. (In Chinese)
5. Luo, W. Germany's "Industry 4.0" strategy for China to promote industrial transformation and upgrading. *Ind. Econ. Forum* **2014**, *4*, 52–59. (In Chinese)
6. Li, P.; Wan, J. Industrial Internet development and the deep integration of "two chemical". *J. Chin. Acad. Sci.* **2014**, *2*, 215–222. (In Chinese)
7. Li, Z.; Zhang, T.; Zhang, Q. A review of information-physical fusion system (CPS) research. *Comput. Sci.* **2011**, *9*, 25–31. (In Chinese)
8. Jiang, H. Industrial Internet and the deep "integration" of the two chemical systems in the same way. *China Informatiz.* **2014**, *8*, 11–13. (In Chinese)
9. Hu, H. Germany's "Industry 4.0" inspiration for China's deep integration of two. *China Economic Times*, 4 August 2014. (In Chinese)

10. Industry 4.0 Working Group. Germany's Industry 4.0 strategic plan implementation proposal (above). *Mech. Eng. Her.* **2013**, *8*, 23–33. (In Chinese)
11. Zhang, H.; Liu, F.; Liang, J. Exploring the architecture of green manufacturing and several strategic issues in its implementation. *Comput. Integr. Manuf. Syst.* **1997**, *3*, 11–14. (In Chinese)
12. John, L.; Sampayo, M.; Peas, P. Lean & Green on Industry 4.0 Context-Contribution to Understand L&G Drivers and Design Principles. *Int. J. Math. Eng. Manag. Sci. Plus Mangey Ram* **2021**, *6*, 1214–1229.
13. Mittal, V.K.; Sindhwani, R.; Kalsariya, V.; Salroo, F.; Sangwan, K.S.; Singh, P.L. Adoption of Integrated Lean-Green-Agile Strategies for Modern Manufacturing Systems. *Procedia CIRP* **2017**, *61*, 463–468. [CrossRef]
14. Ribeiro, T.B.A.; Ferreira, L.M.D.F.; Magalhães, V.S.M.; GarridoAzevedo, S. Analysis of the impact of lean and green practices in manufacturing companies: An exploratory study. *IFAC-Pap. OnLine* **2022**, *55*, 2419–2424. [CrossRef]
15. Galeazzo, A.; Furlana, A.; Vinelli, A. Lean and green in action: Interdependencies and performance of pollution prevention projects. *J. Clean. Prod.* **2014**, *85*, 191–200. [CrossRef]
16. Dües, C.M.; Tan, K.H.; Lim, M. Green as the new lean: How to use lean practices as a catalyst to greening your supply chain. *J. Clean. Prod.* **2013**, *40*, 93–100. [CrossRef]
17. Wiengarten, F.; Fynes, B.; Onofrei, G. Exploring synergetic effects between investments in environmental and quality/lean practices in supply chains. *Supply Chain. Manag.-Int. J.* **2013**, *18*, 148–160. [CrossRef]
18. Duarte, S.; Cruz-Machado, V. Investigating lean and green supply chain linkages through a balanced scorecard framework. *Int. J. Manag. Sci.* **2015**, *10*, 20–29. [CrossRef]
19. Cherrafi, A.; El Fezazi, S.; Govindan, K.; Garza-Reyes, J.A.; Mokhlis, A.; Benhida, K. A framework for the integration of Green and Lean Six Sigma for superior sustainability performance. *Int. J. Prod. Res.* **2017**, *55*, 4481–4515. [CrossRef]
20. Sobral, M.C.; Jabbour, A.B.L.S.; Jabbour, C.J.C. Green benefits from adopting lean manufacturing: A case study from the automotive sector. *Environ. Qual. Manag.* **2013**, *22*, 65–72. [CrossRef]
21. Melnyk, S.A.; Sroufe, R.; Montabon, F. How does management view environmentally responsible manufacturing? *Prod. Inventory Manag. J.* **2001**, *42*, 55–70.
22. Melnyk, S.A.; Sroufe, R.P.; Montabon, F.L.; Hinds, T.J. Green MRP: Identifying the material and environmental impacts of production schedules. *Int. J. Prod. Res.* **2001**, *39*, 1559–1573. [CrossRef]
23. Handfield, R.B.; Walton, S.V.; Seegers, L.K.; Melnyk, S.A. "Green" value chain practices in the furniture industry. *J. Oper. Manag. Vol.* **1997**, *15*, 293–315. [CrossRef]
24. Chen, C. Design for the Environment: A Quality-Based Model for Green Product Development. *Manag. Sci.* **2001**, *47*, 250–263. [CrossRef]
25. Garza-Reyes, J.A. Lean and Green—A systematic review of the state-of-the-art literature. *J. Clean. Prod.* **2015**, *102*, 18–29. [CrossRef]
26. Zhu, X.; Zhang, H.; Jiang, Z. Application of green-modified value stream mapping to integrate and implement lean and green practices: A case study. *Int. J. Comput. Integr. Manuf.* **2020**, *33*, 716–731. [CrossRef]
27. Zhu, X.; Zhang, H. Construction of Lean-green coordinated development model from the perspective of personnel integration in manufacturing companies. *Proc. Inst. Mech. Eng. Part B J. Eng. Manuf.* **2020**, *234*, 1460–1470. [CrossRef]
28. Zhu, X.; Zhang, H. A lean green implementation evaluation method based on fuzzy analytic net process and fuzzy complex proportional assessment. *Int. J. Circuits Syst. Signal Process.* **2020**, *14*, 646–655.
29. Bc, A.; Lmsca, B.; Ppps, B.; Emfa, B. Simulation-based analysis of catalyzers and trade-offs in Lean & Green manufacturing. *J. Clean. Prod.* **2020**, *242*, 118411–118436.
30. Wei, D.; Sin, Y.; Ding, S.; Sue, L.; Anao, A.; Chee, P.; Ponnambalam, S.G.; LoongLam, H. Enhancing the Adaptability: Lean and Green Strategy towards the Industry Revolution 4.0. *J. Clean. Prod.* **2020**, *273*, 122870.
31. Bhattacharya, A.; Nand, A.; Castka, P. Lean-green integration and its impact on sustainability performance: A critical review. *J. Clean. Prod.* **2019**, *236*, 117697.1–117697.16. [CrossRef]
32. Cherrafi, A.; Elfezazi, S.; Chiarini, A.; Mokhlis, A.; Benhida, K. The integration of lean manufacturing, six sigma and sustainability: A literature review and future research directions for developing a specific model. *J. Clean. Prod.* **2016**, *139*, 828–846. [CrossRef]
33. Cherrafi, A.; Elfezazi, S.; Garza-Reyes, J.A.; Benhida, K.; Mokhlis, A. Barriers in Green Lean implementation: A combined systematic literature review and interpretive structural modelling approach. *Prod. Plan. Control.* **2017**, *3*, 1–14. [CrossRef]
34. Chiarini, A. Sustainable manufacturing-greening processes using specific Lean Production tools: An empirical observation from European motorcycle component manufacturers. *J. Clean. Prod.* **2014**, *85*, 226–233. [CrossRef]
35. Garza-Reyes, J.A.; Villarreal, B.; Kumar, V.; Ruiz, P.M. Lean and Green in the Transport and Logistics Sector—A Case Study of Simultaneous Deployment. *Prod. Plan. Control.* **2016**, *27*, 1221–1232. [CrossRef]
36. Jabbour, C.J.C.; de Sousa Jabbour, A.B.L.; Govindan, K.; Teixeira, A.A.; de Souza Freitas, W.R. Environmental management and operational performance in automotive companies in Brazil: The role of human resource management and lean manufacturing. *J. Clean. Prod.* **2013**, *47*, 129–140. [CrossRef]
37. Nadeem, S.P.; Garza-Reyes, J.A.; Leung, S.C.; Cherrafi, A.; Anosike, T.; Lim, M.K. Lean manufacturing and environmental performance e exploring the impact and relationship. In Proceedings of the IFIP International Conference on Advances in Production Management Systems (APMS 2017): Advances in Production Management Systems. The Path to Intelligent,

Collaborative and Sustainable Manufacturing, Hamburg, Germany, 3–7 September 2017; Springer: Berlin/Heidelberg, Germany, 2017; pp. 331–340.

38. Reyes, G.; Arturo, J. Green lean and the need for Six Sigma. *Int. J. Lean Six Sigma* **2015**, *6*, 226–248. [CrossRef]
39. Verrier, B.; Rose, B.; Caillaud, E. Lean and Green strategy: The Lean and Green House and Maturity deployment model. *J. Clean. Prod.* **2016**, *116*, 150–156. [CrossRef]
40. King, A.A.; Lenox, M.J. Lean and Green? An Empirical Examination of the Relationship between Lean Production and Environmental Performance. *Prod. Oper. Manag.* **2001**, *10*, 244–256. [CrossRef]
41. Bergmiller, G.G.; McCright, P.R. Are lean and green programs synergistic? In Proceedings of the Industrial Engineering Research Conference, Norcross, GA, USA, 30 May–3 June 2009; pp. 1155–1160.
42. EPA. The Lean and Environment Toolkit Available in: United States Environmental Protection Agency [EB/OL]. 2013; [2019-9-23]. Available online: https://www.epa.gov/sites/production/files/201310/docu-ments/leanenvirotoolkit.pdf (accessed on 15 October 2022).
43. Silva, S.A.S.; Medeiros, C.F.; Vieira, R.K. Cleaner production and PDCA cycle: Practical application for reducing the Cans Loss index in a beverage company. *J. Clean. Prod.* **2017**, *150*, 324–338. [CrossRef]
44. Abreu, M.F.; Alves, A.C.; Moreira, F. Lean-Green models for eco-efficient and sustainable production. *Energy* **2017**, *137*, 846–853. [CrossRef]
45. Herron, C.; Hicks, C. The transfer of selected lean manufacturing techniques from Japanese automotive manufacturing into general manufacturing (UK) through change agents. *Robot. Comput.-Integr. Manuf.* **2008**, *24*, 524–531. [CrossRef]
46. Kassinis, G.; Vafeas, N. Environmental performance and plant closure. *J. Bus. Res.* **2009**, *62*, 484–494. [CrossRef]
47. Thanki, S.; Govindan, K.; Thakkar, J. An investigation on lean-green implementation practices in Indian SMEs using analytical hierarchy process (AHP) approach. *J. Clean. Prod.* **2016**, *135*, 284–298. [CrossRef]
48. Chauhan, G.; Singh, T.P. Measuring parameters of lean manufacturing realization. *Meas. Bus. Excell.* **2012**, *16*, 57–71. [CrossRef]
49. Hines, P.; Holwe, M.; Rich, N. Learning to Evolve: A Review of Contemporary Lean Thinking. *Int. J. Oper. Prod. Manag.* **2004**, *24*, 994–1011. [CrossRef]
50. Manikas, A.S.; Kroes, J.R. The relationship between lean manufacturing, environmental damage, and firm performance. *Lett. Spat. Resour. Sci.* **2018**, *11*, 1–15. [CrossRef]
51. Ye, F.; Zhang, J. Relationship between green supply chain management drivers, green design and performance. *Sci. Res.* **2010**, *28*, 1230–1239. (In Chinese)
52. Shaw, S.; Grant, D.B.; Mangan, J. Developing environmental supply chain performance measures. *Benchmarking Int. J.* **2010**, *17*, 320–339. [CrossRef]
53. Haslinda, A.; Fuong, C.C. The Implementation of ISO 14001 Environmental Management System in Manufacturing Firms in Malaysia. *Editor. Board* **2010**, *6*, 100–107.
54. Zhu, Q. An empirical study on the influence model of green supply chain management dynamics/pressure. *J. Dalian Univ. Technol. (Soc. Sci. Ed.)* **2008**, *29*, 6–12. (In Chinese)
55. Gandhi, N.S.; Thanki, S.J.; Thakkar, J.J. Ranking of drivers for integrated lean-green manufacturing for Indian manufacturing SMEs. *J. Clean. Prod.* **2018**, *171*, 675–689. [CrossRef]
56. Cabral, I.; Grilo, A.; Cruz-Machado, V. A decision-making model for lean, agile, resilient and green supply chain management. *Int. J. Prod. Res.* **2012**, *50*, 4830–4845. [CrossRef]
57. Reis, L.V.; Kipper, L.M.; Velásquez, F.D.G.; Hofmann, N.; Frozza, R.; Ocampo, S.A.; Hernandez, C.A.T. A model for Lean and Green integration and monitoring for the coffee sector. *Comput. Electron. Agric.* **2018**, *150*, 62–73. [CrossRef]
58. Tang, Y.; Zou, S. Multi-level gray evaluation of enterprise technology innovation capability. *Sci. Technol. Prog. Countermeas.* **1999**, *16*, 3. (In Chinese)
59. Zheng, C.H.D.; He, J.N.S.; Chen, T. Research on the evaluation of enterprise technology innovation capability. *China Soft Sci.* **1999**, *10*, 3. (In Chinese)
60. Basílio, M.P.; Pereira, V.; Costa, H.G.; Santos, M.; Ghosh, A. A Systematic Review of the Applications of Multi-Criteria Decision Aid Methods (1977–2022). *Electronics* **2022**, *11*, 1720. [CrossRef]
61. Wang, Y.; Luo, Y.; Hua, Z. On the extent analysis method for fuzzy AHP and its applications. *Eur. J. Oper. Res.* **2008**, *186*, 735–747. [CrossRef]
62. Opricovic, S.; Tzeng, G.H. Compromise solution by MCDM methods: A comparative analysis of VIKOR and TOPSIS. *Eur. J. Oper. Res.* **2004**, *156*, 445–455. [CrossRef]

*Article*

# Reservoir Advanced Process Control for Hydroelectric Power Production

**Silvia Maria Zanoli** [1,*], **Crescenzo Pepe** [1], **Giacomo Astolfi** [2] **and Francesco Luzi** [2]

1    Dipartimento di Ingegneria dell'Informazione, Università Politecnica delle Marche, Via Brecce Bianche 12, 60131 Ancona, Italy

2    Alperia Green Future, 60015 Falconara Marittima, Italy

*    Correspondence: s.zanoli@univpm.it

**Abstract:** The present work is in the framework of water resource control and optimization. Specifically, an advanced process control system was designed and implemented in a hydroelectric power plant for water management. Two reservoirs (connected through a regulation gate) and a set of turbines for energy production constitute the main elements of the process. In-depth data analysis was carried out to determine the control variables and the major issues related to the previous conduction of the plant. A tailored modelization process was conducted, and satisfactory fitting performances were obtained with linear models. In particular, first-principles equations were combined with data-based techniques. The achievement of a reliable model of the plant and the availability of reliable forecasts of the measured disturbance variables—e.g., the hydroelectric power production plan—motivated the choice of a control approach based on model predictive control techniques. A tailored methodology was proposed to account for model uncertainties, and an ad hoc model mismatch compensation strategy was designed. Virtual environment simulations based on meaningful scenarios confirmed the validity of the proposed approach for reducing water waste while meeting the water demand for electric energy production. The control system was commissioned for the real plant, obtaining significant performance and a remarkable service factor.

**Keywords:** model predictive control; reservoir; hydroelectric power plant; modelization; forecast; advanced process control; regulation gate manipulation; water resources management; process control; process optimization

Citation: Zanoli, S.M.; Pepe, C.; Astolfi, G.; Luzi, F. Reservoir Advanced Process Control for Hydroelectric Power Production. *Processes* **2023**, *11*, 300. https://doi.org/10.3390/pr11020300

Academic Editors: Jie Zhang and Olympia Roeva

Received: 29 November 2022
Revised: 5 January 2023
Accepted: 12 January 2023
Published: 17 January 2023

## 1. Introduction

In hydroelectric power plants, maximizing the efficiency of water exploitation is a major challenge for a profitable energy production. This challenge is part of the more general context aimed at minimizing water waste [1–3]. Over the past decades, hydropower equipment has been optimized to achieve high performance, availability, and flexibility; these improvements contribute to the energy transition [4–7]. Hydroelectric power plants may include different sub-processes, e.g., water collection reservoirs, regulation gates, intakes, rivers, sand traps, turbines, floodgates. Digitalization is playing a key role in improving the efficiency of hydropower plants at different levels of the control/automation hierarchy [8]. In this context, advanced process control (APC) systems [9], optimization algorithms [10], and Industry 4.0 [11] can represent strategic solutions. In order to design APC systems and optimization algorithms for non-standard processes and to apply Industry 4.0 principles, suitable cross-fertilization procedures are needed so as to adapt the smart solutions to the considered case studies [12,13]. Adapting APC, high-level optimization, and Industry 4.0 solutions developed in industrial plants to the field of hydropower requires a methodological approach that takes into account the different hardware and software architectures. APC systems can represent software solutions for improving the conduction of hydropower plants and standardizing control operations in the field. In this way, plant operators can play a role at the supervisory level. The benefits offered by APC systems increase as the complexity of the process increases. The complexity of the process includes several aspects, such as the number of variables involved, the types of relationship between the different variables, and the need for predictive approaches [8,14,15]. Industry 4.0 principles can represent strategic drivers of data exchange between the different levels of the automation hierarchy, while optimization algorithms at high levels of automation are able to take into account overall plant benefits—for example, considering the market prices of electric energy [10]. In the literature, many engineers, researchers, and practitioners have tackled optimization, estimation, and control challenges for hydroelectric power plants.

High-level optimization problems were tackled in [10,16–22]. In [10], a review of optimization algorithms in solving hydro-generation scheduling problems is presented; long-, mid-, and short-term hydro scheduling problems are analyzed, and the target of optimizing the power generation schedule of the accessible hydropower units is presented. Different solutions are detailed, e.g., metaheuristic optimization methods. In [16], the generation scheduling problem, i.e., the unit commitment problem, is investigated through the optimization of the amount of energy that must be provided by each turbine generator. Different versions of the coral reefs optimization algorithm are exploited. In [17], the implementation of reservoir conduction rules using inter-basin water transfer is proposed, and an optimization model based on network flow and particle swarm optimization is adopted for the maximization of the hydroelectric benefits. In [18], the problem of long-term maximization of hydroelectric energy generation from complex multipurpose reservoir systems is solved. The maximization of energy production is the main objective, and genetic algorithms are exploited for the solution of the formulated optimization problems in different scenarios. In [19], a model for a reservoir system is proposed in order to design optimized strategies for the minimization of raw water production cost, the maximization of electrical energy production, and to prevent flood situations. In [20], the authors focus on joint operation and dynamic control of flooding to limit the water level for cascade reservoirs. An effective tradeoff between the flood control and hydropower generation is provided. In [21], a salp swarm algorithm is used to optimize the joint operation of multiple hydropower reservoirs. Multiple strategies combining sine–cosine operators, opposition-based learning mechanisms, and elitism strategies are applied, and the proposed approach is tested through simulations. In [22], optimal operation of reservoirs for power generation plants is proposed. The formulation and implementation of optimal operation schemes are improved by combining the advantages of conventional and optimal operation and using the concept of a warning water level in the operational rules. Calculation under different operating conditions is used to test the developed procedures on a simulated hydropower station.

APC systems for the energy generation devices of hydropower plants are proposed in [23–26]. In [23], a method for controlling the real power delivered by a hydroelectric power plant to a local electrical grid is proposed, based on advanced control techniques, where internal model control and feedforward strategies are combined. Furthermore, a control loop for the frequency correction is added, a tuning method for the controller is proposed, and the designed controller is tested through simulations. In [24], the problem of load distribution between hydraulic units is tackled. The nonlinearity of the hydro turbine characteristics, individual peculiarities of the generation units, and process constraints are taken into account. A field experiment is executed to test the proposed algorithms. In [25], the hydraulic turbine's governing system control problem is analyzed; a nonlinear model is proposed, and Takagi–Sugeno fuzzy linearization and mixed $H_2/H_\infty$ robust control theory are applied to design the controller. The developed APC system is tested through simulations. In [26], modeling and simulation of hydropower plants is assessed in order to test different structures and algorithms for power, frequency, and voltage control. The dynamic and stationary behavior of the hydro units is analyzed in order to implement digital control algorithms.

The high-level optimization algorithms reported in previous studies do not govern the management of the real-time operation of the plants, while the aforementioned APC systems are focused on the control of the energy generation devices. APC systems focused on reservoirs' and tanks' level/volume control are proposed in [27–36] instead. Estimation and control problems are assessed in [27] for water-level control in reservoirs through floodgate manipulation. A nonlinear model predictive control (MPC) with an extended Kalman filter is the solution proposed by the authors, which is tested through simulations. In [28], a system of three interconnected tanks is modeled through Takagi–Sugeno fuzzy models, and a systematic control design procedure is proposed in order to include constraints on the input/output in the formulation. A bank of linear controllers is designed, and linear matrix inequalities are exploited. The proposed procedure is tested through simulations. In [29], a predictive functional control approach is proposed for a three-tank system showing the potential of a predictive controller equipped with an anti-wind-up method through simulations. In [30], an MPC approach is proposed for controlling river systems with water reservoirs, where the main control objective is avoiding the risk of flooding; simulation results are provided for the testing of the designed controller. In [31], five distributed MPC schemes using a hydropower plant benchmark are compared, and specific simulation results are provided; the main objective of the controllers is the coordination of several subsystems over a large geographical area in order to produce the demanded energy while satisfying constraints on water levels and flows. In [32], generalized predictive control is applied to a multivariable model of a pumped-storage hydroelectric power station. The response of the system with constrained predictive control is compared with the existing proportional–integral controller through simulations, showing the benefits provided by MPC. In [33], a predictive control strategy is presented for a process represented by two liquid tanks with a flow control valve. Modelization, control, and disturbance rejection topics are analyzed through simulations. An approach to optimal hydraulic-level tracking based on an inverse optimal controller is proposed in [34], devised with the purpose of regulating power generation rates in a specific hydropower infrastructure. In addition, a neural network is implemented to aid the system in the prediction and management of external perturbations, and the proposed approach is tested through simulations using data collected from the plant throughout a whole year of operation as a tracking reference. A multi-objective MPC approach is presented in [35] for real-time operation of a multi-reservoir system. The approach incorporates the non-dominated sorting genetic algorithm II (NSGA-II), multi-criteria decision-making, and the receding horizon principle to solve a multi-objective reservoir operation problem in real time. The control objectives are to minimize the storage deviations in the reservoirs, to minimize flood risks at a vulnerable downstream location, and to maximize hydropower generation. Tailored simulations are used for the testing of the designed control system. In [36], a six-dimensional nonlinear hydropower system controlled by a nonlinear predictive control method is proposed; a performance index with a terminal penalty function is selected, and numerical experiments are used to test the developed control strategy.

The present paper proposes an APC system aimed at water management for reservoirs in a hydroelectric power plant. Two reservoirs (connected through a regulation gate) and a set of turbines for energy production constitute the main elements of the process. This paper aims to provide holistic knobs and solutions for the assessment of the previously cited aspects of APC systems for hydropower plants. The present paper extends the contents reported in [37], providing additional details and insights on the different phases of the developed project. To the best of the authors' knowledge, in the literature on APC systems in hydropower plants focused on reservoirs' level/volume control, the following aspects have not been explored in depth:

- The procedures for a qualified plant inspection before starting an APC project have not been thoroughly detailed in the literature. The selection, acquisition, storage, and analysis of data play a fundamental role in the plant inspection, together with a detailed study of the plant's devices.

- A detailed feasibility study on the application of MPC for controlling hydropower plants is not present in the literature. Modelization and forecasting need to be assessed in this phase, together with tailored strategies for model mismatch compensation.
- A procedure that defines the MPC constraints, reference trajectories, and tuning parameters in real time based on current and predicted process conditions is not present in the literature on hydropower plants.
- An APC system that takes into account bad data detection, local control loop malfunctions, and lack of efficiency flags in real time is not present in the literature on hydropower plants.
- Smart alarm assessment for hydroelectric power plants is not present in the literature. Smart alarms can represent useful tools during plant conduction, highlighting inefficiency and/or predicting potential problems.

In addition, to the best of the authors' knowledge, projects on hydropower plants that include implementation of real processes that are designed as lasting control applications and not as temporary tests are not widespread. The field application of an APC system designed and tested through virtual environment simulations requires significant reliability and robustness in order to bridge the gap between simulations and field application.

The remainder of this paper is organized as follows: The Section 2 reports the process description, the control specifications, and the data analysis and modelization methods. In addition, the APC design, the field implementation, and the computational framework are described. The Section 3 focuses on data analysis, modelization, forecasting, virtual environment simulations, and field results. Finally, some conclusions and insights for future work are reported.

## 2. Materials and Methods

### 2.1. Process Description, Control Specifications, and Project Definition

The studied process is represented by a hydroelectric power plant located in the Alto Adige region (Italy). Figure 1 shows the geographic characterization of the overall plant. Two artificial water collection reservoirs characterize the process: an upstream reservoir and a downstream reservoir. The downstream reservoir is located in a valley. At the outlet of the downstream reservoir, a tunnel (see Figure 1) takes the water towards a penstock that leads to the power plant (see Figure 2). The electric energy is generated by the rotation of the involved turbines. The water flow between the two reservoirs is controlled through a regulation gate, named the Beikircher gate (see Figure 1). The regulation gate, activated by a butterfly valve, controls the water flow of a pipeline connecting the two reservoirs. The overall process is schematically reported in Figure 3.

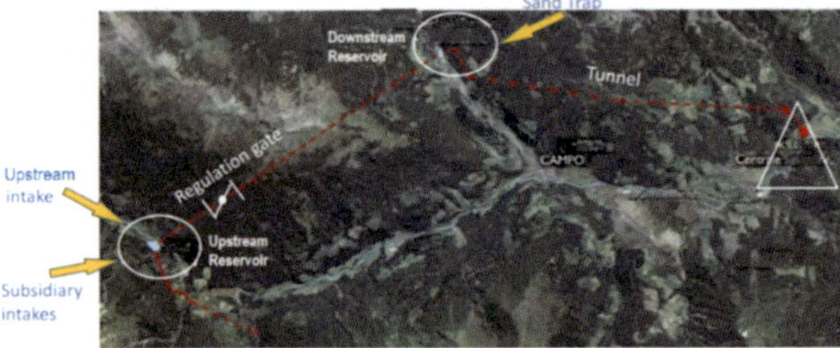


**Figure 1.** Top view of the plant area.

**Figure 2.** Hydroelectric power plant.

**Figure 3.** Schematic representation of the process.

The power plant is characterized by a double group of Francis-type turbines capable of providing an overall efficient power of 22 MW and an average annual electric energy production of 86.81 GWh. In the water catchment area, different rivers are present (see Figure 1). As shown in Figure 3, the reservoirs are characterized by two inlet water flows and one outlet water flow. The reservoirs' level and inlet/outlet water flow rates are measured by suitable sensors. The water flows entering the upstream reservoir consist of the main stream of a river from the intake structure and a set of subsidiary intakes (see Figure 3). Through two side-by-side deicing tanks and two subsequent sand traps, the incoming water from the intake structures flows into the upstream reservoir. The maximum derivable flow rate from the intake structure with clean grids is equal to about 9 m$^3$/s. On the other hand, the subsidiary intakes can provide a maximum flow rate equal to 1.4 m$^3$/s. The two inlet water flows of the upstream reservoir are measured by a level transducer located near the reservoir inlet.

The upstream reservoir was constructed on the right bank of a river and has a maximum length of 280 m and a maximum width of 150 m, with a usable capacity of about 135,500 m$^3$. The bottom level of the reservoir is at 1216 m above sea level (asl), while the overflow level is 1223.60 m asl. The minimum level detectable by the level sensor is 1216.80 m asl; below this level, the operation of the upstream reservoir has to be considered run-of-river. The maximum detectable level is equal to about 1221.85 m. At the outlet of the upstream reservoir, the Beikircher gate regulates the water flow, which enters a tunnel. The length of the tunnel that connects the upstream and downstream reservoirs is equal to 5534 m. The butterfly valve of the regulation gate is controlled by a built-in programmable logic controller (PLC). A hydraulic control unit, placed in a structure near the valve, operates the gate. The flow rate setpoint can be manipulated between 0 and 8 m$^3$/s. According to the plant's needs, the maximum value is typically limited to 7 m$^3$/s. The downstream reservoir was constructed on the right bank of the associated river and has a maximum length of 165 m and a maximum width of 80 m, with a usable capacity of about 52,500 m$^3$. The downstream reservoir's capacity is lower than the upstream reservoir's capacity. The downstream reservoir (see Figure 3), similarly to the upstream reservoir, is characterized by two inlet flows and a single outlet flow. The downstream reservoir's inlet flows are the water flow from the upstream reservoir (regulated by the Beikircher gate) and the intake structure. The intake structure is represented by the water flowing in a gravel reservoir and in a sand trap (see Figure 3). The bottom level of the reservoir is at 1197.20 m asl, while the overflow level is 1203.50 m asl. The minimum level detectable by the level sensor is 1197.40 m asl; below this level, the operation of the downstream reservoir has to be considered run-of-river. The maximum detectable level is equal to about 1202.77 m. At the outlet of the downstream reservoir, the water is conveyed towards the power plant through a tunnel and a penstock. The tunnel is characterized by a length equal to about 7000 m, while the penstock, which consists of a metal pipe, has a length of about 500 m. A jump equal to about 270 m is observed. After passing through the power plant and transferring energy to the turbines, the water flows into a free surface drainage channel, intercepted by two flat gates. The water released by the downstream reservoir is subjected to a flow rate setpoint regulation. The regulation and the flow rate measurement are located not at the outlet of the downstream reservoir, but a few meters downstream. The flow rate regulation is based on the electric energy production plan of the power plant. The electric energy production plan is known a priori with significant confidence. The production plan is sent daily to the managers of the plant and determines how much energy the plant will have to produce hourly during the day. The provided electric power (MW) measurements, together with the related setpoints, are available for the turbines.

Based on tailored plant inspections and plant operators'/managers' interviews/reports, different considerations were made for the process in order to plan the project phases. A list of the manipulated variables (MVs), controlled variables (CVs), and measured disturbance variables (DVs) was obtained [38]. The flow rate setpoint (m$^3$/s) of the regulation gate represents the only MV for the APC system. The CVs are represented by the volume (m$^3$) of the upstream and downstream reservoirs, while the measured DVs are represented by the remaining water inlet/outlet flow rates (m$^3$/s) reported in Figure 3: the upstream reservoir intake, the subsidiary intakes, the downstream reservoir sand trap, and the outlet flow rate from the downstream reservoir. Measured DVs are manipulated by other controllers or related to the natural flow of rivers. All of the reported MVs and DVs were measured, while the CVs were not directly measured. CVs' indirect measurement computation was performed using the measurements of the reservoirs' level.

The previous conduction of the plant was represented by manual and semiautomatic control logics. This conduction was based on empirical laws and experience with the process. The main control specifications were as follows:

- Constrained control of the reservoirs' volume (level). This type of control is usually referred to as zone control [39].
- Avoid water overflow on the reservoirs: the upstream reservoir has a lower priority than the downstream one, since the downstream reservoir is closer to the town.
- Avoid water shortages in the reservoirs: the downstream reservoir has a higher priority, since a lack of water in downstream reservoir could cause a violation of the electric energy production plan of the hydroelectric power plant.

- Compliance with the physical constraints and with the technical operative constraints of the Beikircher regulation gate.

The zone control strategy, resulting from the first specification, is intended for the constrained control of the reservoirs' volume (level), respecting the priorities previously reported in [39]. The constraints can help in avoiding water overflow and water shortages, ensuring the safe conduction of the plant. An efficient zone control is not always achievable in the process under study, because only an MV may be available—for example, the inlet flow rates of the upstream reservoir are not manipulable. The electric energy production plan of the hydroelectric power plant must be respected, because a violation (in excess or in deficit) usually causes an economic penalty for the plant [21]. As explained below, the physical constraints are mainly focused on the Torricelli law [23–25] and on the effective capacity of the plant devices. The technical operative constraints of the Beikircher regulation gate are intended to avoid (if possible) too-frequent control moves in order to minimize the wear damage. In this context, an automatic APC system for real-time control must guarantee an optimal solution for the flow rate setpoint of the regulation gate (MV) under all process conditions. Furthermore, smart alarms highlighting abnormal plant conditions could improve the plant's conduction. In this way, plant operators can play roles at a supervisory level. The control specifications and the proposed control strategy are summarized in Figure 4, while Figure 5 reports the main project phases described below. The accurate definition of the inputs and outputs to be obtained in each project phase represents a critical step.

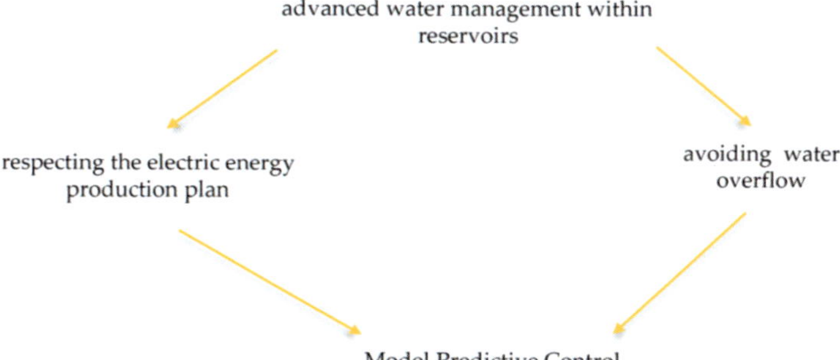


**Figure 4.** Schematic representation of the control specifications and of the proposed control solution.

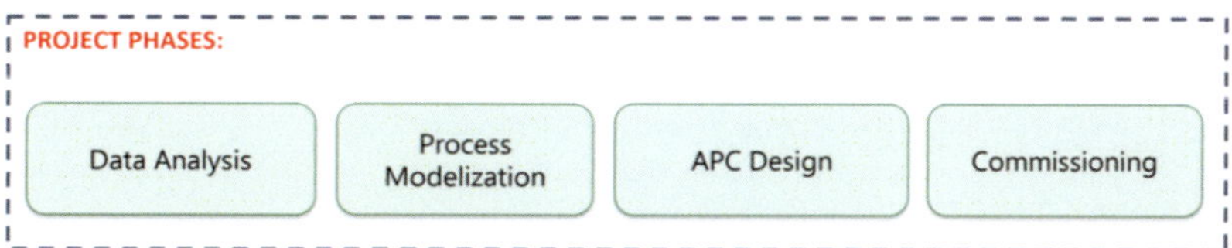


**Figure 5.** Schematic representation of the project management phases.

*2.2. Data Selection, Acquisition, and Storage*

The selection of the process variables to be acquired and stored, together with the definition of the hardware and software architectures to be exploited in real time, was a key phase that required special effort. The plant data were acquired in the field through the PLCs. First, an accurate data selection phase was performed. All MV and measured DV values were selected, together with the reservoirs' level. Also included in the selected data were variables involved in the lower-level control loops—for example, the process variables of the regulation gate and the downstream reservoir outlet flow rate—and the available process variables of the power plant—for example, the electric power of the turbines. Furthermore, a procedure for the supply of the electric energy production plan was defined.

In order to acquire and store the selected data, a suitable architecture was designed (Figure 6). The selected data were acquired from the plant through the PLCs (*PLCs*, Figure 6). Furthermore, the high-level supervisory systems provided the electric energy production plan in advance (*High-Level Supervisory Systems*, Figure 6). A PC server was installed in the plant, and it was connected to the plant's net infrastructure. On the PC server, a supervisory control and data acquisition (SCADA) system was installed and a database was created (*PC Server (SCADA and Database)*, Figure 6). The data selection, acquisition, and storage procedures were implemented on the PC server. Furthermore, a PC client (*Client Control Room*, Figure 6) was installed in the control room of the plant in order to provide selected signal information to the plant's operators, engineers, and managers. Tailored data visualization methods were designed in order to make the provided information user-friendly.

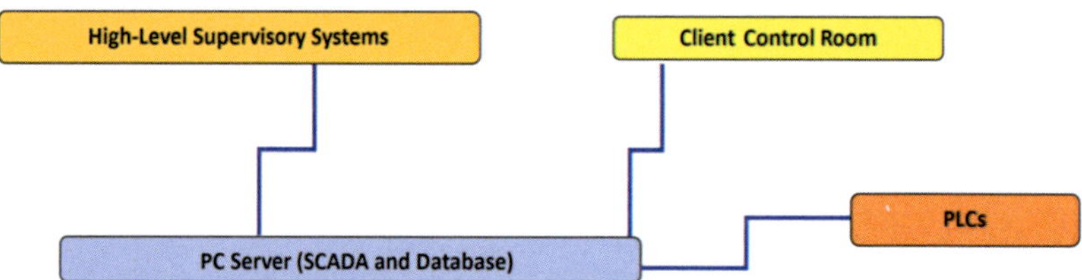


**Figure 6.** Designed architecture for data acquisition and storage.

### 2.3. Data Analysis

Following data selection, acquisition, and storage, data analysis was performed [40–42]. The data analysis sub-phases were difficult to define in the present work. The data analysis phase was divided into three main sub-phases:

- Analysis and processing of process variables and setpoints;
- Performance evaluation of local control loops;
- Assessment related to the electric energy production plan data and the compliance with the electric energy production plan.

The sub-phase consisting in the analysis and processing of process variables and setpoints involved the measurement of the process variables by sensors and the setpoints commanded on the local controllers. The sensors' acquisition/measurement and the PLCs' communication errors/malfunctions were investigated, and the missing data were replaced on the database by the results of tailored regressions. Suitable data preprocessing techniques (e.g., validity limits and spike and freezing checks) were applied in order to detect the bad data, which were discarded. The validity limits and spike and freezing thresholds were tuned based on the sensors' data sheets and the historical data. Furthermore, mobile window filters were used to improve the robustness of the selected measurements. The applied mobile window filters had the following form:

$$m_f(k) = \frac{m(k) + \cdots + m(k - N + 1)}{N} \tag{1}$$

where $k$ is the discrete-time instant, $N$ is the number of samples of the window, $m(\cdot)$ represents the sensor measurements, and $m_f(k)$ is the filtered measurement at instant $k$.

The local control loops' performance evaluation sub-phase consisted of an assessment of the performances of the local controllers of the Beikircher regulation gate and of the downstream reservoir outlet water flow. Some experimental tests were performed consisting of suitable step moves on the gate setpoint, evaluating the rise time, the overshoot, and the settling time. Furthermore, deviation conditions between the setpoints and process variables were investigated [43–45]. In order to motivate the potential abnormal behaviors of the local controllers—and especially of the regulation gate controller—the Torricelli law [23–25] was exploited:

$$Q_{Beik\_MAX}(k) = S_{cond} \cdot \sqrt{2 \cdot g \cdot \Delta h(k)} \qquad (2)$$

where $k$ is the discrete-time instant, $Q_{Beik\_MAX}$ (m$^3$/s) is the maximum reachable value by the regulation gate's flow rate, $S_{cond}$ (m$^2$) is the pipeline section, $g$ (m/s$^2$) is the acceleration of gravity (constant), and $\Delta h$ (m) is the height of the water level above the reservoir outlet conduct. Equation (2) is derived from the Bernoulli equation [23–25]. Table 1 reports the values of the parameters involved in Equation (2) for the regulation gate's flow rate. The involved pipeline connects the upstream reservoir to the regulation gate, so the water height has to be considered with respect to the level of the upstream reservoir outlet. The minimum level detectable by the upstream reservoir's level sensor was taken into account, i.e., 1216.80 m (see Table 1). If the upstream reservoir's level ($h_C$ in Table 1) is greater than or equal to about 1218.35 m asl, a flow rate of up to 8 m$^3$/s can be required on the regulation gate. If the upstream reservoir's level is greater than or equal to about 1218 m, a maximum flow rate of up to 7 m$^3$/s can be required on the regulation gate. On the other hand, if the upstream reservoir's level is lower than the computed thresholds, the physically reachable flow rate setpoint on the regulation gate decreases (see Table 1). For these reasons, as explained in Section 2.7, Equation (2) was also used for real-time modifications of the MV upper constraints.

**Table 1.** Parameters for the application of the Torricelli law to the regulation gate's flow rate.

| Parameter | Value |
| --- | --- |
| $S_{cond}$ | 1.45 (m$^2$) |
| $g$ | 9.81 (m/s$^2$) |
| $\Delta h$ | $h_C - 1216.80$ (m) |

The electric energy production plan data were evaluated in order to verify the implemented data-exchange procedure between the SCADA and the high-level supervisory systems (see Figure 6). Furthermore, an in-depth verification of the compliance of the defined electric energy production plan was performed.

The previously mentioned data analysis procedures were customized in order to be implemented in the real-time APC system. A module was designed, named *Bad Detection, Data Conditioning, and DV Prediction* module which, among its functions, includes an ad hoc bad data detection algorithm together with an algorithm that performs data filtering on mobile windows. Furthermore, the local control loops are checked for malfunction and compliance with the electric energy production plan is verified within this module. An overall *data analysis reliability flag* results from the aforementioned checks. This flag is exploited by the APC system (see Section 2.7); in this way, bad data detection, local control loop malfunctions, and f inefficient conditions are included in the real-time implementation of the APC system.

To the best of the authors' knowledge, the proposed methods for data selection, acquisition, storage, and analysis represent an innovation in the literature on APC systems for hydropower plants. Not using accurate methods of data selection, acquisition, storage, and analysis may represent a missing key prerequisite for designing a robust APC system.

### 2.4. Modelization

In order to design an MPC solution for the process under consideration, the modelization is a fundamental requirement, because MPC techniques strictly depend on the goodness of the obtained process model. A linear modelization approach, based on first-principles equations [21,24,46,47] and empirical data-based time delay identification [48], was adopted. The resulting continuous-time model was as follows:

$$\frac{d}{dt}w_C(t) = Q_{ingr\_C}(t) + Q_{prese\_sus}(t) - Q_{Beik}(t) \tag{3}$$

$$\frac{d}{dt}w_V(t) = Q_{dissab}(t) + Q_{Beik}(t-43) - Q_{gall\_Sar}(t-3) \tag{4}$$

where $t$ is the continuous time variable (min), $w_C$ (m$^3$) and $w_V$ (m$^3$) are the upstream and downstream reservoirs' water volumes, respectively, and $Q_{Beik}$ (m$^3$/s) is the regulation gate flow rate setpoint. $Q_{ingr\_C}$ (m$^3$/s) is the upstream reservoir's intake flow rate, $Q_{prese\_sus}$ (m$^3$/s) is the upstream reservoir's subsidiary intakes' flow rate, $Q_{dissab}$ (m$^3$/s) is the downstream reservoir's sand trap inlet flow rate, and $Q_{gall\_Sar}$ (m$^3$/s) is the downstream reservoir's outlet flow rate (see Figure 3). In Equations (3) and (4), note the sign of each term—the inlet flow rates have a positive sign, while the outlet ones have a negative sign. Furthermore, note that the flow rate of the regulation gate has an immediate effect on the upstream reservoir, while its action on the downstream reservoir is delayed (delay equal to 43 min). Finally, it should be noted that due to the regulation and flow rate sensors' location, a delay (3 min) is also present in the outlet flow rate of the downstream reservoir (see Section 2.1). For this reason, the resulting process model is a MIMO process with time delays on the inputs (i.e., MVs and DVs). The empirical data-based time-delay identification phase was executed by performing suitable step test procedures on the regulation gate's flow rate setpoint (MVs) and from data analysis on the downstream reservoir's outlet flow rate (DVs) [48]. Equations (3) and (4) consider the regulation gate setpoint. In fact, the dynamics of the lower-level controller were negligible with respect to the adopted controller's sampling time (equal to 60 s).

The reservoirs' water volume dynamic behavior was modeled through Equations (3) and (4). Since the reservoirs' field data are level measurements, an ad hoc volume-level conversion was investigated and implemented. Equations (3) and (4), enriched with the aspects reported in Section 2.6, were recast in order to obtain a continuous-time state-space model. The state-space description provides the dynamics as a set of coupled first-order differential equations in a set of internal variables (state variables), together with a set of algebraic equations that combine the state variables into physical output variables [49]. Subsequently, a discretization procedure was performed, using a zero-order hold and a sample time equal to 60 s, and time delays were included in the process dynamics [49,50]. In this way, the following discrete-time state-space model was obtained:

$$x(k+1) = Ax(k) + B_u u(k) + B_d d(k) \quad y(k) = Cx(k) + v(k) \tag{5}$$

where $k$ is the discrete-time instant, $x$ is the state vector, $u$ is the MV vector (scalar), $d$ is the DV vector that acts on the state, $y$ is the output vector, $v$ is an unmeasured DV vector which acts on the output, and $A$, $B_u$, $B_d$, and $C$ are matrices of suitable dimensions [39,49,50]. $d$, i.e., the state DVs vector, includes the measured DVs reported in Equations (3) and (4), along with the additional fictitious DVs added for model mismatch compensation (see Section 2.6). $v$, i.e., the output DV vector, includes the unmeasured disturbances added for model mismatch compensation (see Section 2.6).

In Equations (3) and (4), the upstream and downstream reservoirs' volume is considered. In order to exploit the reservoirs' level feedback, a volume-level relationship was formulated. Poor information on the shape and geometry of the reservoirs was available, so an estimation of the volume-level relationships was obtained. Based on known volume-level pairs, different mathematical laws were tested and compared, e.g., nonlinear, linear, and piecewise linear laws. The best results were obtained using the following piecewise linear law:

$$w_x = w_1 + (h_x - h_1)\cdot(w_2 - w_1)/(h_2 - h_1) \tag{6}$$

where $h_x$ (m) is the level value to be converted into the volume $w_x$ (m$^3$), while $(w_1, h_1)$ and $(w_2, h_2)$ are known volume-level pairs. The volume-level pairs were provided by the plant managers and covered the entire operating range of the reservoirs.

### 2.5. Forecasting of the Measured DVs

An MPC solution needs accurate CV predictions. On the other hand, CV predictions depend on the predictions of the measured DVs. Thus, the predictions of the measured DVs represent an additional significant aspect and are difficult to address for MPC purposes [40,50–52]. Using Equations (3) and (4), the measured DVs were reported, where $Q_{ingr\_C}$ (m$^3$/s) is the upstream reservoir's intake flow rate, $Q_{prese\_sus}$ (m$^3$/s) is the upstream reservoir's subsidiary intakes flow rate, $Q_{dissab}$ (m$^3$/s) is the downstream reservoir's sand trap inlet flow rate, and $Q_{gall\_Sar}$ (m$^3$/s) is the downstream reservoir's outlet flow rate (see Figure 3). Future values of the flow rates $Q_{ingr\_C}$, $Q_{prese\_sus}$, and $Q_{dissab}$ are unknown. Even though their flow rates are relatively constant or slowly vary for most of the time, there are nevertheless periods where significant variations are observed—for example, due to unexpected and unpredictable maneuvers on the pipelines (see Section 3). In these periods, the DVs' behavior significantly affected the model performances in term of future predictions. For this reason, Equation (1) was used to filter the values, testing and comparing different lengths of the window. The tuning phase was a critical step. Effective values were obtained in each operating condition to be considered over the whole MPC prediction horizon $H_p$ (see Section 2.7). With regard to the downstream reservoir's outlet flow rate, i.e., $Q_{gall\_Sar}$ (m$^3$/s), a correlation analysis with the total provided electric power (MW) was executed using the stored historical data. This analysis was motivated by the fact that the water flow to be sent to the power plant depends on the electric energy to be produced; as previously explained, the electric energy production plan is known in advance. The following relationship was obtained:

$$Q_{gall\_Sar} \cong P_{tot}/2.1 \tag{7}$$

where $P_{tot}$ (MW) is the total power. Exploiting Equation (7) and the electric energy production plan, reliable predictions of $Q_{gall\_Sar}$ (measured DV) were obtained.

The computation of the measured DV predictions was performed using the *Bad Detection, Data Conditioning, and DVs Prediction* module (see Sections 2.3 and 2.7).

### 2.6. Model Mismatch Compensation

The model described in Section 2.4 takes into account the inlet and/or outlet flows that resulted from the plant inspection. The flows considered were the measured flows used in the previous conduction of the process. From the early stages of the validation phase of the model described in Section 2.4 with field data, a significant model mismatch was encountered in many situations. This constituted an unexpected difficulty. A moderate model mismatch could be justified by the accuracy of the flow and level sensors and the possible presence of leaks. Furthermore, the presence of rainfall and inaccuracies in the volume-level conversion (and vice versa) were further causes of possible model mismatch, but an accurate data analysis made it clear that additional and more significant causes of uncertainty were present in some periods. The extent of the model mismatch was accounted for by the presence of unknown and unmeasured inflows and/or outflows. Uncertainties must be evaluated and taken into account to ensure acceptable control performance. In order to include a real-time model mismatch compensation and smoothing method, the combination of two strategies was formulated. First, Equations (3) and (4) were modified as follows:

$$\frac{d}{dt}w_C(t) = Q_{ingr_C}(t) + Q_{prese_{sus}}(t) - Q_{Beik}(t) + Q_{loss,\ w_C}(t) \tag{8}$$

$$\frac{d}{dt}w_V(t) = Q_{dissab}(t) + Q_{Beik}(t-43) - Q_{gall\_Sar}(t-3) + Q_{loss,\ w_V}(t) \tag{9}$$

where $Q_{loss,\,w_C}$ (m$^3$/s) and $Q_{loss,\,w_V}$ (m$^3$/s) represent the DVs related to the fictitious flow rates. With respect to the MVs and to the measured DVs' flow rates, $Q_{loss,\,w_C}$ and $Q_{loss,\,w_V}$ can be characterized by positive or negative signs. Furthermore, as an additional strategy, a $v$ vector was added to the final discrete-time state-space model (see Equation (5)). The computation of the DVs added for model mismatch compensation was performed using the *Bad Detection, Data Conditioning, and DVs Prediction* module (see Sections 2.3 and 2.7). $Q_{loss,\,w_C}$ and $Q_{loss,\,w_V}$ were observed and computed by evaluating the reservoirs' outlet behavior. For example, $Q_{loss,\,w_C}$ was computed through setting the flow rate setpoint of the regulation gate to zero and evaluating the effective volume decrease of the upstream reservoir. Values computed for $Q_{loss,\,w_C}$ and $Q_{loss,\,w_V}$ were considered to be constant over the whole MPC prediction horizon $H_p$ (see Section 2.7). The $v$ vector values were computed taking into account data for the last 50 min (included the current instant). The update was performed at most every 50 min. For each output, if the difference between the one-step-ahead estimation at the current instant and the measurement was greater than a threshold, only the last measurement was taken into account.

The overall process model was validated based on typical metrics (e.g., goodness-of-fit statistics) and on MPC purposes. The availability of forecasts on water requests by the hydroelectric plant and the goodness of fit of the obtained linear process model motivated the choice of an MPC approach (see Section 3 for the modelization results).

To the best of the authors' knowledge, the proposed methods for model mismatch compensation represent an innovation in the literature on APC systems for hydropower plants. A non-adaptive model mismatch compensation strategy could cause delayed correction in the presence of fast, unmeasured disturbance actions.

### 2.7. APC Design

As reported in Figure 5, the APC design phase was executed after the data analysis and modelization steps. Thanks to the modelization and forecasting results (see Section 3), MPC was selected as the control strategy [39,50–52]. The main difficulty faced in the APC design phase was the need to propose a solution that could handle all process conditions. According to the process dynamic behavior and control specifications, a sampling time equal to one minute was defined for the APC system.

Figure 7 reports the schematic representation of the APC system's architecture. At each control instant $k$, plant data and parameters (Figure 7, *plant data and parameters*) were provided by the *SCADA and Database* module (see Figure 6 for further details on this module). Furthermore, the *SCADA and Database* module provides an initial APC status flag (Figure 7, *APC status*) that defines the permission for the APC system to set the MV setpoint for the process. In other words, this flag defines whether the APC system can really be used to operate the plant. For example, if a watchdog communication error is detected in the communication between the SCADA system and the PLCs (see Figure 6), the APC system's conduction is disabled. Plant data and parameters and APC status were processed using the previously defined *Bad Detection, Data Conditioning, and DVs Prediction* module (see Sections 2.3 and 2.6). This module provides smart alarms to the plant (see below). Furthermore, this module performs the checks and the operations described in Sections 2.3 and 2.6, computing an overall *data analysis reliability flag* (see Section 2.3). This flag influences the final APC status flag, which is provided by the module together with the conditioned plant data and the prediction of the DVs (see Figure 7). For example, if a bad condition is detected on a plant measurement, the APC status flag is used to inhibit the APC system's actions. Some of the outputs computed by the *Bad Detection, Data Conditioning, and DVs Prediction* module were provided to the *MPC Parameters Selector* module. This module, based on the current and predicted process conditions, defines the MPC constraints, reference trajectories, and tuning parameters in real time (see below). The outputs computed by the *MPC Parameters Selector* module are provided to the *MPC* module (see Figure 7). The *MPC* module, based on a receding horizon strategy (see below), computes the MV value to be applied to the plant (Figure 7, $u(k)$). Furthermore, smart alarms are also provided by the *MPC* module (see below).

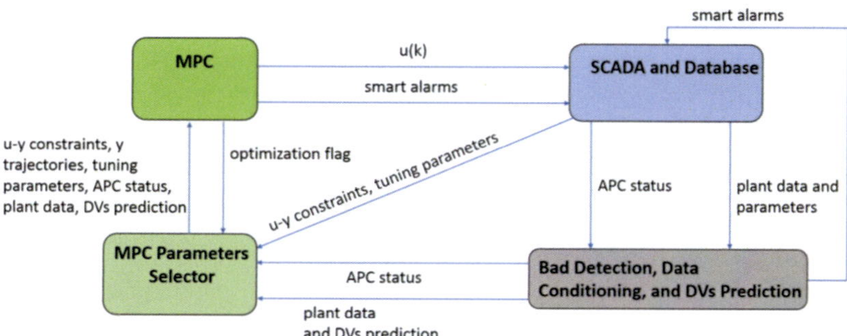


**Figure 7.** Schematic representation of the APC system's architecture.

A detailed analysis was performed based on the obtained process model and the DVs' forecasting in order to define a reliable prediction horizon $H_p$. The selected prediction horizon was equal to 130 min. No move-blocking strategies [53] were implemented, and the control horizon $H_u$ was set equal to $H_p$. The proposed MPC strategy is based on a quadratic programming (QP) problem. The quadratic cost function to be minimized is as follows:

$$V(k) = \sum_{j=0}^{H_u-1} \|\hat{u}(k+j|k)\|^2_{\mathcal{S}(j)} + \sum_{j=0}^{H_u-1} \|\Delta\hat{u}(k+j|k)\|^2_{\mathcal{R}(j)} + \sum_{j=1}^{H_p} \|\hat{y}(k+j|k) - r(k+j|k)\|^2_{Q(j)} + \rho \cdot \varepsilon^2(k) \quad (10)$$

subject to the following linear constraints:

$$\begin{aligned}
&\text{i. } lb_{du}(k) \leq \Delta\hat{u}(k+j|k) \leq ub_{du}(k), \; j = 0, \ldots, H_u - 1 \\
&\text{ii. } lb_u(k) \leq \hat{u}(k+j|k) \leq ub_u(k), \; j = 0, \ldots, H_u - 1 \\
&\text{iii. } lb_{h,yj}(k) - \gamma_{lbh,yj} \cdot \varepsilon(k) \leq \hat{y}_j(k+i|k) \leq ub_{h,yj}(k) + \gamma_{ubh,yj} \cdot \varepsilon(k), \\
&\qquad j = 1, 2; \; i = H_{wj}, \ldots, H_p \\
&\text{iv. } lb_{s,yj}(k) - \gamma_{lbs,yj} \cdot \varepsilon(k) \leq \hat{y}_j(k+i|k) \leq ub_{s,yj}(k) + \gamma_{ubs,yj} \cdot \varepsilon(k), \\
&\qquad j = 1, 2; \; i = H_{wj}, \ldots, H_p \text{v. } \varepsilon(k) \geq 0
\end{aligned} \quad (11)$$

In Equation (10), $\|\cdot\|$ represents the Euclidean norm, $\hat{u}$ and $\hat{y}$ are the predictions of the MVs and the CVs, respectively, and $\Delta\hat{u}$ represents the future control moves on the MVs. $\hat{u}$ and $\hat{y}$ are parametrized based on the known information up to the current control instant $k$ and on $\Delta\hat{u}$ terms [39]. Within the known information up to the current control instant $k$, the DV predictions are included. The $r$ terms are the reference trajectories on the CVs. The MVs' magnitude and moves are penalized over the control horizon in Equation (10), while the CVs' tracking errors are penalized on the prediction horizon. The suitable positive semidefinite matrices $\mathcal{R}$, $\mathcal{S}$, and $Q$ can weight the described terms. In Equation (11), $lb_{du}$, $ub_{du}$, $lb_u$, and $ub_u$ define the MVs constraints over the control horizon. The MVs' constraints are *hard* constraints, i.e., they can never be violated. On the other hand, two groups of CVs constraints were included in the formulation: The first group is represented by the terms $lb_{h,y}$ and $ub_{h,y}$ in Equation (11). These constraints are initially set as *hard* constraints, i.e., the related $\gamma$ terms are equal to zero in Equation (11); then, based on the process conditions, they can be converted to *soft* constraints (see below). These CV constraints refer to the reservoir volume constraints associated with the minimum and maximum volumes (i.e., water shortage and water overflow). A second group is represented by the $lb_{s,y}$ and $ub_{s,y}$ terms in Equation (11); these constraints are defined based on the process conditions and are always *soft* constraints; these constraints are always tighter with respect to the first group. Their relaxation is allowed through a slack variable $\varepsilon$. The slack variable $\varepsilon$ is included in the constraints (see Equation (11)) through suitable $\gamma$ coefficients, while its introduction in the cost function Equation (10) is performed through a positive coefficient $\rho$ [54].

In order to meet the specifications reported in Section 2.1, the downstream reservoir volume (CV) was set with a greater priority with respect to the upstream reservoir volume (CV); the $\gamma$ coefficients related to the associated *soft* constraints in Equation (11) were used for this purpose. In Equation (11), $H_{wj}$ represents the first prediction instant where the associated CV can be constrained; its definition is based on the obtained process model considering the MV time delays. The decision variables were included in the $\Delta\hat{u}$ and $\varepsilon$ terms. The QP problem was solved through the MATLAB quadprog solver [55]. At each control instant $k$, the *MPC Parameters Selector* module of Figure 7 considers the upper MV constraints provided by the SCADA system taking into account the Torricelli law (see Section 2.3) for their potential modification.

At predetermined hours of the day (typically every six hours starting from midnight), the *MPC Parameters Selector* module computes a long-range prediction of the reservoirs' volume up to the next prediction time instant. When exploiting a defined lower volume threshold for each reservoir, a potential water shortage indication is given. This indication can be exploited as smart alarm and for the setup of the MPC problem reported in Equations (10) and (11). If a water shortage condition is predicted at the current control instant, none of the *soft* CVs are considered in Equation (11), and all of the MVs' weights are zeroed in Equation (10). However, the $lb_{h,y}$ and $ub_{h,y}$ constraints are maintained in Equation (11). Furthermore, a reference trajectory is assigned to the reservoirs' volume—the upstream reservoir's volume tracks its *hard* lower constraint, while the downstream reservoir's volume tracks its *hard* upper constraint. In this way, the best action to fill the downstream reservoir is guaranteed through the introduction of reference trajectories. If a water shortage condition is not predicted at the current control instant, the *MPC Parameters Selector* module evaluates the DVs' prediction and, in particular, the prediction of the $Q_{gall\_Sar}$ flow rate in order to check whether there will be electric energy production on the prediction horizon. If no production is detected and the downstream reservoir volume is lower than a defined threshold (*max*), the *MPC Parameters Selector* module defines a zone control from the MPC formulation as shown in Equations (10) and (11); the $Q$ matrix weights are zeroed in Equation (10). In Equation (11), $lb_{h,y}$ and $ub_{h,y}$ are considered to be *hard* constraints, while $lb_{s,y}$ and $ub_{s,y}$ are not. In this way, an optimal solution can be sought in order to guarantee the transit of the only needed water from the upstream reservoir to the downstream reservoir. On the other hand, if production is detected, the $lb_{s,y}$ and $ub_{s,y}$ *soft* constraints are added, but the $S$ matrices' weights are zeroed in Equation (10) by the *MPC Parameters Selector* module, in order to avoid minimizing the regulation gate opening.

If the *MPC* module finds a solution (whether in cases of water shortage or not), the first term of the computed control sequence $\hat{u}(k + j - 1|k)$—i.e., $\hat{u}(k|k) = u(k)$—is sent to the plant. If the *MPC* module does not find a solution (i.e., infeasibility), a suitable *optimization flag* is computed by the *MPC* module and provided to the *MPC Parameters Selector* module (see Figure 7). The *optimization flag* reports the cause of the failure—the MPC formulation needs to be adjusted, and a new MPC problem is solved to find a solution. If the current volume of the upstream reservoir violates its $ub_{s,y}$ constraint—i.e., an overflow condition is likely to occur for the upstream reservoir—and the downstream reservoir's current volume is no greater than its *soft* upper constraint, the *hard* upper constraint of the upstream reservoir's volume is set as *soft*. Furthermore, two conditions are distinguished, depending on the violation of the *soft* lower constraint of the downstream reservoir level. If the violation takes place, the same solution for the water shortage condition is adopted. If the condition is not verified, the CVs' constraints are not considered, and the upstream reservoir volume tracks its *soft* lower constraint while the downstream reservoir volume tracks its *soft* upper constraint. Furthermore, the MVs' weights are not zeroed in Equation (10). In this way, the best action to avoid wasted water is guaranteed through the introduction of suitable reference trajectories.

If the second MPC attempt fails or the previous conditions are not satisfied, a heuristic law is applied to adjust the MPC formulation in order to find a solution. The heuristic law takes into account the current downstream reservoir volume and computes the desired

MV target. This computation is performed by the *MPC Parameters Selector* module, which suitably processes the constraints of the MV, taking into account the desired target. A range $(min, max)$ is defined for the downstream reservoir volume, represented by a lower and an upper threshold. If the downstream reservoir's volume is greater than a defined threshold $(max)$, a zero value is desired for the MV; the *MPC Parameters Selector* module suitably processes the constraints of the MV, taking into account the desired target. Otherwise, if the lower threshold is violated, the MV target must be equal to the allowed maximum value. Finally, if the downstream reservoir's volume is within the defined range, the following equation is used:

$$u_{target}(k) = ub_u(k) \cdot \left( 1 - \frac{w_V(k) - min}{max - min} \right) \tag{12}$$

where $min$ and $max$ are the defined thresholds, $ub_u$ is the upper constraint of the MV (see Equation (11)), and $w_V(k)$ is the current downstream reservoir volume obtained thanks to the volume-level relationship reported in Equation (6).

Through the aforementioned procedural steps, the APC system can properly handle the feasibility issues associated with the MPC optimization problem. Moreover, if the optimizer does not find a solution on the first or second attempt, the proposed heuristic law allows the APC system to efficiently control the process.

A set of smart alarms was designed in order to improve the reliability of the proposed APC system. Smart alarms were computed and sent by the *Bad Detection, Data Conditioning, and DVs Prediction* module and by the *MPC* module. Smart alarms computed by the *Bad Detection, Data Conditioning, and DVs Prediction* module refer to the aspects reported in Section 2.3—for example, a smart alarm on the detected missed compliance with respect to the electric energy production plan. Examples of smart alarms sent by the *MPC* module refer to the different checks described above, e.g., water overflow or shortage prediction.

To the best of the authors' knowledge, the proposed APC system, which takes into account bad data detection, local control loop malfunctions, and lack of efficiency flags used in real time, represents an innovation in the literature on hydropower plants. Additional novelties are represented by the proposed *MPC Parameters Selector* module reported in Figure 7 and by the designed smart alarms.

### 2.8. Computational Framework and Field Implementation

The computational framework exploited for all of the project phases—excluding the commissioning phase—was represented by a laptop computer with the following specifications: Intel(R) Core(TM) i8-3840QM CPU with 3 GHz HDD. A MATLAB environment was used for data analysis, modelization, DV forecasting, and virtual environment simulations [55]. The MATLAB Identification Toolbox, MATLAB Control System Toolbox, and MATLAB Optimization Toolbox were exploited for process identification and controller synthesis. The MATLAB functions for scatterplots were also exploited. Furthermore, a MATLAB environment was also used for the project maintenance in order to analyze the APC system's performance and its key performance indicators (KPIs) [55]. For the commissioning phase (i.e., field implementation), the architecture reported in Figure 6, enriched with the schematic representation of Figure 7, was exploited.

### 3. Results and Discussion

The project phases reported in Figure 5 were implemented in order to target the APC system commissioning at the real plant. The commissioning was executed in December 2019, and different upgrades of the controller were carried out during the project maintenance. The APC system obtained Industry 4.0 compliance certification, thanks to the architecture reported in Figure 6 and the functional aspects depicted in Figure 7. Figure 8 depicts some pages of the graphical user interface (GUI) of the APC system. In the first panel, a schematic representation of the plant is reported, together with the available process measurements; note the regulation gate between the two reservoirs. In the second panel, the APC system's main variables are reported, together with their current values and the related constraints/parameters. Moreover, the flags that define the initial APC status flag (Figure 7, *APC status*) can be noted; these flags refer to the overall status of the APC application and to the Beikircher regulation gate (see the top-left side of Figure 8). Furthermore, the electric energy production plan (red line) and the produced electric energy are shown in a customized display.

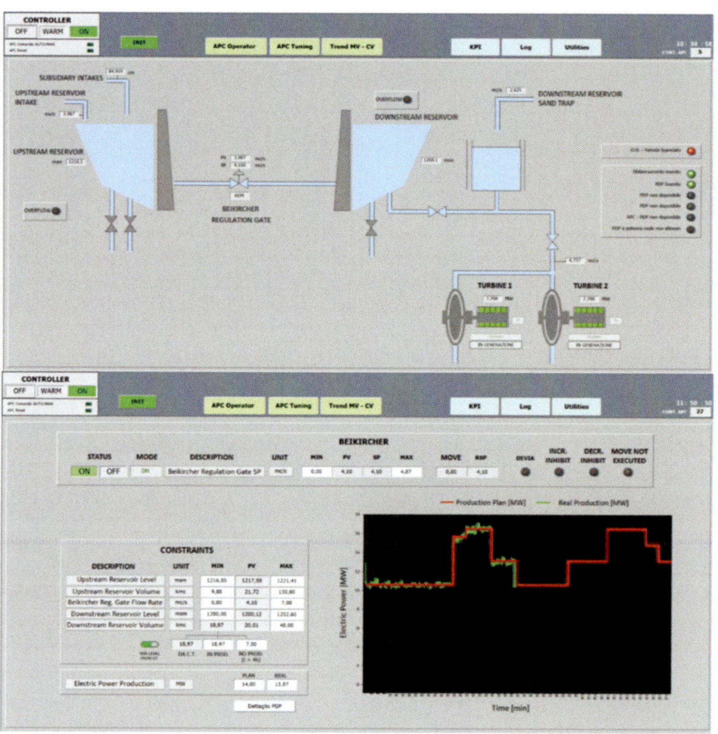

**Figure 8.** Results: panels of the GUI of the APC system.

### 3.1. Data Analysis Results

As explained in Section 2.3, data analysis represents one of the key phases of the present work. Significant data analysis results were obtained; in the following, three remarkable examples of the data analysis results are reported. Analyzing the upstream reservoir's intake flow rate process variable ($Q_{ingr\_C}$ (m$^3$/s)), unexpected behaviors were observed. Figure 9 reports an example of this behavior (one day of data). For forecasting purposes, the filtering procedures reported in Sections 2.3 and 2.6 were used, and a filtered process variable was obtained using a window of $N = 20$ samples, i.e., 20 min (see the red line in Figure 9). Analyzing the performance of the previous conduction of the plant, represented by manual and semiautomatic control logics, abnormal behaviors of the regulation gate's flow rate control loop were observed. In Figure 10, two plots are reported (one day of data): the first plot represents the flow rate setpoint (blue) and the process variable (green) of the regulation gate, while the upstream reservoir level (red) is depicted in the second plot. As can be noted, sometimes the control loop of the regulation gate does not present reliable tracking performances. In fact, persistent deviation between the setpoint and the process variable can be observed. Analyzing the upstream reservoir level (red line) in Figure 10, and based on the theoretical analysis of the Torricelli law reported in the previous sections, the cause of the abnormal behavior of the control loop was deduced. Under the previous conduction, the Torricelli law was not used for real-time conditioning of the MV upper constraint. This affected the performances of the control loop of the regulation gate, and it was not an optimal condition for running an APC system at a higher control hierarchy level.

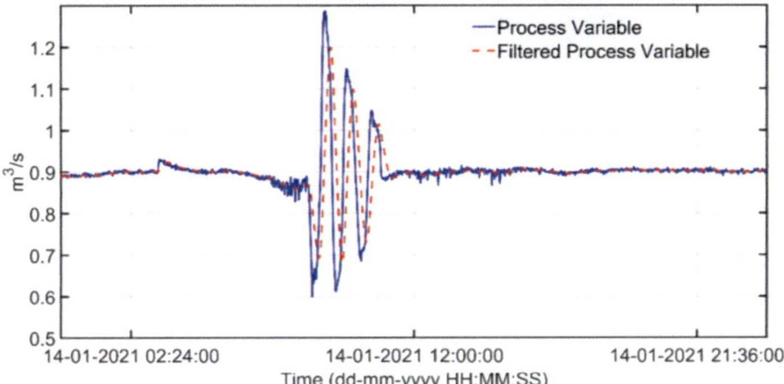

**Figure 9.** Data analysis results: upstream reservoir intake process variable and filtered process variable.

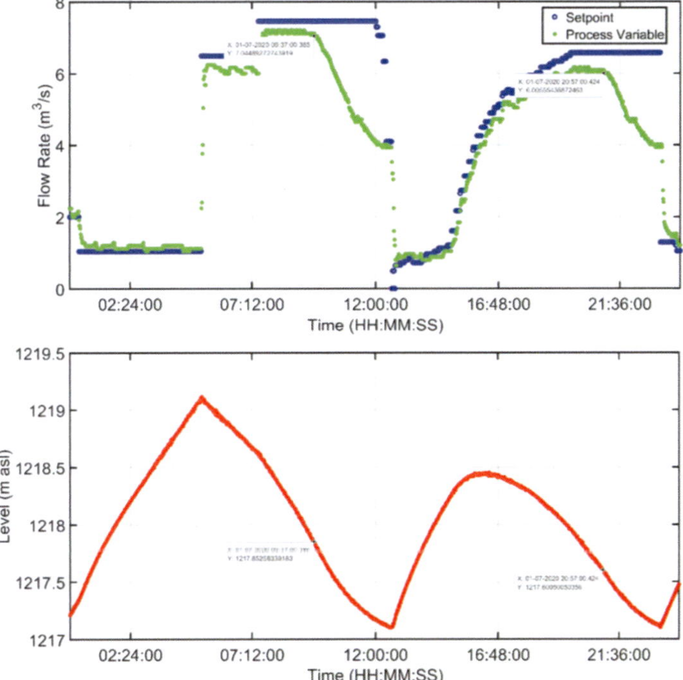

**Figure 10.** Data analysis results: Beikircher regulation gate control loop variables and upstream reservoir level.

Analyzing the compliance of the real electric energy production with respect to the electric energy production plan, some abnormal conditions were observed through data analysis. One day of data are reported in Figure 11; the setpoint (red line) and the process variable (blue line) of the electric power production are depicted, together with the planned electric power (green line). As can be noted, the electric energy production plan is not respected, because an anticipation of about 80 min is observed. As detailed in the previous sections, reliable forecasting of the electric energy production is a fundamental requirement for the APC system and for avoiding penalties. A smart alarm was introduced in the APC system for the non-compliance with the electric energy production plan, and the *data analysis reliability flag* (see Sections 2.3 and 2.7) was exploited for the inhibition of the APC system in this critical condition.

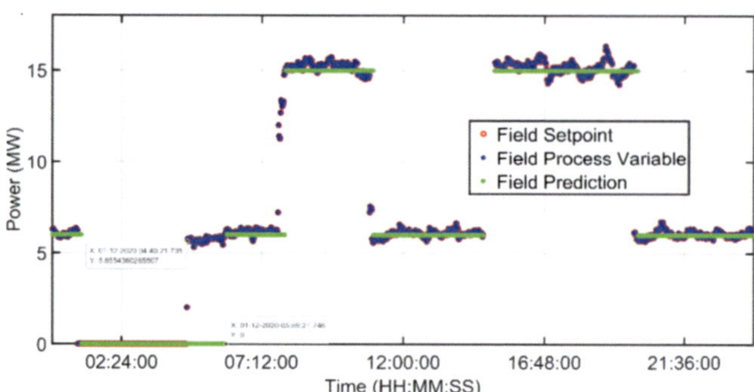

**Figure 11.** Data analysis results: electric energy production plan and real production.

*3.2. Modelization, Forecasting of Measured DVs, and Model Mismatch Compensation Results*

As previously described, the selection of the MPC strategy was a consequence of the fact that the modelization, measured DV forecasting, and model mismatch compensation results were remarkable. A significant example of the measured DVs' forecasting performances is reported in Figure 12, where the reliability of Equation (7) is illustrated, i.e., the achieved relationship between the downstream reservoir's outlet flow rate and the electric energy power. Two months of data are reported in two different plots: power–flow rate pairs are represented in blue, while a yellow line depicts the obtained linear law. The known terms of the linear law were neglected.

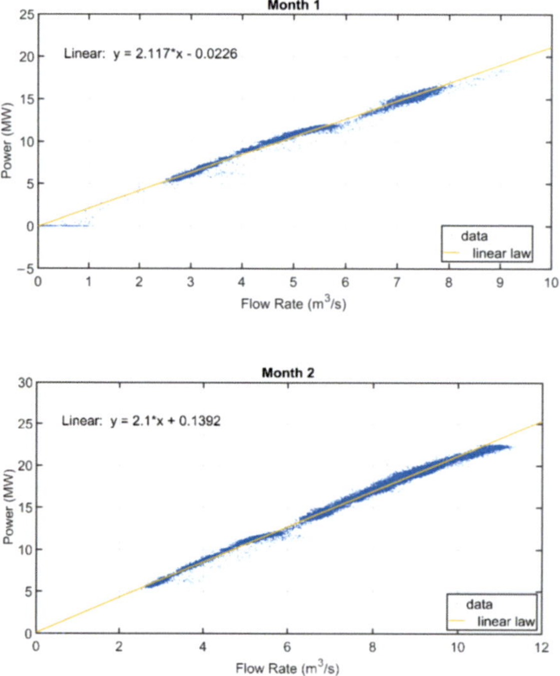

**Figure 12.** Measured DVs' forecasting results: electric power production plan and downstream reservoir outlet flow rate (two months of data).

An example of the cause of the introduction of the time delays in Equation (4) is reported in Figure 13, where a step test experiment is presented. The figure shows the delayed effect of the gate opening ($Q_{Beik}$ term in Equation (4)) on the downstream reservoir level. The downstream reservoir level is shown in the first plot, while the second plot reports the downstream reservoir's inlet–outlet flow rates: the green line is the flow rate at the Beikircher gate ($Q_{Beik}$), the light blue is the intake from the sand trap ($Q_{dissab}$), and the black and red lines represent the scheduled water flow request from the hydroelectric plant and its effective flow rate value ($Q_{gall\_Sar}$), respectively. In the selected period, only the inlet flow rate controlled by the regulation gate was varied, while the others were largely constant. The variation was performed at about 10:41, causing a slope change on the downstream reservoir level after about 43 min, at 11:24.

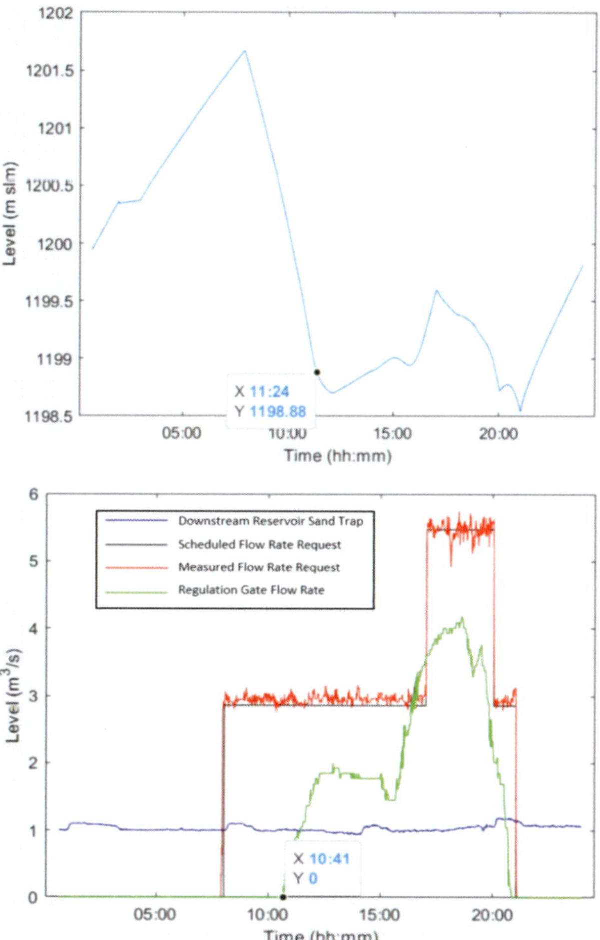


**Figure 13.** Modelization results: time delay between the Beikircher regulation gate's flow rate and the downstream reservoir level.

Figures 14 and 15 show the motivation behind the adopted model mismatch compensation strategy. Here, the model performances on the upstream and downstream reservoirs' levels are depicted for two different days. Blue lines represent the process variables, while orange lines represent the model results. The first plot of the figures represents the model's performances without model mismatch compensation; as can be noted, the model diverges.

In the second plot, a model mismatch compensation strategy is added. The $v$ vector term's computation (see Section 2.6) is reported through a dark yellow line, and the benefits of the modification can be observed. Finally, Figure 16 reports an example of the performance of the upstream reservoir level model. The blue line indicates the field process variable, while the red line represents the model's performance exploiting the available data on input–output flow rates. The one-hour-ahead prediction (exploiting the measured DVs' forecasting) computed at 15:18 on the selected day is depicted by an orange line; as can be observed, the behaviors of the orange and red lines are similar.

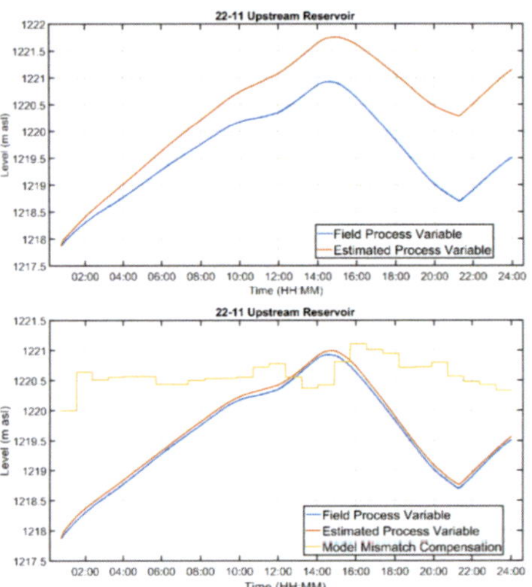

**Figure 14.** Model mismatch compensation results: upstream reservoir level with and without model mismatch compensation.

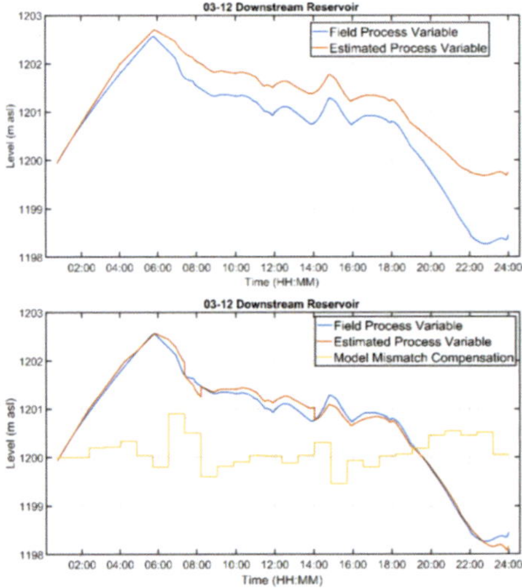

**Figure 15.** Model mismatch compensation results: downstream reservoir level with and without model mismatch compensation.

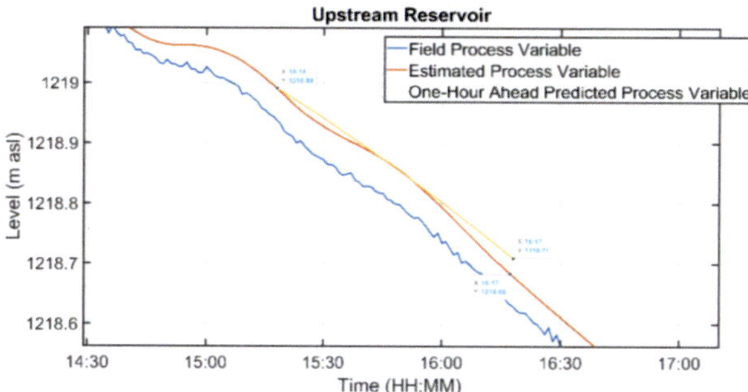


**Figure 16.** Modelization results: upstream reservoir level (blue line), predicted level with available data on input–output flow rates (red line), and one-hour-ahead prediction (orange line).

### 3.3. Virtual Environment Simulation Results

Before the installation of the APC system at the real plant, tailored virtual environment simulations were performed in order to test the developed controller. Figure 17 reports an example of one-day simulation results. A model mismatch between the simulated plant and the internal model of the controller was added in order to effectively test all of the proposed APC system, including the model mismatch compensation aspect. In the first plot of Figure 17, the measured disturbances are reported. The upstream reservoir's subsidiary intakes are reported in green, the upstream reservoir's intake is depicted in light blue, and the downstream reservoir's sand trap is represented by a blue line. Furthermore, the current data of the downstream reservoir's outlet flow rate are represented in red, and the values assumed for it by the conversion of the production plan are depicted by a black line. The second and the third plots of Figure 17 refer to the upstream and downstream reservoirs, respectively. Black lines represent the *hard* constraints, cyan lines represent the *soft* constraints, and light blue lines represent the real level, while orange lines represent the modeled level. When infeasibility is detected, the orange lines are replaced by red lines with a greater width. The dark yellow line represents the values computed for the $v$ term of Equation (5), while the purple horizontal lines represent the water shortage prediction indication. The Beikircher regulation gate's flow rate setpoint (MV) is represented by a purple line.

During the simulated day, two electric energy production periods are present (see the black and red lines in the first plot of Figure 17). In the first part of the simulation, no water shortage condition was predicted and no production requests were predicted; the *soft* constraints were not present in the MPC formulation (note the absence of cyan lines in the first part of the reservoirs plots in Figure 17). The controller did not perform any move on the MV, leaving it at 0 m$^3$/s and guaranteeing an acceptable behavior at the level of the two reservoirs. At about 05:50, the production request started on the prediction horizon. In this condition, without any indication of water shortage, *soft* constraints were present in the MPC formulation (note the presence of cyan lines in the middle part of the reservoir plots in Figure 17). The controller started to move the MV when, at about 08:00, it predicted a violation of the downstream reservoir level's *soft* constraint. The alternation of non-productive and productive periods repeated between 12:00 and 16:00. From about 16:45 to about 20:00, a water shortage condition was detected. According to the rules reported in Section 2.7, the MPC formulation was adapted in order to send as much water as possible to the downstream reservoir. Furthermore, at about 17:00, the infeasibility condition was verified and the heuristic law ensured reliable behavior of the controller. When the water shortage condition disappeared, the MPC formulation was updated and the MV was moved to 0 m$^3$/s in order to avoid a useless anticipated transit of the water from the upstream reservoir to the downstream reservoir.

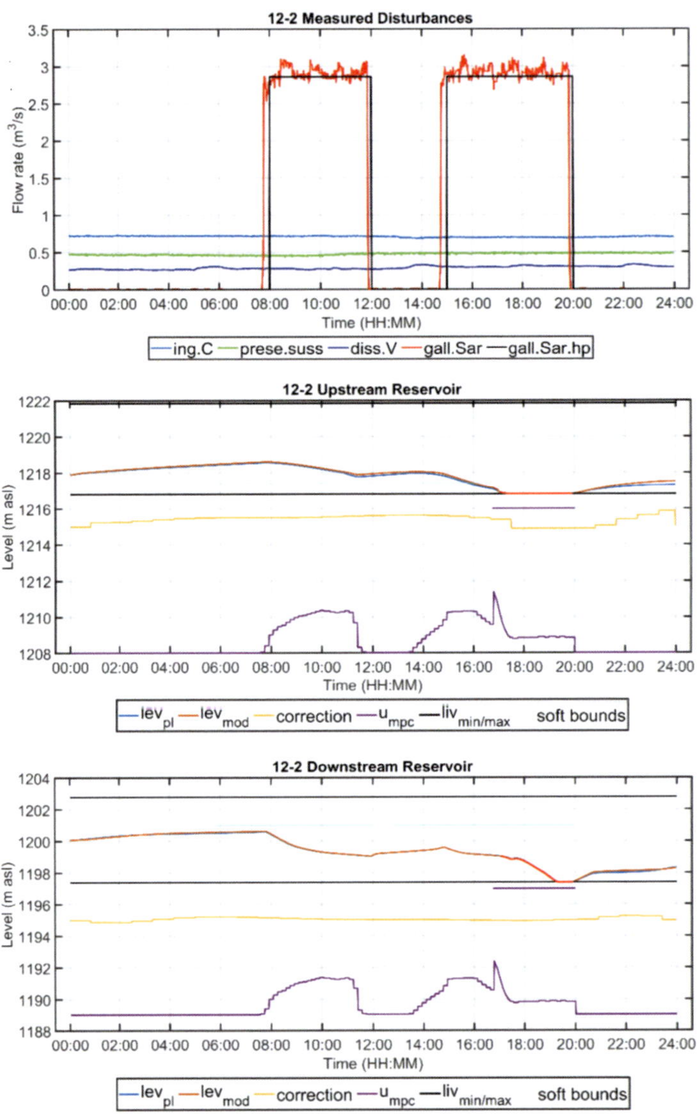

**Figure 17.** Virtual environment results: measured disturbances, upstream reservoir, and downstream reservoir.

*3.4. Field Results*

Using the developed field architecture, the field results were suitably stored in order to allow an accurate evaluation of the APC system's performance. An example of the field performance of the developed APC system is reported in Figure 18. In the first plot, the electric energy production plan is depicted (green line), together with the setpoint (red line) and the process variable (blue line). The upstream and downstream reservoirs' levels are depicted by blue lines in the second and third plots, together with their lower and upper *soft* constraints (red lines). Finally, the Beikircher regulation gate's flow rate setpoint and process variable are reported in Figure 18 (red line and blue line, respectively), together with the upper and lower *hard* constraints (red lines).

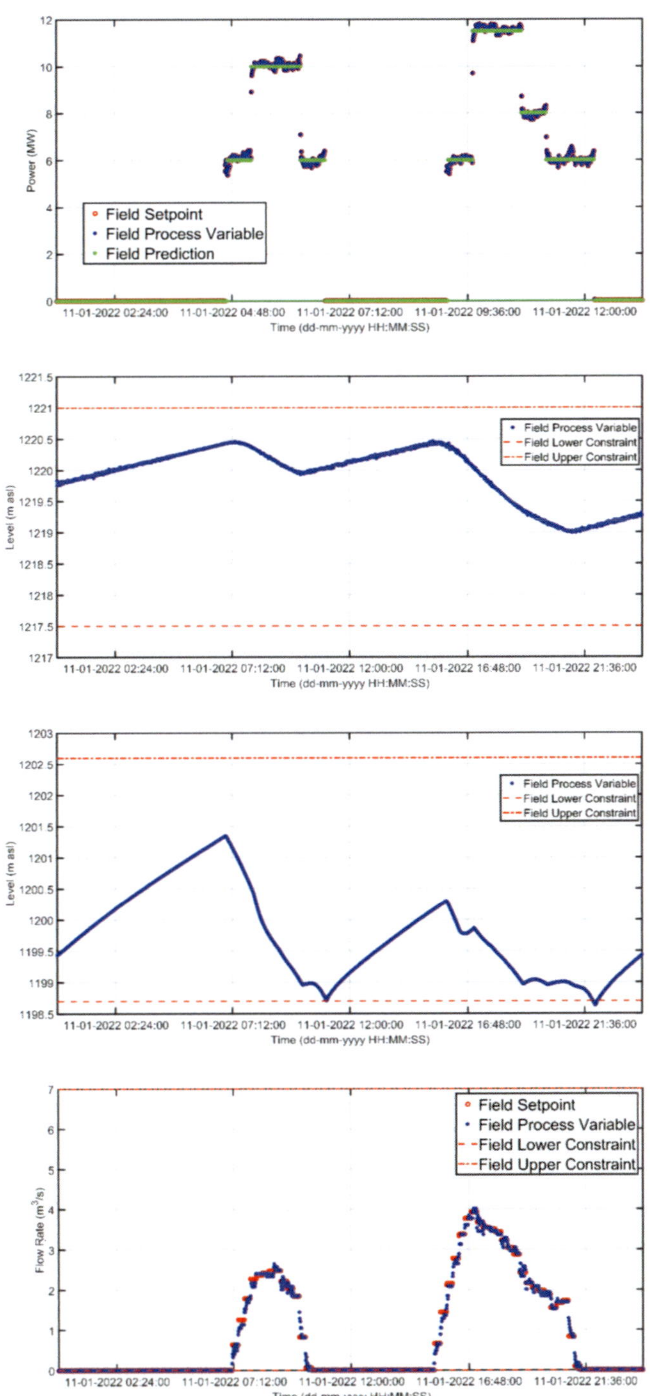

**Figure 18.** Field results: electric energy production, upstream reservoir, downstream reservoir, and Beikircher regulation gate.

Two production periods were present on the considered day, and the electric energy production plan was always respected. The upstream reservoir's level was always within the defined constraints, while the downstream reservoir's level violated its lower constraint under two conditions of the day. However, the maximum observed violation was less than 2% of the difference between the upper and the lower constraints, i.e., a negligible violation. It should be noted that the regulation gate's flow rate setpoint was not increased in advance in the first production period, because the APC system predicted that the water volume already present in the downstream reservoir was adequate; on the other hand, in the second production period, the regulation gate's setpoint was increased in advance due to the lower starting point of the downstream reservoir level.

## 4. Conclusions

An APC system based on an MPC strategy for hydroelectric power plants is proposed in the present paper. A hydroelectric power plant located in Italy was selected as a case study. Two reservoirs (connected through a regulation gate) and a set of turbines for energy production constituted the main elements of the process. Insights into the project phases—i.e., data analysis, modelization, controller design, and field implementation— were provided. In particular, an assessment on data selection, acquisition, storage and analysis was presented, together with a feasibility study on MPC applications for controlling hydropower plants. Furthermore, two modules were introduced in the classic MPC architecture in order to enhance the soundness and the reliability of the proposed solution. These modules cover different functions, e.g., real-time bad data detection and real-time definition of the controller parameters based on the current and predicted process conditions, together with the computation of smart alarms. The proposed APC system represents a reliable control tool, as proven by the high performances and the remarkable service factor obtained on the real plant. The service factor is the percentage of time for which the APC system is fully in service. The service factor after about three years from commissioning is higher than 90%. The results of this KPI prove the robustness of the proposed solution in terms of performances in real-time control.

Future works will focus on the further improvement of the modelization and controller synthesis phases. Furthermore, the studies will be aimed at obtaining high-level supervisors in order to further enhance the plant's benefits.

**Author Contributions:** Conceptualization, S.M.Z. and C.P.; Data Curation, S.M.Z., C.P., F.L., and G.A.; Formal Analysis, S.M.Z. and C.P.; Investigation, S.M.Z. and C.P.; Methodology, S.M.Z. and C.P.; Software, S.M.Z., C.P., and F.L.; Validation, S.M.Z. and C.P.; Visualization, S.M.Z., C.P., and F.L.; Writing–Original Draft S.M.Z. and C.P.; Writing–Review and Editing, S.M.Z. and C.P. All authors have read and agreed to the published version of the manuscript.

**Funding:** This research received no external funding.

**Data Availability Statement:** Not applicable.

**Conflicts of Interest:** The authors declare no conflict of interest.

## References

1. Agenda 2030. Available online: https://unric.org/it/agenda-2030/ (accessed on 7 November 2022).
2. UNDP. Available online: https://www.undp.org/ (accessed on 7 November 2022).
3. PNRR. Available online: https://www.mise.gov.it/index.php/it/pnrr (accessed on 7 November 2022).
4. Hydropower Europe. Available online: https://hydropower-europe.eu/ (accessed on 7 November 2022).
5. Kougias, I.; Aggidis, G.; Avellan, F.; Deniz, S.; Lundin, U.; Moro, A.; Muntean, S.; Novara, D.; Pérez-Díaz, J.I.; Quaranta, E.; et al. Analysis of emerging technologies in the hydropower sector. *Renew. Sustain. Energy Rev.* **2019**, *113*, 109257. [CrossRef]
6. Kougias, I. *Hydropower Technology Development Report 2020*; EUR 30510 EN; Publications Office of the European Union: Luxembourg, 2020. [CrossRef]
7. Ramos, H.M.; Carravetta, A.; Nabola, A.M. New Challenges in Water Systems. *Water* **2020**, *12*, 2340. [CrossRef]
8. Yang, W. *Hydropower Plants and Power Systems—Dynamic Processes and Control for Stable and Efficient Operation*; Springer: Cham, Switzerland, 2019. [CrossRef]

9.  AIChE. Available online: https://www.aiche.org/resources/publications/cep/2016/june/understand-advanced-process-control (accessed on 7 November 2022).
10. Thaeer Hammid, A.; Awad, O.I.; Sulaiman, M.H.; Gunasekaran, S.S.; Mostafa, S.A.; Manoj Kumar, N.; Khalaf, B.A.; Al-Jawhar, Y.A.; Abdulhasan, R.A. A Review of Optimization Algorithms in Solving Hydro Generation Scheduling Problems. *Energies* **2020**, *13*, 2787. [CrossRef]
11. Bundesministerium für Wirtschaft und Klimaschutz. Available online: https://www.plattform-i40.de/ (accessed on 7 November 2022).
12. Zanoli, S.M.; Pepe, C.; Rocchi, M. Control and optimization of a cement rotary kiln: A model predictive control approach. In Proceedings of the 2016 Indian Control Conference (ICC), Hyderabad, India, 4–6 January 2016. [CrossRef]
13. Zanoli, S.M.; Pepe, C.; Orlietti, L.; Barchiesi, D. A Model Predictive Control strategy for energy saving and user comfort features in building automation. In Proceedings of the 2015 19th International Conference on System Theory, Control and Computing (ICSTCC), Cheile Gradistei, Romania, 14–16 October 2015. [CrossRef]
14. Munoz-Hernandez, G.A.; Mansoor, S.P.; Jones, D.I. *Modelling and Controlling Hydropower Plants*; Springer: London, UK, 2013. [CrossRef]
15. *Handbook of Water Resources Management: Discourses, Concepts and Examples*; Springer: Cham, Switzerland, 2021. [CrossRef]
16. Marcelino, C.G.; Camacho-Gómez, C.; Jiménez-Fernández, S.; Salcedo-Sanz, S. Optimal Generation Scheduling in Hydro-Power Plants with the Coral Reefs Optimization Algorithm. *Energies* **2021**, *14*, 2443. [CrossRef]
17. Passos de Aragão, A.; Teixeira Leite Asano, P.; de Andrade Lira Rabêlo, R. A Reservoir Operation Policy Using Inter-Basin Water Transfer for Maximizing Hydroelectric Benefits in Brazil. *Energies* **2020**, *13*, 2564. [CrossRef]
18. Bakanos, P.I.; Katsifarakis, K.L. Optimizing Current and Future Hydroelectric Energy Production and Water Uses of the Complex Multi-Reservoir System in the Aliakmon River, Greece. *Energies* **2020**, *13*, 6499. [CrossRef]
19. Westerhoff, T.; Scharaw, B. Model based management of a reservoir system. In Proceedings of the 1999 European Control Conference (ECC), Karlsruhe, Germany, 31 August–3 September 1999. [CrossRef]
20. Chen, J.; Guo, S.; Li, Y.; Liu, P.; Zhou, Y. Joint Operation and Dynamic Control of Flood Limiting Water Levels for Cascade Reservoirs. *Water Resour. Manag.* **2012**, *27*, 749–763. [CrossRef]
21. Qiu, H.; Zhou, J.; Chen, L.; Zhu, Y. Multiple Strategies Based Salp Swarm Algorithm for Optimal Operation of Multiple Hydropower Reservoirs. *Water* **2021**, *13*, 2753. [CrossRef]
22. Zhang, Y.; Wu, J.; Yu, H.; Ji, C. Formulation and Implementation of Short-Term Optimal Reservoir Operation Schemes Integrated with Operation Rules. *Water* **2019**, *11*, 944. [CrossRef]
23. Ungureşan, M.L.; Mureşan, V.; Abrudean, M.; Vălean, H.; Clitan, I.; Bondici, C.; Puşcaşiu, A.; Fanca, A.; Stan, O. Advanced control of a hydroelectric power plant. In Proceedings of the 2017 21st International Conference on System Theory, Control and Computing (ICSTCC), Sinaia, Romania, 19–21 October 2017. [CrossRef]
24. Kazantsev, Y.V.; Glazyrin, G.V.; Khalyasmaa, A.I.; Shayk, S.M.; Kuparev, M.A. Advanced Algorithms in Automatic Generation Control of Hydroelectric Power Plants. *Mathematics* **2022**, *10*, 4809. [CrossRef]
25. Li, L.; Qian, J.; Zou, Y.; Tian, D.; Zeng, Y.; Cao, F.; Li, X. Optimized Takagi–Sugeno Fuzzy Mixed H2/H∞ Robust Controller Design Based on CPSOGSA Optimization Algorithm for Hydraulic Turbine Governing System. *Energies* **2022**, *15*, 4771. [CrossRef]
26. Vinatoru, M. Monitoring and control of hydro power plant. *IFAC Proc. Vol.* **2007**, *40*, 44–55. [CrossRef]
27. Zhou, W.; Thoresen, H.M.; Glemmstad, B. Application of Kalman filter based nonlinear MPC for flood gate control of hydropower plant. In Proceedings of the 2012 IEEE Power and Energy Society General Meeting, San Diego, CA, USA, 22–26 July 2012. [CrossRef]
28. Namazov, M.; Alili, A. Design of stable Takagi Sugeno fuzzy control system for three interconnected tank system via LMIs with constraint on the output. *IFAC-PapersOnLine* **2018**, *51*, 721–726. [CrossRef]
29. Arnold, C.; Aissa, T.; Lambeck, S. Implicit Regulator Calculation for Regular MIMO-Systems with Predictive Functional Control Demonstrated at a Three Tank System. *IFAC-PapersOnLine* **2014**, *47*, 5375–5380. [CrossRef]
30. Breckpot, M.; Agudelo, O.M.; De Moor, B. Flood Control with Model Predictive Control for River Systems with Water Reservoirs. *J. Irrig. Drain. Eng.* **2013**, *139*, 532–541. [CrossRef]
31. Maestre, J.M.; Ridao, M.A.; Kozma, A.; Savorgnan, C.; Diehl, M.; Doan, M.D.; Sadowska, A.; Keviczky, T.; De Schutter, B.; Scheu, H.; et al. A comparison of distributed MPC schemes on a hydro-power plant benchmark. *Optim. Control. Appl. Methods* **2015**, *36*, 306–332. [CrossRef]
32. Munoz-Hernandez, G.A.M.; Jones, D. MIMO Generalized Predictive Control for a Hydroelectric Power Station. *IEEE Trans. Energy Convers.* **2006**, *21*, 921–929. [CrossRef]
33. Essahafi, M. Model Predictive Control (MPC) Applied to Coupled Tank Liquid Level System. *arXiv* **2014**, arXiv:1404.1498. [CrossRef]
34. Perez-Villalpando, M.A.; Gurubel Tun, K.J.; Arellano-Muro, C.A.; Fausto, F. Inverse Optimal Control Using Metaheuristics of Hydropower Plant Model via Forecasting Based on the Feature Engineering. *Energies* **2021**, *14*, 7356. [CrossRef]
35. Myo Lin, N.; Tian, X.; Rutten, M.; Abraham, E.; Maestre, J.M.; van de Giesen, N. Multi-Objective Model Predictive Control for Real-Time Operation of a Multi-Reservoir System. *Water* **2020**, *12*, 1898. [CrossRef]
36. Zhang, R.; Chen, D.; Ma, X. Nonlinear Predictive Control of a Hydropower System Model. *Entropy* **2015**, *17*, 6129–6149. [CrossRef]
37. Zanoli, S.M.; Pepe, C.; Astolfi, G.; Luzi, F. Model Predictive Control for Hydroelectric Power Plant Reservoirs. In Proceedings of the 2022 23rd International Carpathian Control Conference (ICCC), Sinaia, Romania, 29 May–1 June 2022. [CrossRef]

38. Zanoli, S.M.; Pepe, C.; Rocchi, M.; Astolfi, G. Application of Advanced Process Control techniques for a cement rotary kiln. In Proceedings of the 2015 19th International Conference on System Theory, Control and Computing (ICSTCC), Cheile Gradistei, Romania, 14–16 October 2015. [CrossRef]
39. Maciejowski, J.M. *Predictive Control with Constraints*; Prentice-Hall, Pearson Education Limited: Harlow, UK, 2002.
40. Kelleher, J.D.; Tierney, B. *Data Science*; MIT Press: Cambridge, MA, USA, 2018.
41. Archdeacon, T. *Correlation and Regression Analysis: A Historian's Guide*; University of Wisconsin Press: Madison, WI, USA, 1994.
42. Navidi, W. *Probabilità e Statistica per L'ingegneria e le Scienze*; McGraw-Hill Education: New York, NY, USA, 2006.
43. Aström, K.J.; Hägglund, T. *PID Controllers: Theory, Design, and Tuning*; ISA: Research Triangle Park, NC, USA, 1995.
44. Shinskey, F.G. *Process Control Systems: Application, Design, and Tuning*; McGraw-Hill Professional Publishing: New York, NY, USA, 1996.
45. Magnani, G.; Ferretti, G.; Rocco, P. *Tecnologie dei Sistemi di Controllo*; McGraw-Hill: Milan, Italy, 2007.
46. Ramos, V.S.; Sena, H.J.; Fileti, A.M.F.; Silva, F.V. Teaching Multivariable Model Predictive Control in a Laboratory Scale Three-Tank Process. *Chem. Eng. Trans.* **2017**, *57*, 1579–1584. [CrossRef]
47. Join, C.; Sira-Ramírez, H.; Fliess, M. Control of an uncertain three-tank system via on-line parameter identification and fault detection. *IFAC Proc. Vol.* **2005**, *38*, 251–256. [CrossRef]
48. Ljung, L. *System Identification: Theory for the User*; Prentice-Hall PTR: Upper Saddle River, NJ, USA, 1999.
49. De Schutter, B. Minimal state-space realization in linear system theory: An overview. *J. Comput. Appl. Math.* **2000**, *121*, 331–354. [CrossRef]
50. Bemporad, A.; Morari, M.; Ricker, N.L. *Model Predictive Control Toolbox User's Guide*; MathWorks: Natick, MA, USA, 2015.
51. Camacho, E.F.; Bordons, C. *Model Predictive Control*; Springer: London, UK, 2007. [CrossRef]
52. Rawlings, J.B.; Mayne, D.Q.; Diehl, M.M. *Model Predictive Control: Theory and Design*; Nob Hill Publishing, 2020. Available online: http://www.nobhillpublishing.com/mpc-paperback/index-mpc.html (accessed on 10 November 2022).
53. Cagienard, R.; Grieder, P.; Kerrigan, E.C.; Morari, M. Move blocking strategies in receding horizon control. *J. Process Control.* **2007**, *17*, 563–570. [CrossRef]
54. Zanoli, S.M.; Pepe, C.; Rocchi, M. Cement rotary kiln: Constraints handling and optimization via model predictive control techniques. In Proceedings of the 2015 5th Australian Control Conference (AUCC), Gold Coast, QLD, Australia, 5–6 November 2015. Available online: https://ieeexplore.ieee.org/document/7361950 (accessed on 15 November 2022).
55. MathWorks. Available online: https://it.mathworks.com/ (accessed on 15 November 2022).

 *processes*

*Article*

# Study on the Influencing Factors of the Response Characteristics of the Slide Valve-Type Direct-Acting Relief Valve with External Orifice

Huiyong Liu [1,*] and Qing Zhao [2]

[1] Department of Mechanical Design, School of Mechanical Engineering, Guizhou University, Guiyang 550025, China
[2] Department of Water Resources and Hydropower Engineering, College of Civil Engineering, Guizhou University, Guiyang 550025, China
* Correspondence: heartext@163.com

**Abstract:** The slide valve-type direct-acting relief valve with external orifice (SVTDARVWEO) is widely used in hydraulic systems, and its response characteristics are influenced by key factors. It is of great significance to carry out research on the influencing factors of the response characteristics of the SVTDARVWEO. The working principle of the SVTDARVWEO is analyzed in the present study. The simulation model of the SVTDARVWEO is established using AMESim. The influence of the orifice diameter, viscosity coefficient, valve element mass, spring stiffness, oil seal length, and valve element diameter on the response characteristics of the SVTDARVWEO is studied. The results show that: (1) The smaller the orifice diameter is, the smaller the oscillation frequency, amplitude and maximum overshoot of pressure, flowrate, displacement and velocity are. (2) When the viscosity coefficient is 50 N/(m/s), 55 N/(m/s) and 60 N/(m/s), the pressure, flowrate, displacement and velocity oscillate periodically, but the amplitude of the oscillation decreases gradually, and the oscillation frequency is 250 Hz. When the viscosity coefficient is 60 N/(m/s), the pressure, flowrate, displacement and velocity will reach their respective stable values earlier. (3) When the valve element mass is 0.01 kg, 0.015 kg and 0.02 kg, the pressure, flowrate, displacement and velocity oscillate periodically, but the amplitude of oscillation decreases gradually. When the valve element mass is 0.01 kg, the pressure, flowrate, displacement and velocity will reach the stable value earlier. (4) The smaller the spring stiffness is, the greater the maximum overshoot of pressure, flowrate, displacement and velocity is, and the higher the number of oscillations to reach the stable value are, in addition to more time being required. (5) With the increase in oil seal length, the maximum overshoot of pressure and velocity, stability value of displacement also increase correspondingly. (6) With the increase in the valve element diameter, the stable value of pressure decreases, and the oscillation frequency of pressure, flowrate, displacement and velocity increase, but the oscillation amplitude decreases.

**Keywords:** SVTDARVWEO; AMESim; response characteristics; influencing factors

**Citation:** Liu, H.; Zhao, Q. Study on the Influencing Factors of the Response Characteristics of the Slide Valve-Type Direct-Acting Relief Valve with External Orifice. *Processes* **2023**, *11*, 397. https://doi.org/10.3390/pr11020397

Academic Editor: Sergey Y. Yurish

Received: 23 December 2022
Revised: 21 January 2023
Accepted: 22 January 2023
Published: 28 January 2023

## 1. Introduction

The direct-acting relief valve (DARV) relies on the pressure oil in the hydraulic system to directly act on the valve element to balance it with the spring force, so as to control the opening and closing of the overflow port in order to keep the hydraulic pressure of the controlled system or circuit constant, and to realize the functions of pressure stabilization, pressure regulation or pressure limitation. The application of the DARV mainly includes constant pressure overflow, safety protection, back-pressure generation, remote pressure regulation and multi-stage pressure control. (1) Constant pressure relief. Figure 1 is a typical inlet throttle speed-control system. When the system pressure is lower than the opening pressure of the DARV, the DARV will close. At this time, the system pressure depends on the load. When the system pressure reaches the set value of the DARV, the DARV is normally open and the system pressure is limited. When the load speed change of the actuator causes the flowrate change, the regulating function of the DARV keeps the system pressure essentially constant, and overflows the excess oil back to the tank, thus realizing constant pressure overflow. (2) Security protection. Figure 2 shows the parallel throttling speed-control circuit with a constant displacement pump. Figure 3 shows the volumetric throttling speed-control circuit with a variable displacement pump. Figure 4 shows the volumetric speed-control circuit with a variable/constant displacement pump/motor. In the above three circuits, the DARV is often used as a safety valve to prevent overloading of the system. Under normal conditions, the DARV is normally closed. When the system pressure is too high due to fault, abnormal load and other reasons, the DARV opens to overflow in order to protect the safety of the entire hydraulic system. (3) Back pressure generation. As shown in Figure 5, by connecting the DARV-3 to the oil return path of the actuator 4, a certain oil return resistance is created to improve the motion smoothness of the actuator. (4) Remote pressure regulation. As shown in Figure 6, by connecting the DARV-3 with the remote control port of the pilot relief valve and adjusting the pressure of the DARV-3, the pilot relief valve can be remotely regulated within the set pressure range. (5) Multi-stage pressure control. Figure 7 is a typical three-stage pressure control circuit. The DARV-2 and DARV-3 are connected with the remote control port of the pilot relief valve through a three-position four-way solenoid directional valve. The multi- stage pressure control of the hydraulic system is realized by switching the different working positions of the solenoid directional valve. The pressure regulating value of the pilot relief valve is set as p1, and the pressure regulating value of DARV-2 and DARV-3 are set as p2 and p3, respectively. When the solenoid directional valve works in the middle position, the working pressure of the system is p $\leq$ p1. When the solenoid directional valve is switched to the left and right positions, the working pressure of the system is p $\leq$ p2 and p $\leq$ p3, respectively.

The structure of the DARV mainly includes slide valve, cone valve, ball valve, etc. The SVTDARVWEO is widely used in medium- and low-pressure hydraulic systems due to its simple structure, easy processing, convenient adjustment and high sensitivity. However, the response characteristics of the SVTDARVWEO are influenced by factors such as the orifice's diameter, viscosity coefficient, valve element mass, spring stiffness, oil seal length, and the valve element's diameter. Therefore, it is of great significance to carry out research on the influencing factors of the response characteristics of the SVTDARVWEO.

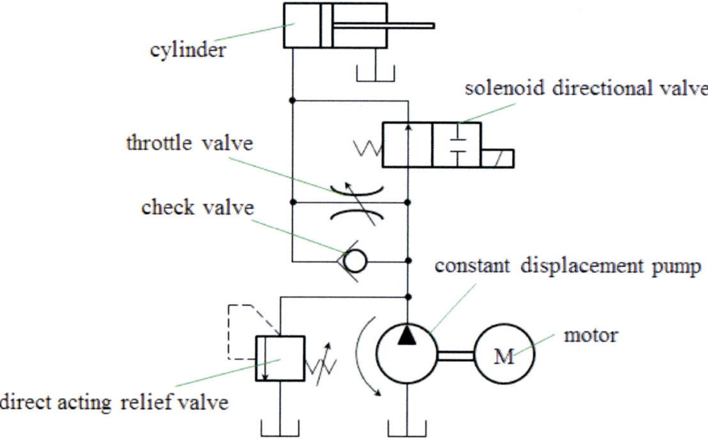


**Figure 1.** Constant pressure relief circuit of inlet throttle speed-control system.

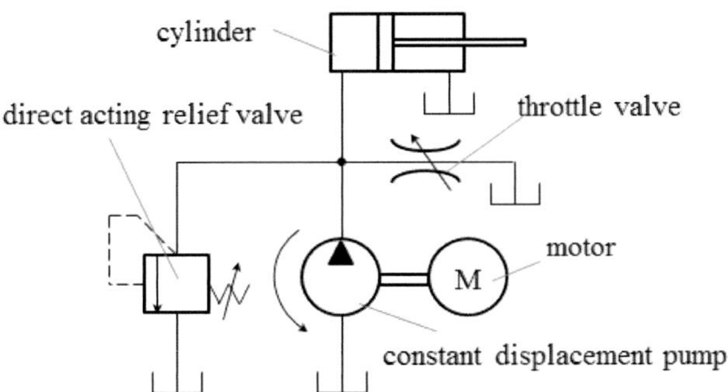


**Figure 2.** The parallel throttling speed-control circuit with constant displacement pump.

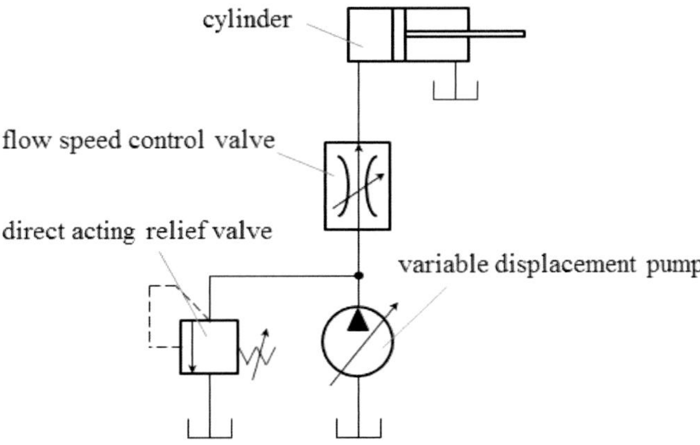


**Figure 3.** The volumetric throttling speed-control circuit with variable displacement pump.

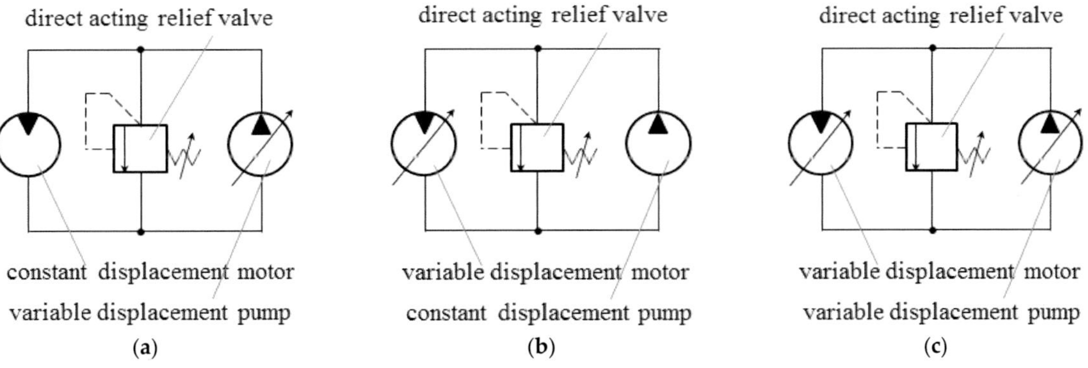


**Figure 4.** The volumetric speed-control circuit with variable/constant displacement pump/motor. (a) VDP—CDM, (b) CDP—VDM, (c) VDP—VDM.

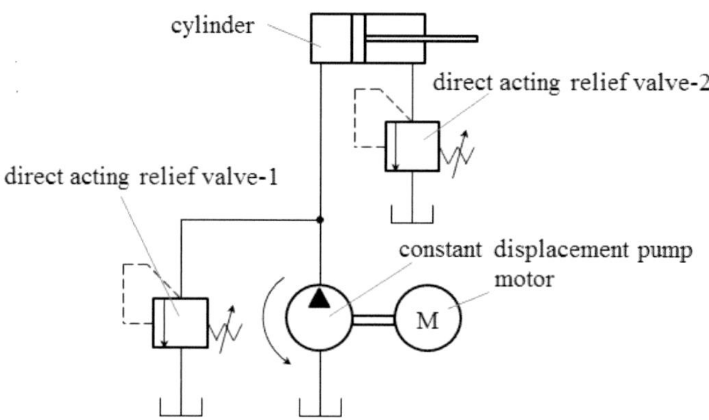


**Figure 5.** The back pressure generation circuit.

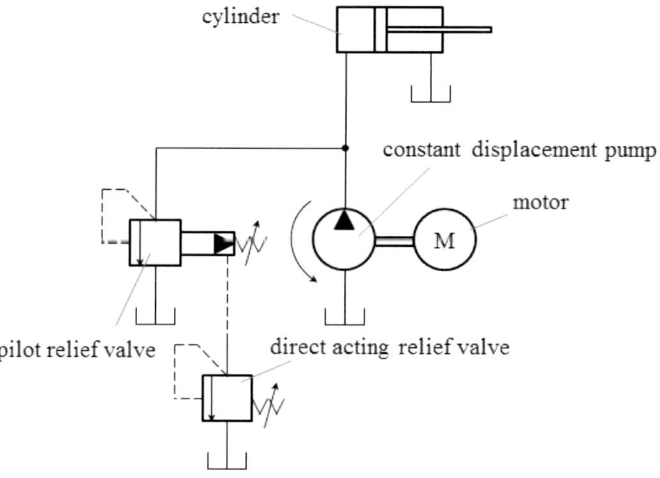


**Figure 6.** The remote pressure regulation circuit.

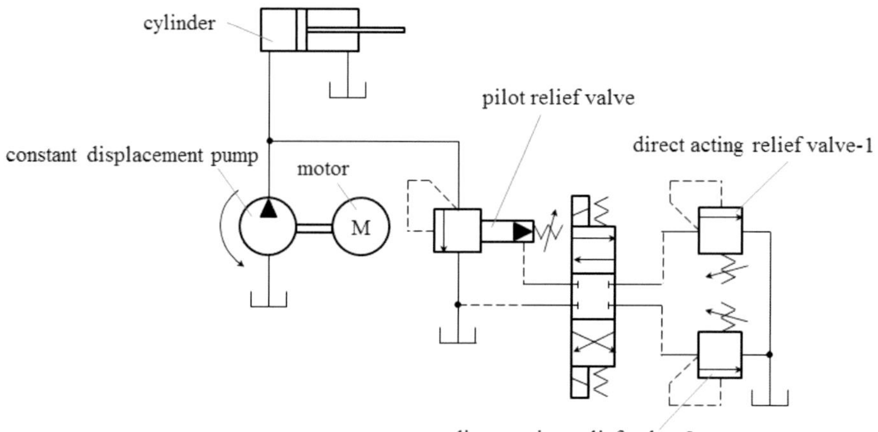


**Figure 7.** The multi-stage pressure control circuit.

In the last few decades, there have been many relevant studies on the direct-acting relief valve. Song et al. [1] developed a numerical model using CFD techniques to investigate the fluid and dynamic characteristics of a direct-operated safety relieve valve (SRV), and examined the comparison of the effects of the design parameters, including the adjusting ring position, vessel volume and spring stiffness. Burhani and Hos [2] addressed the static and dynamic behavior of a direct spring-operated PRV of conical shape in the presence of two-phase non-flashing flow, and recorded the effect of the system parameters, such as spring stiffness and reservoir capacity. Zong et al. [3] established the accuracy of CFD models for the prediction of the flow force exerted on the disk of a direct-operated pressure safety valve in energy system. Kadar et al. [4] presented a delayed oscillator model of pressure relief valves with outlet piping which contained a time delay originated in the pipe length and the velocity of sound in the pipe. Fu et al. [5] designed a direct-acting relief valve with permanent magnet spring to solve the problems of helical compression spring, and established the nonlinear model of air gap-magnetic force to reveal the impact of different configurations and air gap adjustments on magnetic field distribution. Burhani et al. [6] addressed the effect of non-flashing multiphase flow on the dynamic behavior of direct spring operated pressure relief valve, and developed a formula providing an order-of-magnitude estimation of the opening time of direct spring-operated pressure relief valve. Liao et al. [7] developed a two-degree-of-freedom fluid–structure coupling model of a direct-acting relief valve for underwater applications, and carried out parameter optimization with reliability analysis via an optimization closed loop. Zahariea [8] developed a functional diagram to perform numerical analysis of an electromagnetic normally closed direct-acting ball valve with cylindrical seat using the MATLAB/Simscape/SimHydraulics programming language. Wen et al. [9] developed a 2-DOF fluid–structure coupling dynamic model to explain the sudden jump of pressure as the variation of water depth for a direct-acting relief valve used by a torpedo pump as the variation of water depth. Erdodi and Hos [10] proposed two CFD-based methods for the analysis of the fluid forces and valve stability, including steady-state CFD method and dynamic CFD simulations. Liu et al. [11] established the mathematic model of sea water direct-acting relief valve (SDARV), and conducted related dynamic characteristic simulations. Wu et al. [12] established a mathematic model of a direct-operated seawater hydraulic relief valve under deep sea, and conducted stability analysis of the relief valve. Syrkin et al. [13] used the method of phase trajectories to establish the controller's parameters and modes, and obtained the dynamic characteristics providing the absence of self-oscillations at the expense of additional damping of the shut-off and regulating elements. Raeder et al. [14] proposed an approach involving numerical simulation of non-stationary 3D gas dynamics, which enables one to determine a spatial structure of flow in a direct-acting safety valve and its quantitative characteristics (pressure, density, velocity, temperature). Sohn [15] investigated the fluid dynamics of a spring-loaded-type safety valve operated with steam through computational fluid dynamics (CFD), and analyzed the opening process by running the total ten-step simulations of lift level from 0 to 100%. Dempster et al. [16] questioned the accuracy of the scaling approach, examined the influence of the effect of built-up pressure in the discharge region of the safety relief valve, and investigated the problem theoretically using a CFD technique via the commercial code FLUENT. Zong et al. [17] performed the numerical and experimental investigation on a direct-operated pressure safety valve to deeply explore the mechanism of the discontinuities. Suzuki and Urata [18] developed a balanced-piston-type water hydraulic relief valve, which focused on preventing cavitation and improving the static characteristics and stability. Bazsó and Hős [19] presented some detailed experimental results on the static and dynamic behavior of a hydraulic pressure relief valve with a poppet valve body. Hős et al. [20] derived a model of an in-service direct-spring pressure relief valve, which coupled low-order rigid body mechanics for the valve to one-dimensional gas dynamics within the pipe. Hős, Bazsó and Champneys [21] developed a mathematical model of a spring-loaded pressure relief valve connected to a reservoir of compressible fluid via a single, straight pipe. Hős et al. [22] carried out the study of gas-service direct-spring pressure relief valves connected to a tank via a straight pipe by deriving a reduced-order model for predicting oscillatory instabilities such as valve flutter and chatter. Hyunjun, Dawon, and Sanghyun [23] used a multi-objective genetic algorithm to carry out the optimum design of direct spring-loaded pressure relief valve in water distribution system. Hyunjun et al. [24] explored the optimization of a direct spring-loaded pressure relief valve (DSLPRV) to consider the instability issue of a valve disk and the surge control for a pipeline system. Lei et al.'s [25] study concerned the flow model and dynamic characteristics of a direct spring-loaded poppet relief valve. Kim et al. [26] presented design concepts to improve the disadvantages of conventional direct-acting relief valves such as the low pressure precision, the low allowable flow rate, and unstable chattering phenomena. Dimitrov and Krstev [27] examined experimentally and theoretically the transients in hydraulic systems with direct-operated pressure relief valves, and determined the coefficient of hydrodynamic force acting on the valve poppet.

The novelty and significance of this work is that it establishes the AMESim simulation model of the SVTDARVWEO, and analyzes the influencing factors (such as the orifice's diameter, viscosity coefficient, valve element mass, spring stiffness, oil seal length, and valve element diameter) of the response characteristics of the SVTDARVWEO systematically and comprehensively, provides theoretical basis for the design and manufacturing of the SVTDARVWEO, and offers a reference for the design and manufacturing of other valves, including pneumatic valves and hydraulic valves, helping manufacturers to reduce the development costs and shorten the development cycle.

The rest of this paper is organized as follows. In Section 2, the working principle of the SVTDARVWEO is analyzed. In Section 3, the AMESim simulation model of the SVTDARVWEO is established. The influence of the orifice's diameter, viscosity coefficient, valve element mass, spring stiffness, oil seal length, valve element diameter on the response characteristics of the SVTDARVWEO is analyzed in Section 4. Finally, some conclusions are drawn in Section 5.

## 2. Working Principle of the SVTDARVWEO

Figure 8 is the structural diagram of the SVTDARVWEO, the typical feature of which is that the orifice is not on the valve element but external to the valve element. The basic working principle of the SVTDARVWEO is as follows: the pressure oil of the hydraulic system flows to the inlet of the SVTDARVWEO (port $P$) through a pipe, and then it is divided into three paths: the first path acts on the ring with an area of $A_2$ at the upper part of the valve element (pressure: $p_2$); the second path acts on the ring with an area of $A_3$ at the lower part of the valve element (pressure: $p_3$); and the third path acts on the circular bottom with an area of $A_4$ at the lower part of the valve element (pressure: $p_4$) through thin pipes and orifice. In addition, the top of the valve element is affected by the pressure $p_1$, and the effective area is $A_1$. The valve element is jointly affected by the following forces: the hydraulic force $F_1$ ($F_1 = p_1 A_1$) acting on the top of the valve element (area: $A_1$); the hydraulic force $F_2$ ($F_2 = p_2 A_2$) acting on the upper part of the valve element (area: $A_2$); the hydraulic force $F_3$ ($F_3 = p_3 A_3$) acting on the lower part of the valve element (area: $A_3$); the hydraulic force $F_4$ ($F_4 = p_4 A_4$) acting on the bottom of the valve element (area: $A_4$); the mass force $G(G = mg)$; the spring force $F_s(F_s = k(x_0 + x))$; and the viscous friction force $f(f = \mu A \frac{v}{h})$. When the resultant force of the hydraulic forces acting on the valve element 3 is greater than the spring force, mass force, viscous friction occurs between the valve element and the valve body, and the valve port is opened, allowing the oil to overflow to the oil tank through the port T.

It can be seen from Figure 8 that $A_1 = A_4$, $A_2 = A_3$; therefore, the values of $F_2$ and $F_3$ are equal and opposite, which can offset each other. In addition, since $p_1$ is connected to port T, $p_1 = 0$ and $F_1 = 0$. According to the above analysis, the resultant force of hydraulic forces acting on the valve element 3 is $F_4$. As long as $(F_4 \pm G) > (F_s + f)$ is satisfied, the valve port can be opened and the oil overflows to the oil tank.

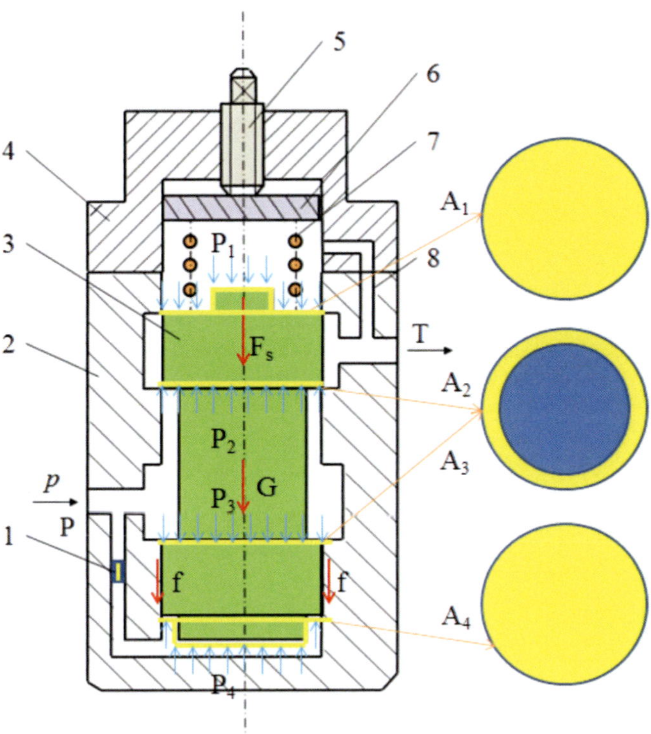



**Figure 8.** The structural diagram of the SVTDARVWEO.

## 3. AMESim Simulation Model of the SVTDARVWEO

With the development of fluid mechanics, hydraulic transmission, modern control theory, and other related disciplines in addition to the the rapid development of computer technology, hydraulic simulation technology and software are becoming an increasingly mature and become a powerful tool for designers of hydraulic components and systems. At present, hydraulic simulation software mainly includes MATLAB, EASY5, FluidSIM, HyPneu, DSHplus, HOPSAN, Automation Studio, 20-sim, AMESim, etc. Compared with other hydraulic simulation software, such as the familiar MATLAB, the biggest advantage of AMESim is that it does not need to establish the mathematical model of the system. The establishment, expansion or change of the simulation model is carried out through the graphical user interface, which frees users from numerical simulation algorithms, time-consuming programming, and tedious mathematical modeling, so as to focus on the design of the physical system itself in the engineering project, and modeling and simulation analysis can be conducted directly without special learning of programming language. Another advantage of AMESim is that it has a variety of simulation methods, such as batch-processing simulation, discontinuous continuous simulation, steady-state simulation, dynamic simulation, etc. It can dynamically switch the integration algorithm and adjust the integration step according to the characteristics of the system model at different simulation times, thus improving the stability of the system model and ensuring the accuracy of the simulation results.

AMESim is an advanced engineering system simulation modeling environment based on bond graph, integrating component libraries in mechanical, hydraulic, pneumatic, control, thermal, electrical and magnetic fields. Component libraries in different fields can be connected to each other, providing a complete platform for system engineering design, enabling users to build complex multidisciplinary simulation system models on the same platform and, on this basis, to conduct in-depth simulations and analysis, as well as to study the steady and dynamic performance of any component or system on this platform. HCD (hydraulic component design) is the library of hydraulic component designs in AMESim, which enables users to build sub models of any component from very basic modules, and to further study the steady and dynamic performance of the component, thus greatly enhancing the function of AMESim.

Based on the structure and working principle of the SVTDARVWEO, the AMESim simulation model of the valve is established using the hydraulic component design library (HCD), one-dimensional mechanical library (Mechanical), and standard hydraulic library (Hydraulic), as shown in Figure 9. In this model, the thick solid line represents the oil pressure action surface, and the arrow represents the pressure action's direction. The four pressure action surfaces in the model correspond to the four pressure action surfaces of the valve element. The position of the orifice in the model corresponds to the position of the orifice in the structural diagram. The following basic assumptions were made when building the AMESim simulation model:

(1)   The operating temperature and ambient temperature do not change;
(2)   The physical and chemical properties of the working medium do not change;
(3)   The working medium is not polluted;
(4)   There is no geometric shape error between valve element and valve body;
(5)   The radial fit clearance between valve element and valve body is equal;
(6)   There is no assembly error between spring and valve element;
(7)   The foundation is stable without vibration;
(8)   The working environment is the earth, and the gravity acceleration remains unchanged, g = 9.8 m/s$^2$;
(9)   No internal and external leakage;
(10)  The influence of gravity can be ignored;
(11)  All parts will not deform during operation.

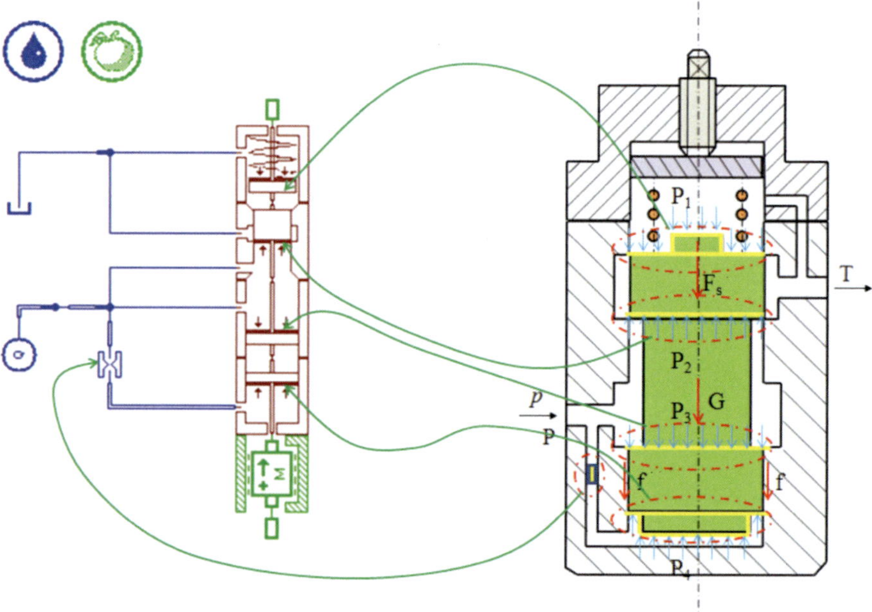


**Figure 9.** AMESim simulation model of the SVTDARVWEO.

## 4. Results and Discussion

Basic parameters: operating temperature $T = 40\,^{\circ}\text{C}$, density $\rho = 850\ \text{kg/m}^3$, bulk modulus $K = 17,000$ bar, absolute viscosity $v = 51$ cP, flowrate $Q = 10$ L/min. By changing the value of each parameter, the influence of each parameter on the response characteristics of the SVTDARVWEO is studied.

### 4.1. Influence of Orifice Diameter on Response Characteristics

The simulation parameters are shown in Table 1. The values of viscosity coefficient, valve element mass, spring stiffness, oil seal length, and valve element diameter remain constant, and the orifice's diameter is set as 1 mm, 2 mm, 3 mm and 4 mm, respectively. The influence of the orifice's diameter on the response characteristics of the SVTDARVWEO is studied.

**Table 1.** Simulation parameters—group 1.

| Orifice Diameter | Viscosity Coefficient | Valve Element Mass | Spring Stiffness | Oil Seal Length | Valve Element Diameter |
|---|---|---|---|---|---|
| (mm) | (N/(m/s)) | (kg) | (N/mm) | (mm) | (mm) |
| 1~4 | 100 | 0.01 | 1 | 0 | 10 |

The pressure response characteristics with orifice diameters of 1 mm~4 mm are shown in Figure 10a. The final stable value of the pressure is 10.02 bar. When the orifice diameters are 1 mm, 2 mm, 3 mm and 4 mm, the maximum pressure overshoot is about 4.608 bar, 2.712 bar, 2.353 bar and 2.26 bar, respectively. When the orifice's diameter is 1 mm, the oscillation frequency is the lowest, about 167 Hz. When the orifice's diameter is 2 mm, the oscillation frequency is about 200 Hz. When the orifice's diameter is 3 mm and 4 mm, the oscillation frequency is close to about 250 Hz. In addition, when the orifice's diameter is 1 mm, 2 mm, 3 mm and 4 mm, it needs to oscillate to about 0.06 s before the pressure can reach the stable value.

The flowrate response characteristics with an orifice's diameter of 1 mm~4 mm are shown in Figure 10b. The final stable value of the flowrate is 10 L/min, and the flowrate before 0.004 s is 0 L/min. After 0.004 s, the flowrate oscillates to about 0.06 s before it becomes stable. When the orifice's diameter is 1 mm, the oscillation frequency is the lowest, about 167 Hz. When the orifice's diameter is 2 mm, the oscillation frequency is about 200 Hz. When the orifice's diameter is 3 mm and 4 mm, the oscillation frequency is close to about 250 Hz. When the orifice's diameter is 2 mm, the maximum flowrate overshoot is the highest, about 7.171 L/min. The maximum flowrate overshoot of orifice diameters of 1 mm, 3 mm and 4 mm is 6.646 L/min, 6.558 L/min and 6.37 1 L/min, respectively.

The displacement response characteristics of the valve with an orifice's diameter of 1 mm~4 mm are shown in Figure 10c. The final stability value of the displacement is 0.156 mm, the displacement before 0.004 s is 0 mm, and the vibration after 0.004 s is about 0.06 s before it becomes stable. When the orifice's diameter is 1 mm, the oscillation frequency is the lowest, about 167 Hz. When the orifice's diameter is 2 mm, the oscillation frequency is about 200 Hz. When the orifice's diameter is 3 mm and 4 mm, the oscillation frequency is close to about 250 Hz. When the orifice's diameter is 1 mm, the maximum displacement overshoot is the lowest, about 0.107 mm. The maximum displacement overshoot of the orifice diameters of 2 mm, 3 mm and 4 mm is close, being 0.125 mm, 0.122 mm and 0.12 mm, respectively.

The velocity response characteristics of the valve element with an orifice's diameter of 1 mm~4 mm are shown in Figure 10d. The final stable value of the valve element velocity is 0 m/s. The valve element velocity before 0.004 s is 0 m/s. After 0.004 s, the valve element velocity oscillates to about 0.06 s and is 0 m/s again. When the orifice's diameter is 1 mm, the oscillation frequency is the lowest, about 167 Hz. When the orifice's diameter is 2 mm, the oscillation frequency is about 200 Hz. When the orifice's diameter is 3 mm and 4 mm, the oscillation frequency is close to about 250 Hz. When the orifice's diameter is 1 mm, the peak velocity of the valve element is the lowest, about 0.13 m/s. The peak velocity of the valve element with orifice diameters of 2 mm, 3 mm and 4 mm is close, being 0.187 m/s, 0.192 m/s and 0.193 m/s, respectively.

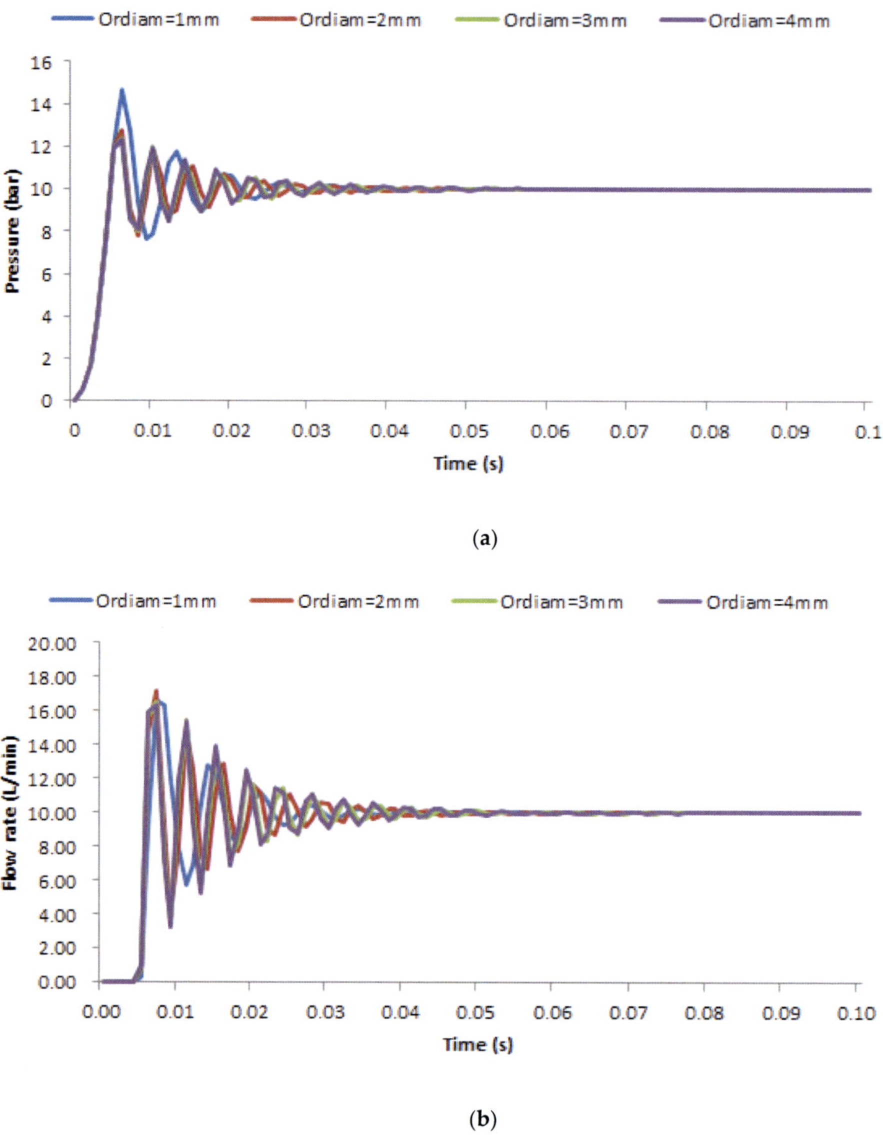

(a)

(b)

**Figure 10.** *Cont.*

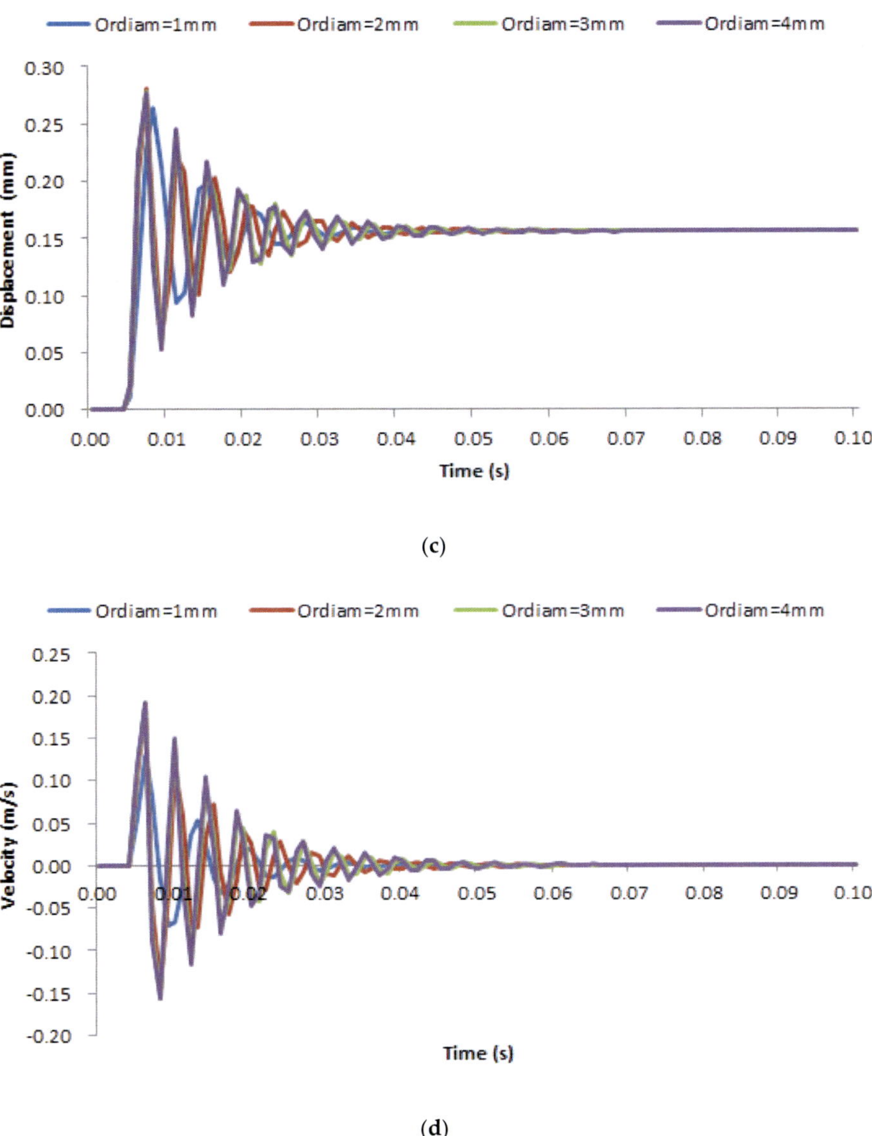

(c)

(d)

**Figure 10.** (**a**) The pressure response characteristics—orifice diameter: 1 mm~4 mm. (**b**) The flowrate response characteristics—orifice diameter: 1 mm~4 mm. (**c**) The displacement response characteristics of the valve element—orifice diameter: 1 mm~4 mm. (**d**) The velocity response characteristics of the valve element—orifice diameter: 1 mm~4 mm.

## 4.2. Influence of Viscosity Coefficient on Response Characteristics

The simulation parameters are shown in Table 2. The values of orifice diameter, valve element mass, spring stiffness, oil seal length, valve element diameter remain constant, and the viscosity coefficient is set as 40 N/(m/s), 45 N/(m/s), 50 N/(m/s), 55 N/(m/s) and 60 N/(m/s), respectively. The influence of the viscosity coefficient on the response characteristics of the SVTDARVWEO is studied.

**Table 2.** Simulation parameters—group 2.

| Orifice Diameter | Viscosity Coefficient | Valve Element Mass | Spring Stiffness | Oil Seal Length | Valve Element Diameter |
|---|---|---|---|---|---|
| (mm) | (N/(m/s)) | (kg) | (N/mm) | (mm) | (mm) |
| 2 | 40~60 | 0.01 | 1 | 0 | 10 |

The pressure response characteristics with the viscosity coefficient of 40 N/(m/s)~60 N/(m/s) are shown in Figure 11a. When the viscosity coefficient is 40 N/(m/s) and 45 N/(m/s), the pressure oscillates periodically, the amplitude of the oscillation does not decrease, and the oscillation frequency is relatively high (333 Hz). When the viscosity coefficient is 50 N/(m/s), 55 N/(m/s) and 60 N/(m/s), the pressure oscillates periodically, but the amplitude of the oscillation gradually decreases, and the oscillation frequency is relatively low (250 Hz). In addition, it can be seen that the larger the viscosity coefficient is, the smaller the amplitude of pressure oscillation for adjacent oscillation periods is. It can be predicted that when the viscosity coefficient is 60 N/(m/s), the pressure will reach the stable value earlier.

The flowrate response characteristics with the viscosity coefficient of 40 N/(m/s)~60 N/(m/s) are shown in Figure 11b. The final stable value of flowrate is 10 L/min, and the flowrate before 0.004 s is 0 L/min. When the viscosity coefficient is 40 N/(m/s) and 45 N/(m/s), the flowrate oscillates periodically, the oscillation amplitude does not decrease, about 10 L/min, and the oscillation frequency is relatively high (333 Hz). When the viscosity coefficient is 50 N/(m/s), 55 N/(m/s) and 60 N/(m/s), the flowrate oscillates periodically, but the amplitude of the oscillation gradually decreases, and the oscillation frequency is relatively low (250 Hz). In addition, it can be seen that the larger the viscosity coefficient is, the smaller the amplitude of flowrate oscillation for adjacent oscillation periods is. It can be predicted that when the viscosity coefficient is 60 N/(m/s), the flowrate will reach the stable value earlier.

The displacement response characteristics of the valve element with the viscosity coefficient of 40 N/(m/s)~60 N/(m/s) are shown in Figure 11c. The displacement before 0.004 s is 0 mm. When the viscosity coefficient is 40 N/(m/s) and 45 N/(m/s), the displacement oscillates periodically, the amplitude of the oscillation does not decrease (about 0.165 mm), and the oscillation frequency is relatively high (333 Hz). When the viscosity coefficient is 50 N/(m/s), 55 N/(m/s) and 60 N/(m/s), the displacement oscillates periodically, but the amplitude of the oscillation gradually decreases, and the oscillation frequency is relatively low (250 Hz). In addition, it can be seen that the larger the viscosity coefficient is, the smaller the amplitude of displacement oscillation for adjacent oscillation periods is. It can be predicted that when the viscosity coefficient is 60 N/(m/s), the displacement will reach the stable value earlier.

The velocity response characteristics of the valve element with the viscosity coefficient of 40 N/(m/s)~60 N/(m/s) are shown in Figure 11d. The velocity before 0.004 s is 0 m/s. When the viscosity coefficient is 40 N/(m/s) and 45 N/(m/s), the velocity oscillates periodically, the amplitude of the oscillation does not decrease (about 0.3 m/s), and the oscillation frequency is relatively high (333 Hz). When the viscosity coefficient is 50 N/(m/s), 55 N/(m/s) and 60 N/(m/s), the velocity oscillates periodically, but the amplitude of the oscillation gradually decreases, and the oscillation frequency is relatively low (250 Hz). In addition, it can be seen that the larger the viscosity coefficient is, the smaller the amplitude of velocity oscillation for adjacent oscillation periods is. It can be predicted that when the viscosity coefficient is 60 N/(m/s), the velocity will reach the stable value earlier.

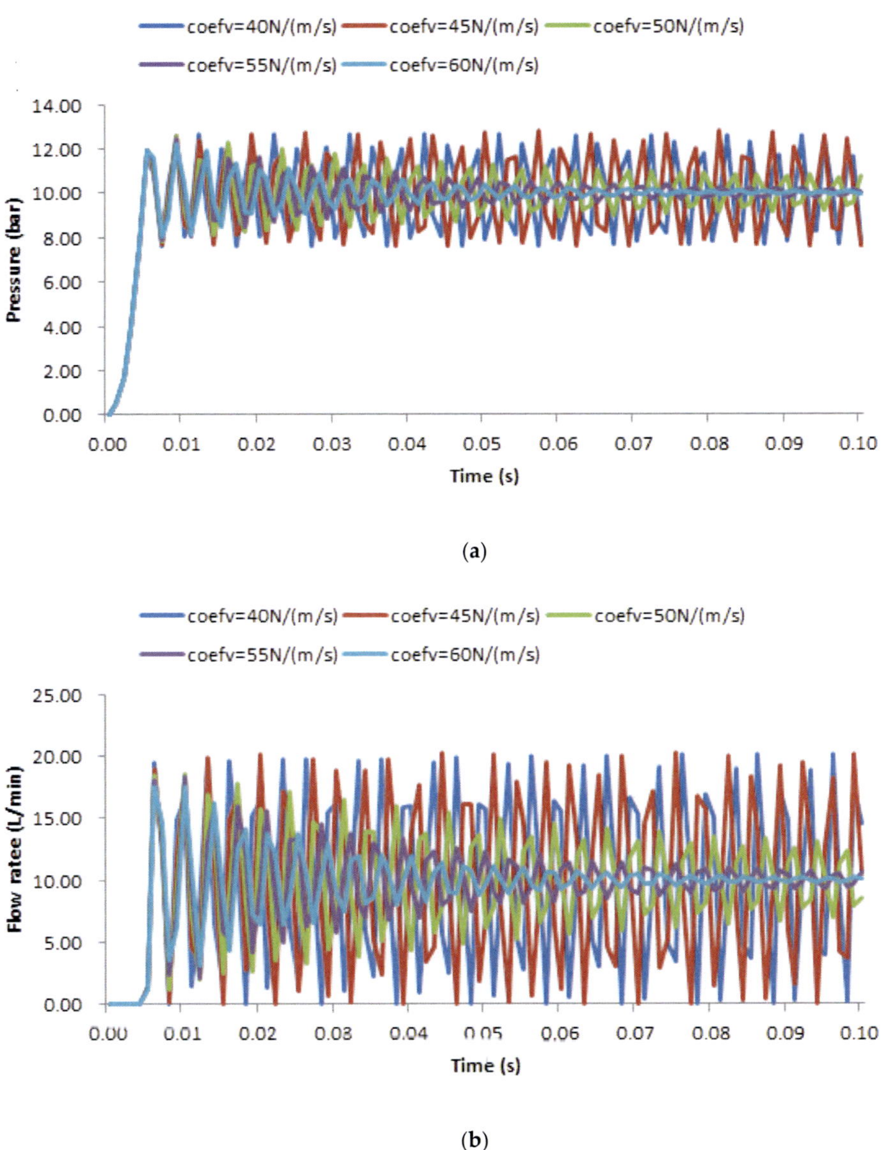



(**a**)



(**b**)

**Figure 11.** *Cont.*

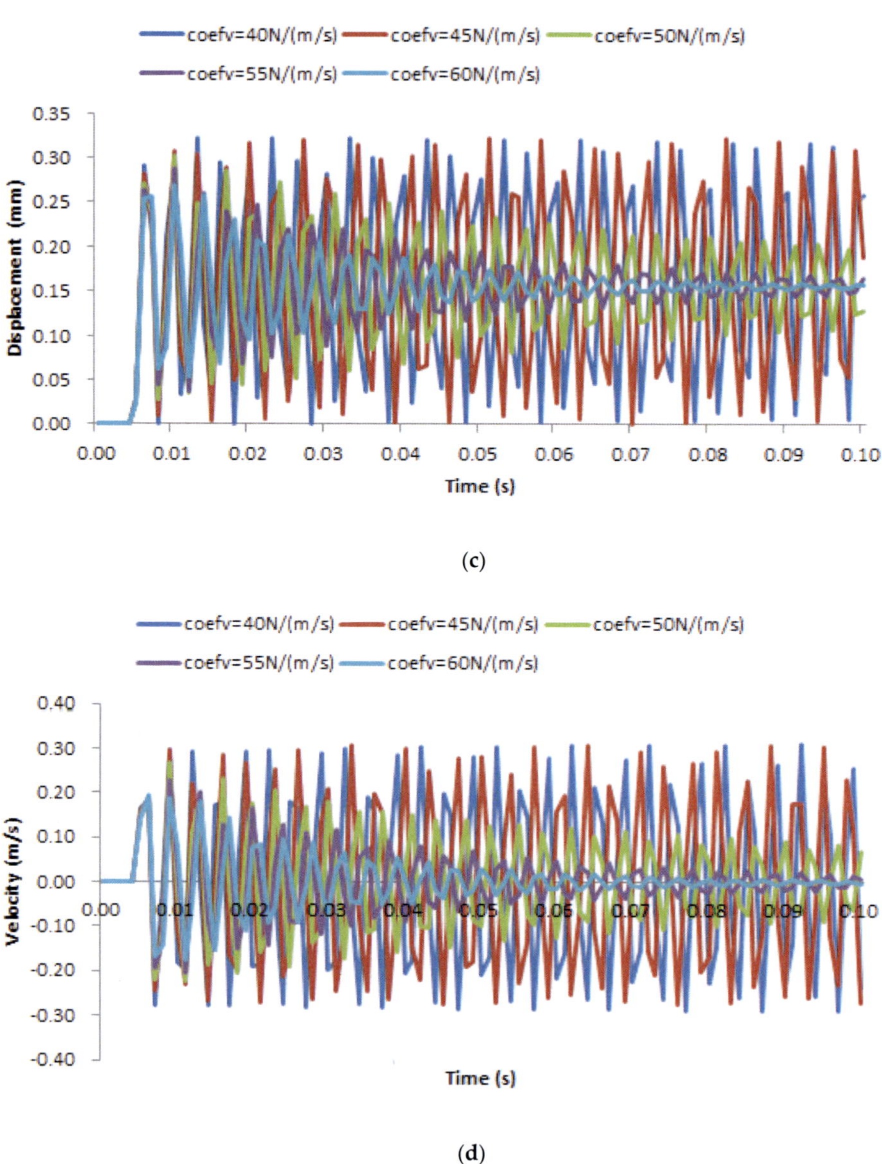

(c)

(d)

**Figure 11.** (**a**) The pressure response characteristics—viscosity coefficient: 40~60 N/(m/s). (**b**) The flowrate response characteristics—viscosity coefficient: 40~60 N/(m/s). (**c**) The displacement response characteristics of the valve element—viscosity coefficient: 40~60 N/(m/s). (**d**) The velocity response characteristics of the valve element—viscosity coefficient: 40~60 N/(m/s).

### 4.3. Influence of Valve Element Mass on Response Characteristics

The simulation parameters are shown in Table 3. The values of orifice diameter, viscosity coefficient, spring stiffness, oil seal length, and valve element diameter remain constant. The valve element mass are set to 0.01 kg, 0.015 kg, 0.025 kg, 0.025 kg and 0.03 kg, respectively. The influence of the valve element mass on the response characteristics of the SVTDARVWEO is studied.

**Table 3.** Simulation parameters—group 3.

| Orifice Diameter | Viscosity Coefficient | Valve Element Mass | Spring Stiffness | Oil Seal Length | Valve Element Diameter |
|---|---|---|---|---|---|
| (mm) | (N/(m/s)) | (kg) | (N/mm) | (mm) | (mm) |
| 2 | 100 | 0.01~0.03 | 1 | 0 | 10 |

The pressure response characteristics of a valve element with a mass of 0.01 kg~0.03 kg are shown in Figure 12a. When the valve element mass is 0.025 kg and 0.03 kg, the pressure oscillates periodically and the amplitude of the oscillation does not decrease (about 3.5 bar). When the valve element mass is 0.01 kg, 0.015 kg and 0.02 kg, the pressure oscillates periodically but the amplitude of oscillation decreases gradually. In addition, it can be seen that the pressure oscillation frequency is about 200 Hz. With the increase in valve element mass, the pressure oscillation frequency will decrease, but not to a large extent. The smaller the valve element mass is, the smaller the amplitude of pressure oscillation for adjacent oscillation periods is. It can be predicted that when the valve element mass is 0.01 kg, the pressure will reach the stable value earlier.

The flowrate response characteristics of a valve element with a mass of 0.01 kg~0.03 kg are shown in Figure 12b. The final stable value of flowrate is 10 L/min, and the flowrate before 0.004 s is 0 L/min. When the valve element mass is 0.025 kg and 0.03 kg, the flowrate rate oscillates periodically and the oscillation amplitude does not decrease (about 10 L/min). When the valve element mass is 0.01 kg, 0.015 kg and 0.02 kg, the flowrate rate oscillates periodically, but the amplitude of oscillation decreases gradually. In addition, it can be seen that the oscillation frequency is about 200 Hz. As the valve element mass increases, the flowrate oscillation frequency will decrease, but the reduction is not significant. The smaller the valve element mass is, the smaller the amplitude of flowrate oscillation for adjacent oscillation periods is. It can be predicted that when the valve element mass is 0.01 kg, the flowrate will reach the stable value earlier.

The displacement response characteristics of a valve element with a mass of 0.01 kg~0.03 kg are shown in Figure 12c. The displacement before 0.004 s is 0 mm. When the valve element mass is 0.025 kg and 0.03 kg, the displacement oscillates periodically and the oscillation amplitude does not decrease, which is about 0.170 mm. When the valve element mass is 0.01 kg, 0.015 kg and 0.02 kg, the displacement oscillates periodically but the amplitude of the oscillation decreases gradually. In addition, it can be seen that the oscillation frequency is about 200 Hz. With the increase in the valve element mass, the displacement oscillation frequency will decrease, but not to a large extent. The smaller the valve element mass is, the smaller the amplitude of displacement oscillation for adjacent oscillation periods is. It can be predicted that when the valve element mass is 0.01 kg, the displacement will reach the stable value earlier.

The velocity response characteristics of a valve element with a mass of 0.01 kg~0.03 kg are shown in Figure 12d. The velocity before 0.004 s is 0 m/s. When the valve element mass is 0.025 kg and 0.03 kg, the velocity oscillates periodically and the amplitude of oscillation does not decrease, which is about 0.25 m/s. When the valve element mass is 0.01 kg, 0.015 kg and 0.02 kg, the velocity oscillates periodically but the amplitude of the oscillation decreases gradually. In addition, it can be seen that the oscillation frequency is about 200 Hz. With the increase in the valve element mass, the velocity oscillation frequency will decrease, but to a small extent. The smaller the valve element mass is, the smaller the amplitude of velocity oscillation for adjacent oscillation periods is. It can be predicted that when the valve element mass is 0.01 kg, the velocity will reach the stable value earlier.

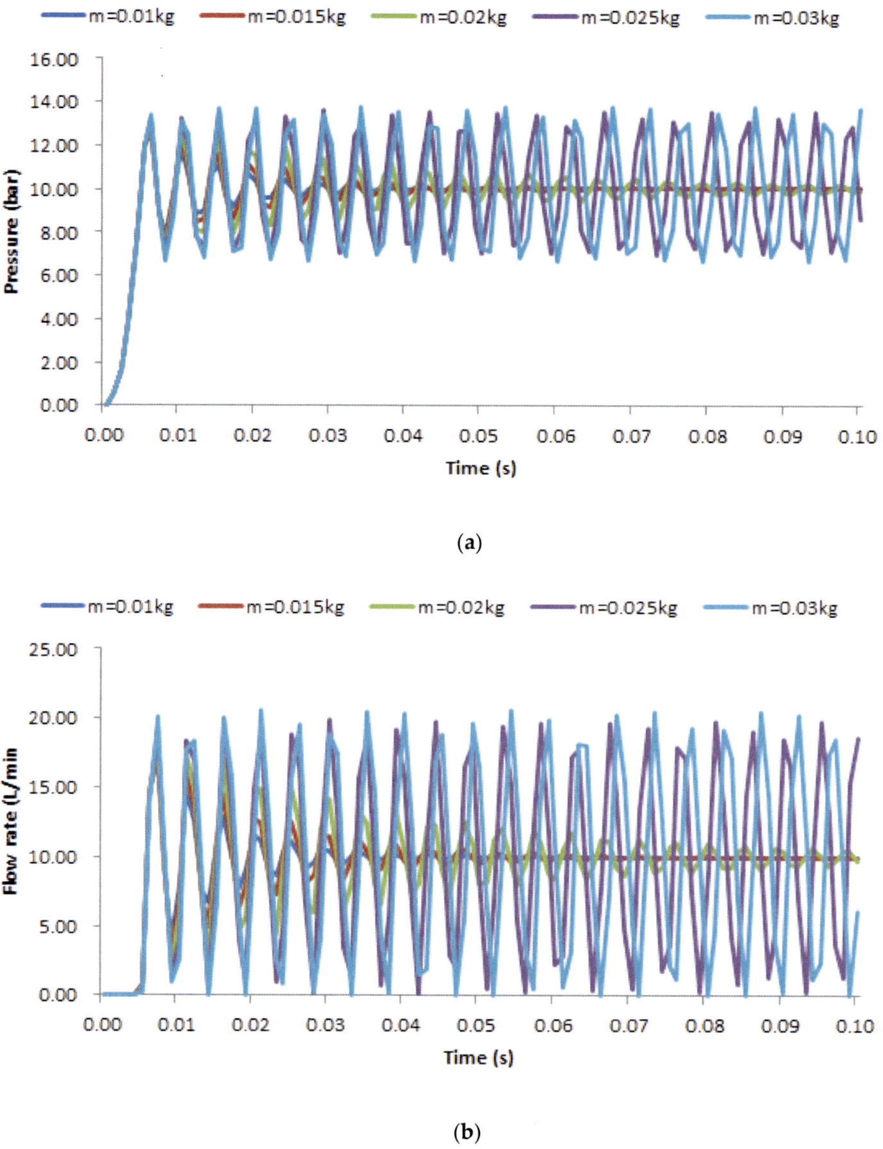



(**a**)



(**b**)

**Figure 12.** *Cont.*

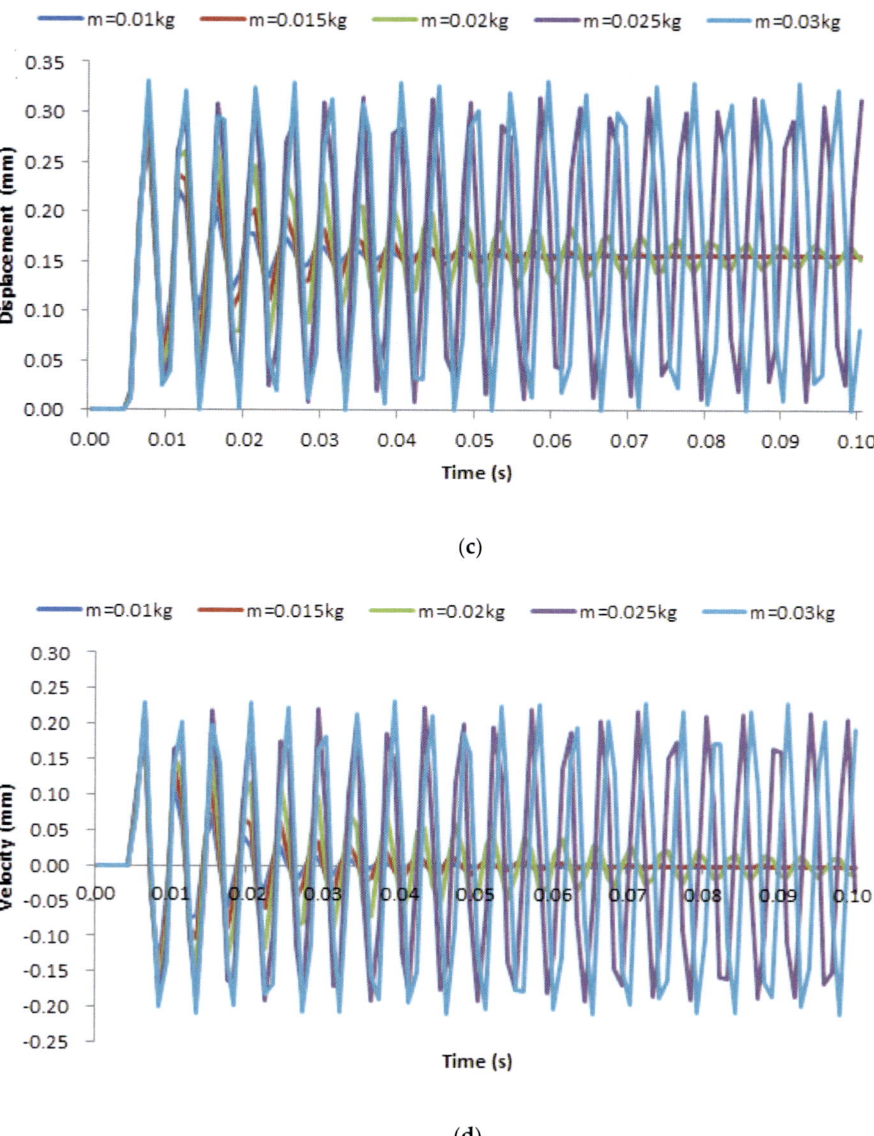



**Figure 12.** (**a**) The pressure response characteristics—valve element mass: 0.01~0.03 kg. (**b**) The flowrate response characteristics—valve element mass: 0.01~0.03 kg. (**c**) The displacement response characteristics—valve element mass: 0.01~0.03 kg. (**d**) The velocity response characteristics—valve element mass: 0.01~0.03 kg.

*4.4. Influence of Spring Stiffness on Response Characteristics*

The simulation parameters are shown in Table 4. The values of orifice diameter, viscosity coefficient, valve element mass, oil seal length, and valve element diameter remain constant, and the spring stiffness is set to 1 N/mm, 10 N/mm, 20 N/mm, 30 N/mm, 50 N/mm, respectively. The influence of spring stiffness on the response characteristics of the SVTDARVWEO is studied.

**Table 4.** Simulation parameters—group 4.

| Orifice Diameter | Viscosity Coefficient | Valve Element Mass | Spring Stiffness | Oil Seal Length | Valve Element Diameter |
|---|---|---|---|---|---|
| (mm) | (N/(m/s)) | (kg) | (N/mm) | (mm) | (mm) |
| 2 | 100 | 0.013 | 1,10,20,30,50 | 0 | 10 |

The pressure response characteristics with a spring stiffness of 1 N/mm~50 N/mm are shown in Figure 13a. The pressure corresponding to different values of spring stiffness oscillates and the oscillation frequency is about 250 Hz. After a certain oscillation, the pressure will eventually reach the stable value. The corresponding pressure stability values for a spring stiffness of 1 N/mm, 10 N/mm, 20 N/mm, 30 N/mm and 50 N/mm are 10.020 bar, 10.197 bar, 10.390 bar, 10.580 bar and 10.950 bar, respectively, and the corresponding maximum pressure overshoots are 2.712 bar, 2.613 bar, 2.504 bar, 2.396 bar and 2.184 bar, respectively. When the spring stiffness is 1 N/mm, the number of oscillations to reach the stable pressure value is the largest and the time required is the longest, but the opening pressure is the lowest. When the spring stiffness is 50 N/mm, the number of oscillations to reach the stable pressure value is the lowest and the time required is the shortest, but the opening pressure is the highest.

The flowrate response characteristics with a spring stiffness of 1 N/mm~50 N/mm are shown in Figure 13b. The flowrate before 0.004 s is 0 L/min, and 0.007 s reaches the maximum flowrate overshoot. The flowrate corresponding to different spring stiffness values oscillates and the oscillation frequency is about 250 Hz. After a certain oscillation, the flowrate will eventually reach the stable value of 10 L/min. The greater the spring stiffness is, the smaller the maximum flowrate overshoot is. The maximum flowrate overshoot corresponding to a spring stiffness of 1 N/mm, 10 N/mm, 20 N/mm, 30 N/mm and 50 N/mm is 7.171 L/min, 6.679 L/min, 6.168 L/min, 5.692 L/min and 4.84 L/min, respectively. When the spring stiffness is 1 N/mm, the number of oscillations is the maximum and the time to reach the stable flowrate value is the longest. When the spring stiffness is 50 N/mm, the number of oscillations to reach the stable flowrate value is the lowest and the time required is the shortest.

The displacement response characteristics with a spring stiffness of 1 N/mm~50 N/mm are shown in Figure 13c. The displacement before 0.004 s is 0 mm, and the maximum displacement overshoot is reached at 0.007 s. The displacement corresponding to different spring stiffness values oscillates and the oscillation frequency is about 250 Hz. After a certain oscillation, the displacement will eventually reach the stable value. The displacement stability values corresponding to a spring stiffness of 1 N/mm, 10 N/mm, 20 N/mm, 30 N/mm and 50 N/mm are 0.156 mm, 0.155 mm, 0.153 mm, 0.152 mm and 0.149 mm, respectively, and the corresponding maximum displacement overshoots are 0.125 mm, 0.114 mm, 0.103 mm, 0.093 mm and 0.076 mm, respectively. When the spring stiffness is 1 N/mm, the number of oscillations is the maximum and the time to reach the stable displacement value is the longest. When the spring stiffness is 50 N/mm, the number of oscillations to reach the stable displacement value is the lowest and the time required is the shortest.

The velocity response characteristics with spring stiffness of 1 N/mm~50 N/mm are shown in Figure 13d. The velocity before 0.004 s is 0 m/s, and the maximum velocity overshoot is reached at 0.006 s. The velocity corresponding to different spring stiffness values oscillates and the oscillation frequency is about 250 Hz. After a certain oscillation, the velocity will eventually reach the stable value of 0 m/s. The maximum velocity overshoots corresponding to spring stiffness of 1 N/mm, 10 N/mm, 20 N/mm, 30 N/mm and 50 N/mm are 0.187 m/s, 0.178 m/s, 0.169 m/s, 0.16 m/s and 0.144 m/s, respectively. When the spring stiffness is 1 N/mm, the number of oscillations is the maximum and the time to reach the stable velocity value is the longest. When the spring stiffness is 50 N/mm, the number of oscillations to reach the stable velocity value is the lowest and the time required is the shortest.

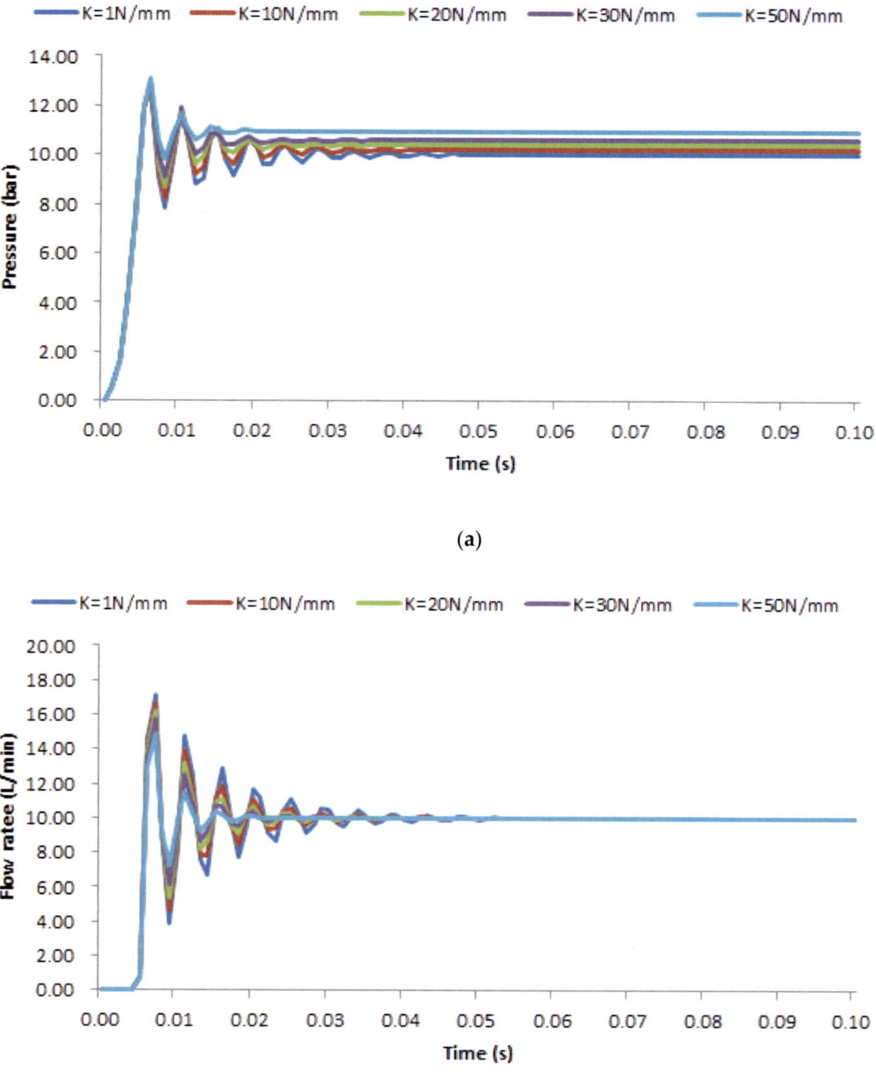



**Figure 13.** *Cont.*

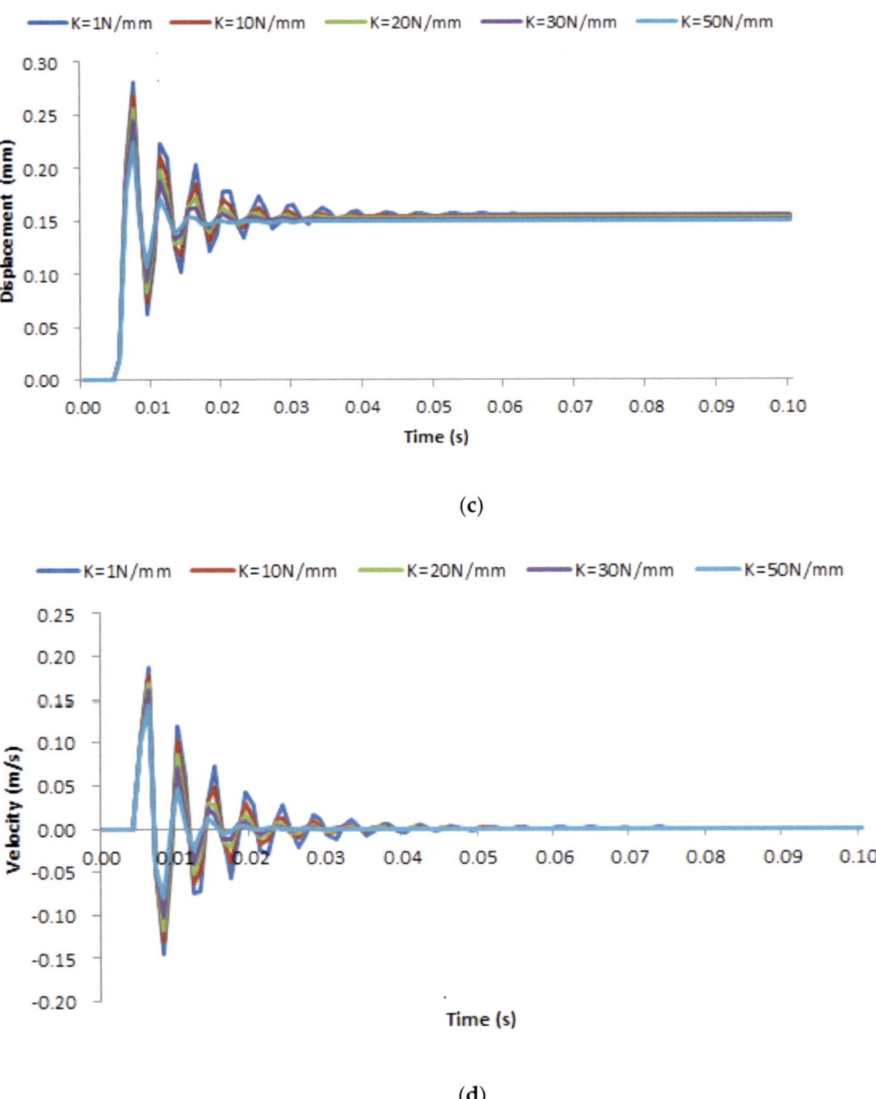



**Figure 13.** (**a**) The pressure response characteristics—spring stiffness: 1~50 N/mm. (**b**) The flowrate response characteristics—spring stiffness: 1~50 N/mm. (**c**) The displacement response characteristics—spring stiffness: 1~50 N/mm. (**d**) The velocity response characteristics—spring stiffness: 1~50 N/mm.

*4.5. Influence of Oil Sealing Length on Response Characteristics*

The simulation parameters are shown in Table 5. The values of orifice diameter, viscosity coefficient, valve element mass, spring stiffness, and valve element diameter remain constant. The oil sealing length is set as 0 mm, 0.5 mm, 1.0 mm, 1.5 mm and 2.0 mm respectively. The influence of oil sealing length on the response characteristics of the SVTDARVWEO is studied.

**Table 5.** Simulation parameters—group 5.

| Orifice Diameter | Viscosity Coefficient | Valve Element Mass | Spring Stiffness | Oil Seal Length | Valve Element Diameter |
|---|---|---|---|---|---|
| (mm) | (N/(m/s)) | (kg) | (N/mm) | (mm) | (mm) |
| 2 | 100 | 0.013 | 1 | 0~2 | 10 |

The pressure response characteristics of the SVTDARVWEO with an oil seal length of 0 mm~2 mm are shown in Figure 14a. The pressure corresponding to different oil sealing lengths oscillates with a frequency of about 200 Hz. After a certain oscillation, the pressure finally reaches the stable value of 10.02 bar. With the increase in oil seal length, the time to reach the maximum pressure overshoot increases correspondingly, and the maximum pressure overshoot also increases correspondingly. The maximum pressure overshoot corresponding to the sealing length of 0 mm, 0.5 mm, 1.0 mm, 1.5 mm and 2.0 mm is 2.712 bar, 7.625 bar, 10.214 bar, 13.707 bar and 13.683 bar, respectively.

The flowrate response characteristics of the SVTDARVWEO with an oil seal length of 0 mm~2 mm are shown in Figure 14b. The flowrate corresponding to different seal oil lengths oscillates with the oscillation frequency of about 200 Hz. After a certain oscillation, the flowrate finally reaches the stable value of 10 L/min. With the increase in oil sealing length, the time for overflow port to generate flowrate and the time for reaching the maximum flowrate overshoot increase accordingly. The maximum flowrate overshoot corresponding to the oil seal length of 0 mm, 0.5 mm, 1.0 mm, 1.5 mm and 2.0 mm is 7.171 L/min, 17.002 L/min, 24.806 L/min, 33.488 L/min and 25.167 L/min, respectively. It is worth noting that the maximum flowrate overshoot does not occur when the maximum oil seal length is 2 mm, but when the oil seal length is 1.5 mm.

The displacement response characteristics of the SVTDARVWEO with an oil seal length of 0 mm~2 mm are shown in Figure 14c. The displacement corresponding to different oil sealing lengths has oscillation, and the oscillation frequency is about 200 Hz. After a certain oscillation, the displacement finally reaches its own displacement stability value. With the increase in oil seal length, the displacement stability value and the time to reach the displacement stability value also increase correspondingly. The displacement stability values corresponding to the sealing length of 0 mm, 0.5 mm, 1.0 mm, 1.5 mm and 2.0 mm are 0.156 mm, 0.656 mm, 1.156 mm, 1.656 mm and 2.156 mm, respectively. With the increase in oil sealing length, the maximum overshoot of displacement increases correspondingly. The maximum overshoot of displacement corresponding to the oil sealing lengths of 0 mm, 0.5 mm, 1.0 mm, 1.5 mm and 2.0 mm is 0.125 mm, 0.313 mm, 0.362 mm, 0.449 mm and 0.516 mm, respectively.

The velocity response characteristics of the SVTDARVWEO with an oil seal length of 0 mm~2 mm are shown in Figure 14d. The velocity corresponding to oil seal lengths oscillates, with an oscillation frequency of about 200 Hz. With the increase in oil seal length, the time to reach the velocity stability value increases correspondingly, and the maximum overshoot of velocity increases correspondingly. The maximum overshoot of velocity corresponding to an oil seal length of 0 mm, 0.5 mm, 1.0 mm, 1.5 mm, and 2.0 mm is 0.187 m/s, 0.493 m/s, 0.602 m/s, 0.78 m/s, and 0.813 m/s, respectively.

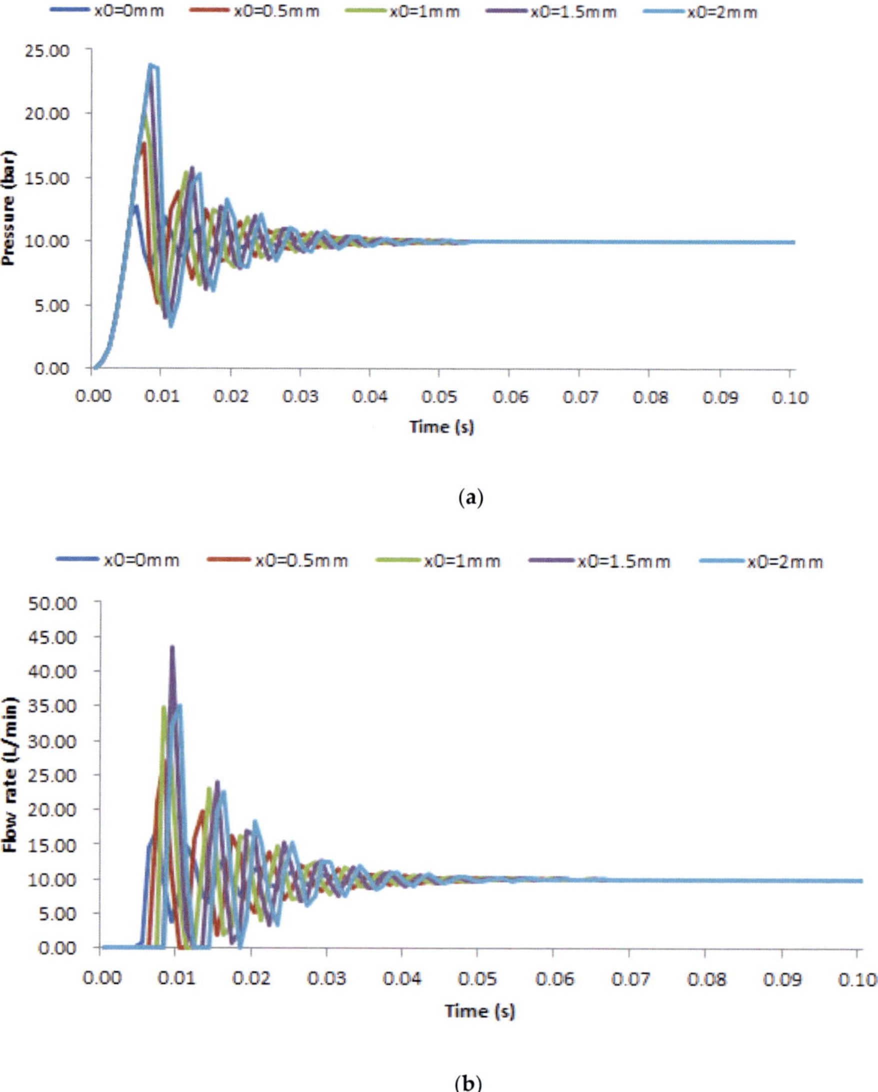

(**a**)

(**b**)

**Figure 14.** *Cont.*

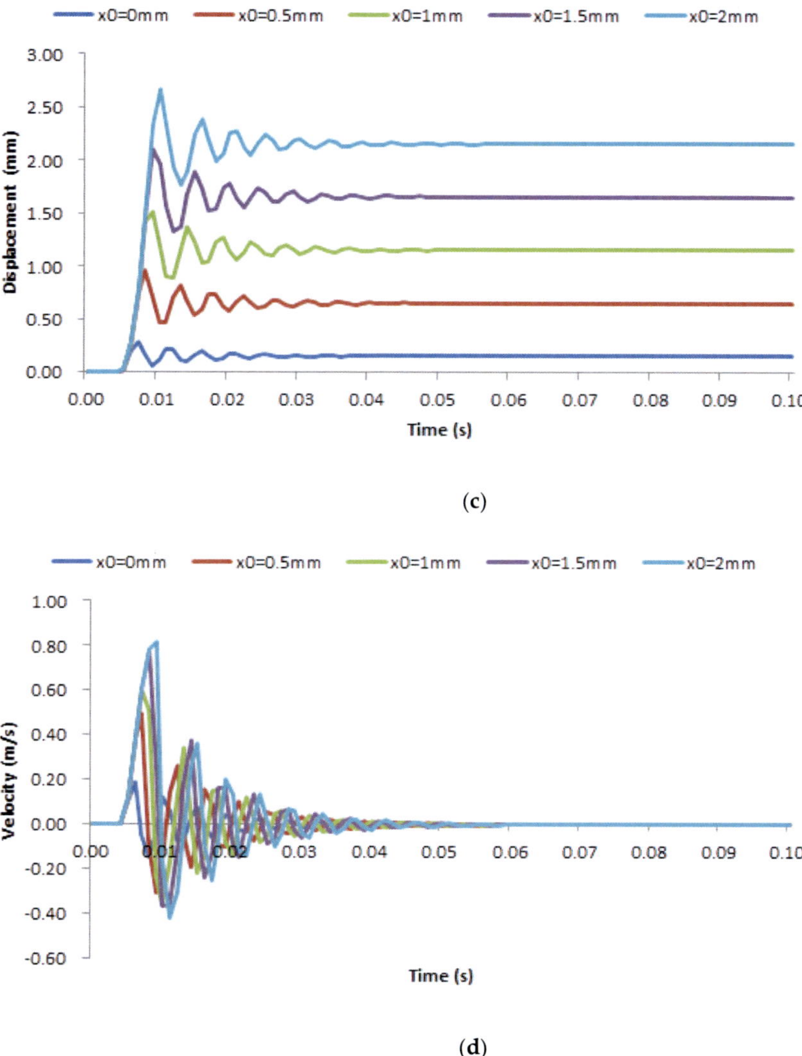

Figure 14. (a) The pressure response characteristics—oil sealing length: 0~2 mm. (b) The flowrate response characteristics—oil sealing length: 0~2 mm. (c) The displacement response characteristics—oil sealing length: 0~2 mm. (d) The velocity response characteristics—oil sealing length: 0~2 mm.

### 4.6. Influence of Valve Element Diameter on Response Characteristics

The simulation parameters are shown in Table 6. The values of orifice diameter, viscosity coefficient, valve element mass, spring stiffness, and oil seal length remain constant. The valve element diameters are set as 10 mm, 11 mm, 12 mm, 13 mm, 14 mm and 15 mm, respectively. The influence of valve element diameter on the response characteristics of the SVTDARVWEO is studied.

The pressure response characteristics of a valve element with a diameter of 10 mm~15 mm are shown in Figure 15a. The pressure corresponding to different valve element diameters oscillates, and the pressure finally reaches its stable value after a certain oscillation. With the increase in valve element diameter, the pressure stability value decreases. The pressure stability values corresponding to the valve element diameters of 10 mm, 11 mm, 12 mm,

13 mm, 14 mm and 15 mm are 10.02 bar, 10.015 bar, 10.011 bar, 10.009 bar, 10.007 bar and 10.006 bar, respectively. In addition, it is easy to see that with the increase in valve element diameter, the pressure oscillation frequency increases and the pressure oscillation amplitude decreases.

**Table 6.** Simulation parameters—group 6.

| Orifice Diameter | Viscosity Coefficient | Valve Element Mass | Spring Stiffness | Oil Seal Length | Valve Element Diameter |
|---|---|---|---|---|---|
| (mm) | (N/(m/s)) | (kg) | (N/mm) | (mm) | (mm) |
| 2 | 100 | 0.013 | 1 | 0 | 10~15 |

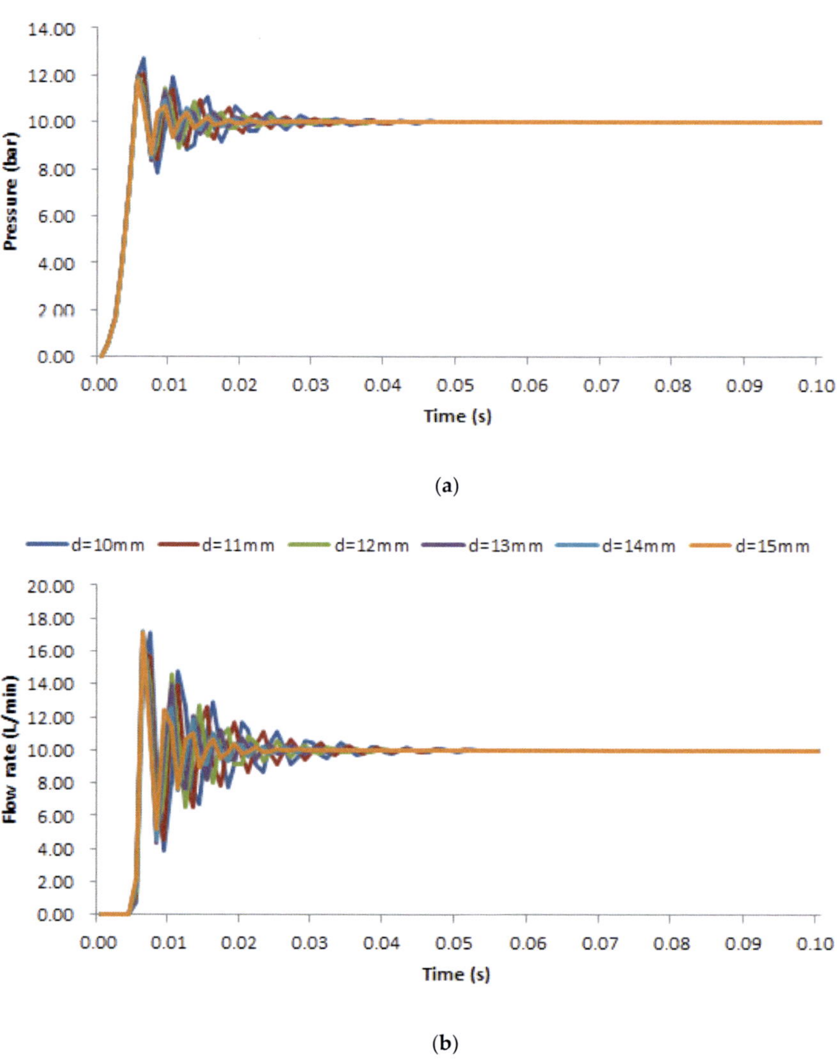


**Figure 15.** *Cont.*

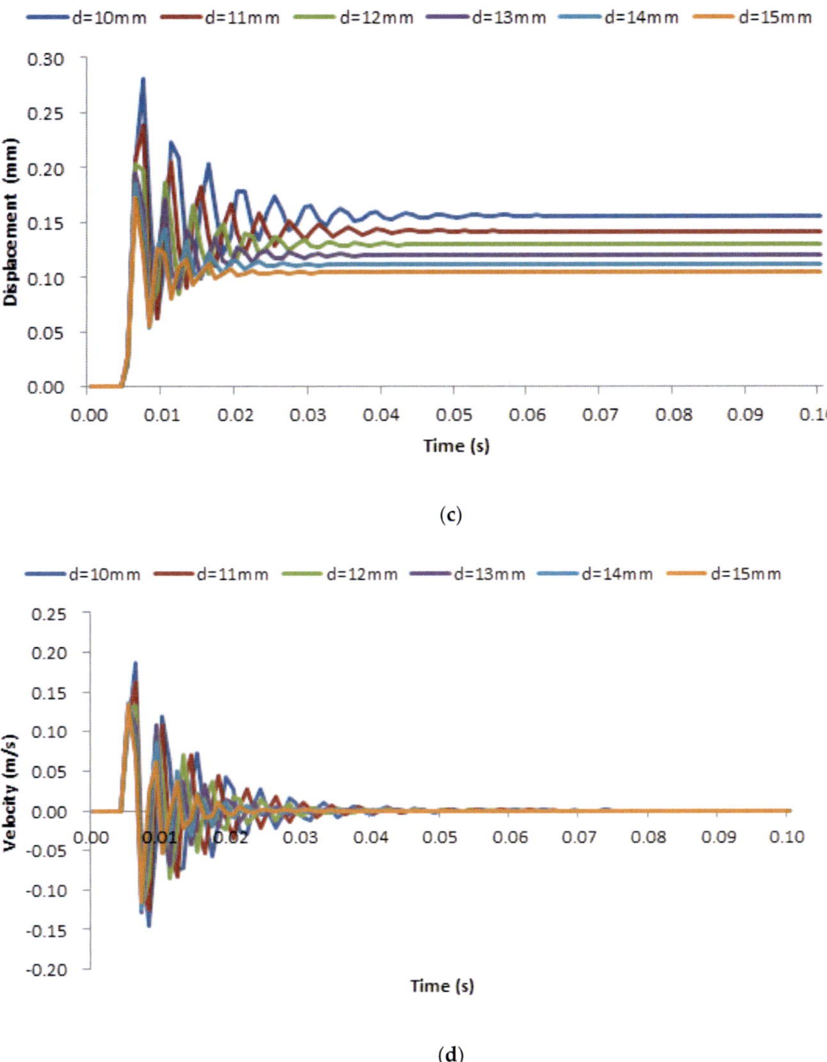



**Figure 15.** (**a**) The pressure response characteristics—valve element diameter: 10~15 mm. (**b**) The flowrate response characteristics—valve element diameter: 10~15 mm. (**c**) The displacement response characteristics—valve element diameter: 10~15 mm. (**d**) The velocity response characteristics—valve element diameter: 10~15 mm.

The flowrate response characteristics of a valve element with a diameter of 10 mm~15 mm are shown in Figure 15b. The flowrate before 0.004 s is 0 L/min, and the flowrate corresponding to different valve element diameters oscillates. After a certain oscillation, the flowrate finally reaches the stable value of 10 L/min. With the increase in valve element diameter, the flowrate oscillation frequency increases and the flowrate oscillation amplitude decreases.

The displacement response characteristics of a valve element with a diameter of 10 mm~15 mm are shown in Figure 15c. The displacement before 0.004 s is 0 mm, and the displacement corresponding to different valve element diameters oscillates. After a certain oscillation, the displacement finally reaches its respective stable value. As the valve element diameter increases, the displacement stability value decreases. The displacement stability values corresponding to the valve element diameters of 10 mm, 11 mm, 12 mm, 13 mm, 14 mm and 15 mm are 0.156 mm, 0.142 mm, 0.130 mm, 0.120 mm, 0.112 mm and 0.104 mm, respectively. With the increase in the valve element diameter, the displacement oscillation frequency increases while the amplitude decreases, and the time for the displacement to reach the stable value also decreases.

The velocity response characteristics of a valve element with a diameter of 10 mm~15 mm are shown in Figure 15d. The velocity before 0.004 s is 0 m/s. The velocity corresponding to different valve element diameters oscillates. After a certain oscillation, the velocity finally reaches the stable value of 0 m/s. With the increase in valve element diameter, the frequency of velocity oscillation increases while the amplitude of oscillation decreases, and the time for the velocity to reach the stable value also decreases.

## 5. Conclusions

Based on the working principle of the SVTDARVWEO, the simulation model of the SVTDARVWEO is established using AMESim. The influence of orifice diameter, viscosity coefficient, valve element mass, spring stiffness, oil seal length, and valve element diameter on the response characteristics of the SVTDARVWEO is analyzed, and the following conclusions are obtained:

(1) The smaller the orifice diameter is, the smaller the oscillation frequency, amplitude and maximum overshoot of pressure, flowrate, displacement, and velocity is. When the orifice diameter is 1 mm, the oscillation frequency is the lowest, about 167 Hz. When the orifice diameter is 2 mm, the oscillation frequency is about 200 Hz. When the orifice diameter is 3 mm and 4 mm, the oscillation frequency is close to about 250 Hz. The flowrate, displacement and velocity before 0.004 s are 0, and the pressure, flowrate, displacement and velocity will oscillate to about 0.06 s to reach individual stable values.

(2) When the viscosity coefficient is 40 N/(m/s) and 45 N/(m/s), the pressure, flowrate, displacement and velocity oscillate periodically, the amplitude of the oscillation does not decrease, and the oscillation frequency is about 333 Hz. When the viscosity coefficient is 50 N/(m/s), 55 N/(m/s) and 60 N/(m/s), the pressure, flowrate, displacement and velocity oscillate periodically, but the amplitude of the oscillation gradually decreases, and the oscillation frequency is about 250 Hz. The flowrate, displacement and velocity before 0.004 s are 0. When the viscosity coefficient is 60 N/(m/s), the pressure, flowrate, displacement and velocity will reach individual stable values earlier.

(3) When the valve element mass is 0.025 kg and 0.03 kg, the pressure, flowrate, displacement and velocity oscillate periodically and the amplitude of oscillation does not decrease. When the valve element mass is 0.01 kg, 0.015 kg and 0.02 kg, the pressure, flowrate, displacement and velocity oscillate periodically, but the amplitude of oscillation decreases gradually. The oscillation frequency is about 200 Hz. As the valve element mass increases, the displacement oscillation frequency will decrease, but to a small extent. The flowrate, displacement and velocity before 0.004 s are 0. When the valve element mass is 0.01 kg, the pressure, flowrate, displacement and velocity will reach individual stable values earlier.

(4) When the spring stiffness is 1 N/mm~50 N/mm, the pressure, flowrate, displacement and velocity corresponding to different spring stiffness values oscillate and the oscillation frequency is about 250 Hz. After the oscillation, the pressure, flowrate, displacement and velocity will eventually reach individual stable values. The greater the spring stiffness is, the smaller the maximum overshoot of pressure, flowrate, displacement and velocity is. The flowrate, displacement and velocity before 0.004 s are 0. When the spring stiffness is 1 N/mm, the pressure, flowrate, displacement and velocity return to individual stable values with the largest number of oscillations and the longest time required. When the spring stiffness is 50 N/mm, the number of

oscillations of pressure, flowrate, displacement and velocity to reach individual stable values is the lowest and the time required is the shortest.

(5) The pressure, flowrate, displacement and velocity corresponding to different oil seal lengths will oscillate, and the oscillation frequency is about 200 Hz. After certain oscillations, the pressure, flowrate, displacement and velocity finally reach individual stable values. With the increase in oil sealing length, the time to reach the maximum overshoot of pressure, the maximum overshoot of pressure, the time to generate flow at the overflow port, and the time to reach the maximum overshoot of flow all increase correspondingly. At the same time, the displacement stability value and the time to reach the displacement stability value increase correspondingly, with the time to reach the velocity stability value and the maximum overshoot of velocity increasing correspondingly.

(6) When the valve element diameter is 10 mm~15 mm, the pressure corresponding to different valve element diameters oscillate, and the pressure, flowrate, displacement and velocity will finally reach individual stable values after certain oscillations. The flowrate, displacement and velocity before 0.004 s are 0. With the increase in valve element diameter, the stable value of pressure decreases, the oscillation frequency of pressure, flowrate, displacement and velocity increases, the oscillation amplitude decreases, and the time for displacement and velocity to reach individual stable values also decreases.

**Author Contributions:** Conceptualization, H.L.; methodology, H.L.; validation, Q.Z.; formal analysis, Q.Z.; investigation, Q.Z.; data curation, H.L.; writing—original draft preparation, H.L.; writing—review and editing, Q.Z.; supervision, Q.Z.; funding acquisition, H.L. All authors have read and agreed to the published version of the manuscript.

**Funding:** This research received no external funding.

**Data Availability Statement:** The data presented in this study are available on request from the corresponding author.

**Acknowledgments:** This work is supported by the National Natural Science Foundation of China [grant number 51365008], and the Joint Foundation of Science and Technology Department of Guizhou Province [grant number Qiankehe LH Zi [2015]7658]. The authors are grateful to them for their support.

**Conflicts of Interest:** The authors declare no conflict of interest.

## References

1. Song, X.; Cui, L.; Cao, M.; Cao, W.; Park, Y.; Dempster, W.M. A CFD analysis of the dynamics of a direct-operated safety relief valve mounted on a pressure vessel. *Energy Convers. Manag.* **2014**, *81*, 407–419. [CrossRef]
2. Burhani, M.G.; Hős, C. An Experimental and Numerical Analysis on the Dynamical Behavior of a Safety Valve in the Case of Two-phase Non-flashing Flow. *Period. Polytech.-Chem. Eng.* **2021**, *65*, 251–260. [CrossRef]
3. Zong, C.Y.; Zheng, F.; Chen, D.; Dempster, W.; Song, X. Computational Fluid Dynamics Analysis of the Flow Force Exerted on the Disk of a Direct-Operated Pressure Safety Valve in Energy System. *J. Press. Vessel. Technol.-Trans. Asme* **2020**, *142*, 011702. [CrossRef]
4. Kadar, F.; Hos, C.; Stepan, G. Delayed oscillator model of pressure relief valves with outlet piping. *J. Sound Vib.* **2022**, *534*, 117016. [CrossRef]
5. Fu, C.Y.; Yang, L.; Si, G.; Li, Y. Design and Mechanical Performance Analysis of Relief Valve with Permanent Magnet Spring. In Proceedings of the IEEE/CSAA International Conference on Aircraft Utility Systems (AUS), Beijing, China, 10–12 October 2016.
6. Burhani, M.G.; Hős, C. Estimating the opening time of a direct spring operated pressure relief valve in the case of multiphase flow of fixed mass fraction in the absence of piping. *J. Loss Prev. Process Ind.* **2020**, *66*, 104169. [CrossRef]
7. Liao, M.L.; Zheng, Y.; Gao, Z.; Song, W. Fluid-structure coupling modelling and parameter optimization of a direct-acting relief valve for underwater application. *Nonlinear Dyn.* **2021**, *105*, 2935–2958. [CrossRef]
8. Zahariea, D. Functional Diagram for Modeling the Electromagnetic Ball Valve with Cylindrical Seat. In Proceedings of the Innovative Manufacturing Engineering Conference (IManE), Chisinau, Moldova, 29–30 May 2014.
9. Song, W.; Yang, C.; Zhang, X.; Li, Y. Mathematical Modelling and Dynamic Analysis of a Direct-Acting Relief Valve Based on Fluid-Structure Coupling Analysis. *Shock. Vib.* **2021**, *2021*, 5581684. [CrossRef]

10. Erdődi, I.; Hős, C. Prediction of quarter-wave instability in direct spring operated pressure relief valves with upstream piping by means of CFD and reduced order modelling. *J. Fluids Struct.* **2017**, *73*, 37–52. [CrossRef]
11. Liu, Y.S.; Ren, X.; Wu, D.; Li, D.; Li, X. Simulation and analysis of a seawater hydraulic relief valve in deep-sea environment. *Ocean. Eng.* **2016**, *125*, 182–190.
12. Wu, S.; Li, C.; Deng, Y. Stability Analysis of a Direct-Operated Seawater Hydraulic Relief Valve under Deep Sea. *Math. Probl. Eng.* **2017**, *2017*, 5676317. [CrossRef]
13. Syrkin, V.V.; Balakin, P.D.; Treyer, V.A. Study on hydraulic direct-acting relief valve. *J. Phys. Conf. Ser.* **2017**, *858*, 12035. [CrossRef]
14. Raeder, T.; Tenenev, V.; Chernova, A.; Koroleva, M. Multilevel Simulation of Direct Operated Safety Valve. In Proceedings of the Ivannikov Ispras Open Conference (ISPRAS), Moscow, Russia, 22–23 November 2018.
15. Sohn, S. A Numerical Analysis of Direct Spring Loaded Type—Steam Safety Valve Using CFD Simulation. In Proceedings of the 22nd International Conference Nuclear Energy for New Europe (NENE), Bled, Slovenia, 9–12 September 2013.
16. Dempster, W.; Taggart, S.; Doyle, C. Limitations in the Use of Pressure Scaling for Safety Relief Valve Design. In Proceedings of the ASME Pressure Vessels and Piping Conference (PVP 2018), Prague, Czech Republic, 15–20 July 2018.
17. Zong, C.Y.; Zheng, F.J.; Song, X.G. Understanding Lift Force Discontinuity of Pressure Safety Valve. In Proceedings of the ASME Pressure Vessels and Piping Conference, San Antonio, TX, USA, 14–19 July 2019.
18. Suzuki, K.; Urata, E. Development of a Direct Pressure-Sensing Water Hydraulic Relief Valve. *Int. J. Fluid Power* **2008**, *9*, 5–13. [CrossRef]
19. Bazsó, C.; Hős, C.J. An experimental study on the stability of a direct spring loaded poppet relief valve. *J. Fluids Struct.* **2013**, *42*, 456–465. [CrossRef]
20. Hős, C.J.; Champneys, A.; Paul, K.; McNeely, M. Dynamic behavior of direct spring loaded pressure relief valves in gas service: Model development, measurements and instability mechanisms. *J. Loss Prev. Process Ind.* **2014**, *31*, 70–81. [CrossRef]
21. Hős, C.; Bazsó, C.; Champneys, A. Model reduction of a direct spring-loaded pressure relief valve with upstream pipe. *IMA J. Appl. Math.* **2015**, *80*, 1009–1024. [CrossRef]
22. Hős, C.J.; Champneys, A.; Paul, K.; McNeely, M. Dynamic behaviour of direct spring loaded pressure relief valves in gas service: II reduced order modelling. *J. Loss Prev. Process Ind.* **2015**, *36*, 1–12. [CrossRef]
23. Kim, H.; Baek, D.; Kim, S. Optimum design of direct spring loaded pressure relief valve in water distribution system using multi-objective genetic algorithm. *J. Korean Soc. Water Wastewater* **2018**, *32*, 115–122. [CrossRef]
24. Kim, H.; Kim, S.; Kim, Y.; Kim, J. Optimization of Operation Parameters for Direct Spring Loaded Pressure Relief Valve in a Pipeline System. *J. Press. Vessel. Technol.* **2018**, *140*, 051603. [CrossRef]
25. Lei, J.; Tao, J.; Liu, C.; Wu, Y. Flow model and dynamic characteristics of a direct spring loaded poppet relief valve. *Proc. Inst. Mech. Eng. Part C J. Mech. Eng. Sci.* **2017**, *232*, 1657–1664. [CrossRef]
26. Kim, S.D.; Kim, J.H. Conceptual Design on a Direct-Operated Relief Valve with High-Precision and Large-Flow. *Proceedings of the Academic Conference of the Active Pressure Construction Machinery Society.* 2021, 52. Available online: https://www.dbpia.co.kr/Journal/articleDetail?nodeId=NODE10755358 (accessed on 1 December 2022).
27. Dimitrov, S.; Krstev, D. Modelling and Simulation of the Transient Performance of a Direct Operated Pressure Relief Valve. *Hidraulica* **2022**, *3*, 75–81.

*processes*


MDPI

*Article*

# Supervision and Control System of the Operational Variables of a Cluster in a High-Pressure Gas Injection Plant

Cristhian Ronceros [1,2,*], José Medina [1], Juan Vásquez [3], Pedro León [1], José Fernández [1] and Estefany Urday [1]

[1] Faculty of Engineering, Private University San Juan Bautista, Ica 11004, Peru
[2] School of Engineering and Applied Sciences, Eastern University, Maturin 6201, Monagas, Venezuela
[3] Mirai Innovations SAC, Ica 11000, Peru
* Correspondence: croncerosm@gmail.com

**Abstract:** The objective of this research was to develop a technological architecture proposal that allows for the supervision and control of the operational parameters of gas injection (flow, temperature, and pressure) in a cluster of a high-pressure gas injection plant. The proposal provides a supervision and control system for the HPGIP I high-pressure gas injection plant that includes instrumentation equipment (transmitters and actuators), a remote terminal unit (RTU) as a control device, and the creation of a control logic as the basis for the development of the SCADA GALBA®, through which the operational variables involved in the process of the gas injection plant can be visualized and controlled, allowing the automatic regulation of the flow of gas that enters the deposits. Automatization of the process allows for the elimination of the average error differential that increases from 2 to 5% when the control valve is opened manually. Currently, the MUC-67 and MUC-68 wells that make up cluster 5 require a control valve opening of 20% and 5%, respectively, and this percentage is directly affected by the average valve opening error when performed manually. In addition, there is a savings of around 40 min in the response time by the operators for the adjustment of the opening or closing parameters of the control valve manually. The proposal allows for the different control actions on the variables or parameters of gas injection present in the clump to be carried out from a control room.

**Keywords:** supervision; control; remote terminal; cluster; high pressure; gas injection

**Citation:** Ronceros, C.; Medina, J.; Vásquez, J.; León, P.; Fernández, J.; Urday, E. Supervision and Control System of the Operational Variables of a Cluster in a High-Pressure Gas Injection Plant. *Processes* 2023, 11, 698. https://doi.org/10.3390/pr11030698

Academic Editor: Sergey Y. Yurish

Received: 28 January 2023
Revised: 13 February 2023
Accepted: 23 February 2023
Published: 26 February 2023



## 1. Introduction

Industrial automation consists of governing the activity and evolution of processes without the continuous intervention of a human operator [1,2]. The main advantages of automation are: replacing human operators in hazardous environments, monotonous tasks, activities involving great physical wear or beyond human capabilities of size, strength, endurance or speed, and economic improvement for companies or society as a whole [3,4].

In this context, the action of replacing a manual process with an automated one in an industry involves relevant progress in two main aspects, production and management, provided that there is an ideal architecture and it is adapted to the processes that allow the elements of the system to work in harmony, fulfilling their activities [5,6]. In this sense, in the oil and gas industry, the different production, control, and maintenance activities depend exclusively on information technologies, and only with them is it possible to achieve an efficient level of operation [7,8]. In this regard [9–11], point out that low-cost automation promotes cost-effective reference architectures and new development approaches to increase the flexibility and efficiency of production operations in the oil industry.

Hydrocarbon industries have been faced with the need to adapt and empower themselves with new technologies to optimize their production processes. For this reason, these industries have focused on keeping crude oil fields operational through techniques that allow them to continue exploiting them and, therefore, guarantee quality standards in the products and services they offer, taking full advantage of the benefits of these advances and being competitive with other companies [12–14]

Control systems are present in the exploration, production, refining, transportation, and distribution processes of the oil industry, providing security and reliability, minimizing work accidents and the use of labor [15–17]. Automation is an essential element in Petroleos de Venezuela (PDVSA), which uses it mainly for the supervision and control of the operational variables of its processes and facilities. PDVSA for the planning of the exploitation process considers the productivity profiles of each oil and gas production field, as well as the requirement of fluid injection for the maintenance of pressure or artificial lift of the well that becomes necessary when the latter loses pressure causing fluctuations in the flow or natural course of extraction.

Extracting the maximum amount of oil from the reservoir is one of the challenges of this industry. The oil industry makes great efforts to develop new technologies to increase the production of residual oil reserves in an economical and environmentally sustainable way [18]. In this sense, in the Eastern Production Executive Directorate (EPED), formed by the operational fields Orocual, Furrial, Jusepin, Carito, and Pirital, they operate through the maintenance of reservoir pressure by means of secondary recovery with gas injection. In the case of the Carito and Pirital fields, as a consequence of having the highest production of crude oil and gas, it is necessary to inject more gas to maintain the pressure and replacement factor, and in this way, guarantee the optimum functionality of these fields.

Among the plants that make up the EPED, there is HPGIP I, which is a secondary recovery plant, which is responsible for the operation of the 1, 2, 3, 4, 5, and 6 blocks in order to ensure the operational conditions contained in the structural design of the plants for their optimal performance, such as the static pressure, suction temperature, and flow of condensed gases according to the handling capacity of the equipment.

The secondary recovery system does not have adequate supervision of the processes [19]; in this sense, the control of the pressures injected into each well is carried out manually, which causes pressure drops and therefore a reduction in the flow of crude oil, generating a reduction in the planned production of crude oil. In HPGIP I, only blocks 1 and 2 operate within its facilities, so the rest are located in a remote geographical area, with a small population and difficulty in access, as is the case of cluster 5. A cluster can be defined as a drilling configuration of wells that are very close on the surface and which, thanks to directional drilling, manage to diversify in the subsoil; it serves to save space, time, costs, and environmental impacts in the drilling process [20–23]

Cluster 5 is made up of injection wells MUC 67 and MUC 68 and does not have a gas injection measurement, nor an automatic control that allows for regulating its flow since it only has a manual action valve. The percentage of opening of the gas flow control valve depends on the conditions of the well (temperature and pressure). These conditions change and this demonstrates that a greater or lesser percentage of opening of the control valve is required. Cluster 5, which is 10 km from the PIGAP 1 plant, makes it difficult to open the valve in a timely manner, as personnel have to travel to the site to open it (currently, the valve is opened manually). Additionally, inadequate injection of gas into the well (more or less than required) not only impacts the oil production of the day but also damages the reservoir for future production.

Depending on the conditions of the well, an additional continuous volume of gas equivalent to the gas/liquid ratio would have to be injected, which would represent the optimal injection rate in the hypothetical case where the injection point coincides with the midpoint of the interval drilled, since the analysis refers to the node corresponding to the flow of the bottom hole pressure of the well. An injection below or above the optimum would increase the flow below the well, reducing its production capacity [24]

In this sense, the objective of the present investigation is to develop an automated system for the supervision and control of operative variables of cluster 5 of the gas injection plant, with the purpose of achieving the control and supervision of the operative variables, with emphasis on the flow that is injected into the reservoirs, with the possibility of regulating the gas injection, thus guaranteeing control in two different states, local and remote, in the valves by configuring them and other control devices. The field complies with PDVSA regulations and rules.

## 2. Methods

This research is based on field research with a descriptive level and feasible project modality. The feasible project consists of the research, elaboration, and development of a viable operational model to solve problems, requirements, or needs of organizations or social groups [25]

The development of the proposal was based on the methodology "Management Guidelines for Capital Investment Projects (MGCIP)" developed by PDVSA. The (MGCIP) contains practical guidelines for the execution of a project in a standardized and orderly manner, so that no detail and/or important step is overlooked, and thus guarantees, with a high degree of confidence, that the projects will be successful and meet the corporation's requirements [26]. The (MGCIP) is structured in five phases for the development and operation of a project within the Venezuelan oil industry, which are visualization, conceptualization, definition, implementation, and operation.

The project only covered the first four phases, since it is an automation proposal so the final activities of the implementation and operation phase are outside the scope of the project. Each of the phases is described below:

Phase I: Visualization

In this phase, operational and technological requirements were identified and analyzed. Among the activities carried out in this phase are the following:

- Description of the production process.
- Description of cluster 5.

Phase II: Conceptualization

In this phase, an evaluation of the technologies to be implemented to monitor the operational parameters and variables of the HPGIP I gas injection cluster 5 was carried out. The activities carried out in this phase were as follows:

- Evaluation and selection of equipment and components.
- Elaboration of the proposal for the system architecture.

Phase III: Definition

The objective of this phase was to develop detailed engineering to carry out the execution of the project. The activities carried out in this phase were as follows:

- Elaboration of the system flow diagram.
- Carrying out the wiring diagram.
- Development of the list of system signals.
- Elaboration of the displays for the interface with the SCADA system.

Phase IV: Implementation

In this phase, we started with the integration of the devices, as well as with the construction of the deployments to later link them with the Guardian of the Alba (GALBA) SCADA. GALBA SCADA is a field data acquisition, supervision, and control system [27,28] made by PDVSA under free software [29]. The activities carried out were:

- Configuring the equipment and devices.
- Connection and testing of the equipment to the data site.
- Construction of the control logic.
- Development of the deployment for the GALBA SCADA.

## 3. Results and Discussion

This section can be divided into subheadings. It should provide a concise and precise description of the experimental results, their interpretation, and the experimental conclusions that can be drawn.

### 3.1. Phase I: Visualize

#### 3.1.1. Description of the Production Process

Secondary recovery methods

Secondary recovery methods lie in the injection of fluids into a reservoir in order to maintain pressure; these substances are injected by certain wells known as injectors, with the displacement of a part of the crude towards the other wells being achieved, and they are producers [20,21]. Generally, the secondary recovery process is used when the natural flow that exerts pressure on the reservoir does not have enough energy to push the oil, and the main flows that are injected for secondary recovery are water and gas.

Secondary recovery process by gas injection

This is a process where gas is injected into a reservoir with the purpose of increasing oil recovery from the reservoir [30,31], as well as controlling oil production and gas conservation; the main objective is to maintain pressure within the reservoir. Gas injection is an impermeable process, which means that it does not mix with the crude oil unless the gas is enriched with hydrocarbons or injected at high pressure [32].

Description of the HPGIP I plant

This plant is a gas injection giant that raises 156 pressures from 84.37 $kg/cm^2$ to 632.76 $kg/cm^2$. The capacity of compression is 157 injections of 1000 million cubic feet per day. It is composed of five modular trains and turbo compressor units, each one mainly made up of a gas turbine model MS-5002-C, NouvoPignone brand of 50,938.33 kw of power, and three centrifugal compressors in series, model BCL 406/B, BCL 305/C, and BCL305/B. Each train requires a volume of six MMPCFD of gas to feed the turbine which provides the required power to the compressors.

In addition, HPGIP has a multiple discharge network that allows the distribution of gas to six clusters, which are made up of seventeen injector wells that help maintain the production of the Punta de Mata Division. On the other hand, it also has auxiliary systems that collaborate for the correct functioning of the facilities and operations to be carried out, which are indicated below:

- Injection network system.
- Relief and ventilation system.
- Fire protection system.
- Instrument air system.
- Air system.
- Oily water drainage system.
- Power supply and electrical distribution system.
- Starting gas system.
- Fuel gas system.
- Process safety system, gas detectors, $CO_2$ unit systems, control room, and UV/IR (alarms).
- Mist injection cooling system (osmosis).

Gas distribution process and injection networks

Gas injection produces enough energy to pressurize the reservoir of low-production wells located in the vicinity of the injection wells. It is important to mention that the injection wells result from the conversion of production wells in a state of abandonment and/or closed due to low production; for this purpose, work must be carried out on the heads and facilities that are part of the structure of the well [18]. Figure 1 shows the diagram of the gas injection process in a cluster.

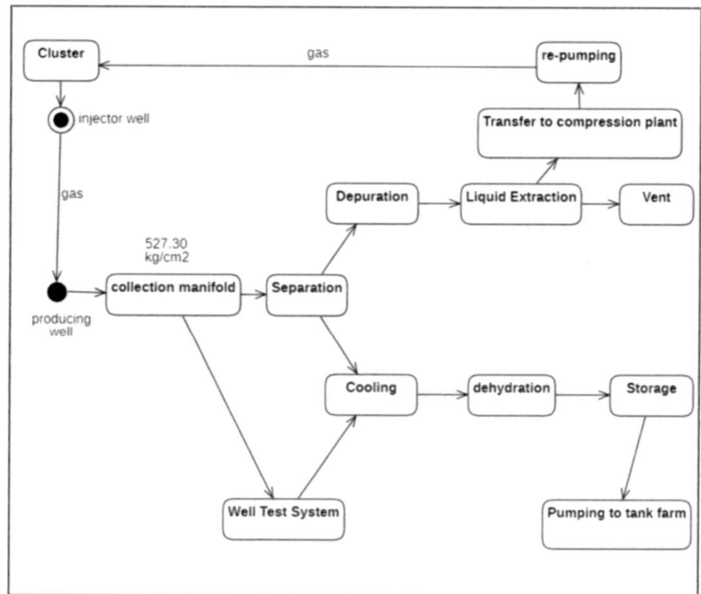


**Figure 1.** Diagram of the gas injection process in a cluster.

The gas from the turbocompressor units enters the discharge manifold with an approximate pressure of 527.30 kg/cm$^2$, for the average injection of 1000 MMPCFD through the injection networks, which are distributed in six clusters that supply gas to seventeen gas injection wells. Next, Table 1 indicates how the wells are distributed with respect to the cluster, with their respective valve opening percentage per well (% choke).

**Table 1.** Well distribution.

| Clusters | % Choke | Well |
| --- | --- | --- |
| Cluster #1 | 50% | MUC-53 |
| | 50% | MUC-59 |
| Cluster #2 | 25% | MUC-54 |
| | 25% | MUC-57 |
| | 25% | MUC-55 |
| | 25% | MUC 60 |
| Cluster #3 | 40% | MUC-56 |
| | 50% | MUC-66 |
| | 40% | MUC-58 |
| | 50% | MUC-65 |
| Cluster #4 | 40% | MUC-91 |
| | 40% | MUC-83 |
| | 40% | MUC-69 |
| Cluster #5 | 20% | MUC-67 |
| | 5% | MUC-68 |
| Cluster #6 | 40% | CRC-24 |
| | 40% | CRC-25 |

Each of the injection wells has an associated injection manifold through which the well can be opened or closed manually, and the inflow through the injection well can also be regulated. In the gas injection network, it must be determined how much energy is required by a reservoir in order to estimate a flow, which is completed by reservoir engineers using an energy recovery factor.

Therefore, if each reservoir requires an amount of energy and there is a failure in one of the turbocompressor units, the person in charge of safeguarding the clusters has the power to close a well and thus the gas will be distributed to the other clusters that require a high gas injection; on the other hand, they have the ability to regulate the flow that enters a reservoir according to the needs of the reservoir at a given time.

### 3.1.2. Description of Cluster 5

In spite of belonging to the distribution of wells of the high-pressure gas injection plant (HPGIP I), cluster 5 is located outside these facilities; it is located in a remote area to the west of the Muscar Operational Complex (MOC), on the Punta de Mata-Maturin National Highway, in the municipality of Ezequiel Zamora. This cluster has two associated injector wells, MUC-67 and MUC 68.

As for the operating architecture, it has a control or telemetry house, which includes antennas and communication radios, which are not operational due to theft of the power lines and transformers. It also has a ball valve used by the well operator to regulate the flow of gas or to control the opening and closing of the well. In the fully open position, the valve is approximately 75% of the pipe size in size. There is a Shaffer safety valve which is intended to carry out the emergency shutdown of the well when the fluid exceeds the preset limit, avoiding the explosion of the system in the event of excess pressure; however, this device is currently out of operation, so the emergency shutdown is performed through the ball valve; a manometer, which measures the gas pressure in the pipeline, which is independent of electrical power because it is purely mechanical.

The orifice plate and the manual flow control valve complete the system; the latter regulates the gas injection and, despite being a rigid and transcendental process, it is worn and corroded due to the action of the natural elements and little intervention and maintenance of the system.

### 3.2. Phase II: Conceptualize

In this phase, the various technologies available on the market are studied in relation to the monitoring and control processes of operational variables.

### 3.2.1. Identify the Equipment and Devices Available on the Market

The control devices, taking into account the brands, models, and characteristics most used by PDVSA to guarantee the homologation of the platform according to the regulations of the Executive Directorate of Automation, Information Technology, and Telecommunications of PDVSA, the governing body of guaranteeing technological services to the Venezuelan oil industry, are the following:

- Controller 1: Linux-based PAC. Inside this device is a 1.33 GHz Intel Atom Z520 Series CPU. This model comes with VGA, USB, RS-232/485, Ethernet modules, and E-8K serial I/O slots designed for a high-performance profile. User programs can be saved to an external storage device, such as a CF card or USB mass storage device. Users can develop applications using the GNU C language. Users can achieve the redundancy function.
- Controller 2: Ideal for oil, gas, pipeline, electrical, and industrial combustion applications. They provide support for applications that require wide temperature ranges and low power consumption, and remote applications powered by solar cells or wind power. Among its main features are the following. An advanced 32-bit processor with integrated real-time multitasking, an operating system (RTOS), eight analog inputs (two with a point-to-point HART interface), two analog outputs, eight discrete inputs, four discrete outputs, and three pulse inputs, an integrated Ethernet, two serial ports (RS232/RS485), and one HART port; a multi-drop interface; the Native protocols include Modbus RTU, Modbus ASCII, and Modbus TCP; 6 M Flash, 1 M RAM, and 32 K Ferroelectric RAM for long-term storage of configured parameters.

- Controller 3: It has an SD card (secure digital) that provides non-volatile storage in which the user program is permanently stored. Key features include: An integrated 1 gigabit (Gb) Ethernet port which provides high-speed motion and I/O control; a screen which makes it easy to diagnose and solve problems; and increased security and new capabilities. The energy storage module avoids the need for a battery. The controller's digitally signed firmware provides an additional layer of security. It provides role-based access control to routines and add-on instructions.

Multivariable transmitter technology: Among the most widely used equipment in PDVSA's operating platform are the following:

- Multivariable Transmitter 1: Capable of three process measurements and flow calculations, all integrated into a single device, it allows for changes in process conditions and provides an accurate reading every time. It is compact in size and easily fits into any system due to its features, with 10 years of stability and features such as flow reading up to 0.65% on turndown, differential pressure reading, and a 200:1 differential pressure ratio. It has the HART communication protocol mounted on a 4–20 mA loop.
- Multivariable Transmitter 2: This makes optimal use of the unique features of the DPharp sensor to give greater insight into processes and its features include $\pm0.04\%$ differential pressure accuracy, $\pm0.1\%$ static pressure accuracy, $\pm0.9\,^{\circ}\text{F}$ external temperature accuracy, 4500 psi MWP, and a HART-type communication protocol. Likewise it has two 4–20 mA loop analog inputs to read two variables, be it differential pressure, static pressure, external temperature, or flow signal.
- Multivariable Transmitter 3: This offers accurate measurement of instantaneous flow, cumulative flow, process pressure, differential pressure, and temperature. As a fully digital transmitter, the product offers a built-in choice of Modbus protocol over serial and/or Ethernet communications. It has an optional analog output: 0–20 mA, serial type protocol: Modbus RTU. It is powered with a voltage of 9 to 30 VDC, and has a differential and absolute pressure of $\pm0.05\%$, temperature accuracy of $\pm0.15\,^{\circ}\text{C}$ ($\pm0.27\,^{\circ}\text{F}$), and Stability of approximately 5 years.

Actuators: The actuators most used by PDVSA are the following:

- Actuator 1: This provides a small to medium thrust output for bracket-mounted applications. It includes operation of built-in butterfly valves and turbine valves, louvers, dampers, and other similar equipment. Its type of control is all/nothing and adjustable position control. Linear and spring return form the actuation.
- Actuator 2: This responds to signals from electronic processes or remote manual adjustments. It allows manual adjustments when there is no power. It is built on the basis of stepper motor technology, allowing it to travel precisely to any position without overshooting. It has a standard 4–20 mA input and output signal.
- Actuator 3: A compact actuator with a quick on–off maneuver and 0–90° rotation to automate the ball, butterfly, and taper plug valves, among others. Among its characteristics are the following: a multivolt power supply of 100–240 V AC (125–320 V DC), 24 V AC/DC, or 400 V tri 50–60 Hz; regulation stops; a digital action command; an on–off action command; manual control in case of emergency; a visual position indicator; 4–20 mA-type proportional control; and maximum rotation under command from 180° to 270°.

Relay 1: They are switching elements between the control system and the actuators and/or sensors, allowing one to connect, separate, control, amplify, or multiply the current or voltage. Among its main features are the type of input voltage: AC/DC and UC; The relay rated current: Max. 10A; the solid-state relay rated current: Max. 10A; Contacts: Max. two normally open changeover contacts; connection technology: a screw, a push-in, and a spring; Bridging: A1, A2, 11, and 14; and it is expandable with logic and time functions in combination with PLC logic, RTU, and other control devices.

### 3.2.2. Identify the Equipment and Devices Available on the Market

After identifying the technologies, we proceeded the evaluation and selection of the most relevant ones for the control process, which are the ones that will make up the control system in cluster 5.

Criteria for the selection of the control device: We proceeded with the identification of the criteria to be considered for the selection of the control device that will later be used for the proposal of the system architecture; among them are:

- Programming: Coding that requires fewer instruction numbers and fewer lines of code. Encoding that contains a structure similar to that of the equipment installed in other facilities.
- Communication: The ability of the control device to achieve communication with field devices.
- Robustness: The ability to adapt and withstand critical environments.
- Technical support and maintenance: The ability to make contact with the manufacturer for the facilitation and availability of materials and user manuals.
- Cost: Economical and accessible.

Criteria for the selection of the field devices

For the selection of the transmitter and actuator, we proceed with the identification of the most relevant criteria; due to the fact that both will be part of the proposal for the system architecture of cluster 5 and are field devices, the same criteria are considered for both among the criteria as follows:

- Security.
- Communication protocols.
- Adaptability.
- Maintainability.
- Cost.
- Margin of error.

Technical evaluation for the selection of devices and equipment for the system architecture. The steps to make a comparative matrix are as follows:

- Each criterion was associated with a corresponding letter, i.e., row A represents criterion A, row B corresponds to criterion B, and so on.
- Subsequently, a weight was assigned to each criterion, selecting the degree of importance of the criterion (1: none; 2: low; 3: medium; 4: high), which is placed in the respective box of the comparison value.
- After emptying the matrix of all the comparisons, the points of each criterion were added, and each of them was weighted, with each having a weight of 1 to 10, where 1 is the minimum and 10 is the maximum.
- In the lower part of the matrix, each of the options studied was placed with a range of numbers from 1 to 5 to indicate whether the technology is appropriate to meet the evaluation criteria, as shown in the Figure 2.

Evaluation of the control device.

According to the comparison made in the table above, it is shown that controller 2 is the one that best meets the proposed requirements, with it obtaining a total of 89.94 points over controllers 1 and 3 with weightings of 78.14 and 87.23, respectively. For the criterion related to programming, controller 2 obtained a score of 24.54 while controller 1 obtained 116.36 points and Controller 3 was the winner of this first criterion with 32.72 points. Because the programming environment of controller 3 is more flexible than the other devices, it is noted that all of the devices comply with the standardized languages in the IEC 1131-3 standard (block diagram, ladder logic, structured text, and instruction list).

**PROJECT:** Supervision and control system of the operational variables of a cluster in a high pressure gas injection plant. A case study of cluster 5 from PIGAP 1

**EVALUATION CRITERIA**

| | |
|---|---|
| A | Programming |
| B | Communication |
| C | Sturdiness |
| D | Technical support and maintenance |
| E | Costs |

**DEGREE OF IMPORTANCE**

| 4. High | 3. Medium | 2. Low | 1. None |
|---|---|---|---|

| OPTIONS MATRIX | | A | B | C | D | E | TOTAL |
|---|---|---|---|---|---|---|---|
| | Weighting results | 9 | 3 | 2 | 2 | 10 | |
| | Weight from 1 to 10 | 8.18 | 2.73 | 1.81 | 1.81 | 10 | |
| 1 | Controller 1 | 2 | 4 | 3 | 3 | 4 | 78.14 |
| | | 16.36 | 10.92 | 5.43 | 5.43 | 40 | |
| 2 | Controller 2 | 3 | 4 | 5 | 3 | 4 | 89.94 |
| | | 24.54 | 10.92 | 9.05 | 5.43 | 40 | |
| 3 | Controller 3 | 4 | 5 | 3 | 3 | 3 | 87.23 |
| | | 32.72 | 13.65 | 5.43 | 5.43 | 30 | |

**Figure 2.** Evaluation matrix for control devices. Yellow Highlight: This is to highlight the highest score achieved in the technical comparison.

Evaluation and selection of the multivariable transmitter.

The results shown in the Figure 3 indicate that multivariable transmitter 1 is the most suitable for the architecture of the control system with a final weight of 107.00 points, while multivariable transmitter 2 and multivariable transmitter 3 had scores of 99.00 and 92.00 points, respectively. Multivariable transmitter 1 stands out in certain criteria such as its adaptability and margin of error, elements that are critical due to the moment of selecting a transmitter; it must have the characteristic of easy adaptation to the system and also provide a reading of the same operational variables, which are necessary to take the corresponding actions regarding the control to be exercised in the system.

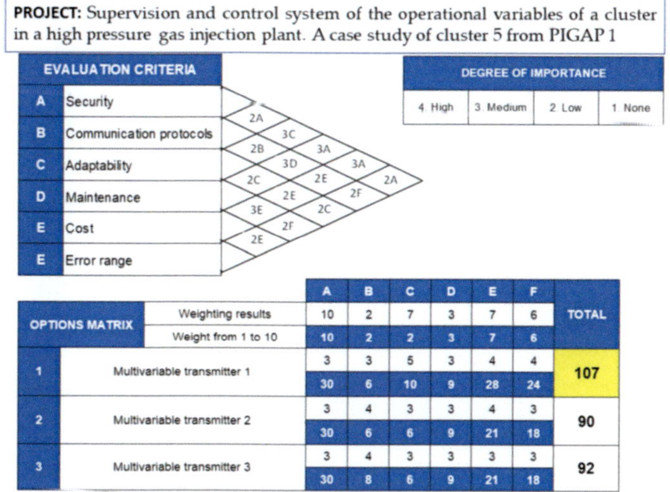


| OPTIONS MATRIX | | A | B | C | D | E | F | TOTAL |
|---|---|---|---|---|---|---|---|---|
| | Weighting results | 10 | 2 | 7 | 3 | 7 | 6 | |
| | Weight from 1 to 10 | 10 | 2 | 2 | 3 | 7 | 6 | |
| 1 | Multivariable transmitter 1 | 3 | 3 | 5 | 3 | 4 | 4 | 107 |
| | | 30 | 6 | 10 | 9 | 28 | 24 | |
| 2 | Multivariable transmitter 2 | 3 | 4 | 3 | 3 | 4 | 3 | 90 |
| | | 30 | 6 | 6 | 9 | 21 | 18 | |
| 3 | Multivariable transmitter 3 | 3 | 4 | 3 | 3 | 3 | 3 | 92 |
| | | 30 | 8 | 6 | 9 | 21 | 18 | |

**Figure 3.** Evaluation matrix for the transmitters. Yellow Highlight: This is to highlight the highest score achieved in the technical comparison.

Evaluation and Selection Actuators

The results obtained show (See Figure 4) that actuator 2 is the most ideal and it is indicated for the architecture of the control system, with it reaching a score of 111.00 points, while actuator 1 and actuator 3 reached scores of 103.00 and 97.00 points, respectively. Actuator 2 stands out in the criteria such as its adaptability and costs, with costs being one of the most important factors when selecting equipment that meets the needs of the system in the same way that it contemplates easy adaptability to the process, remembering that the actuator is the equipment that allows the passage of the gas flow towards the injection well.

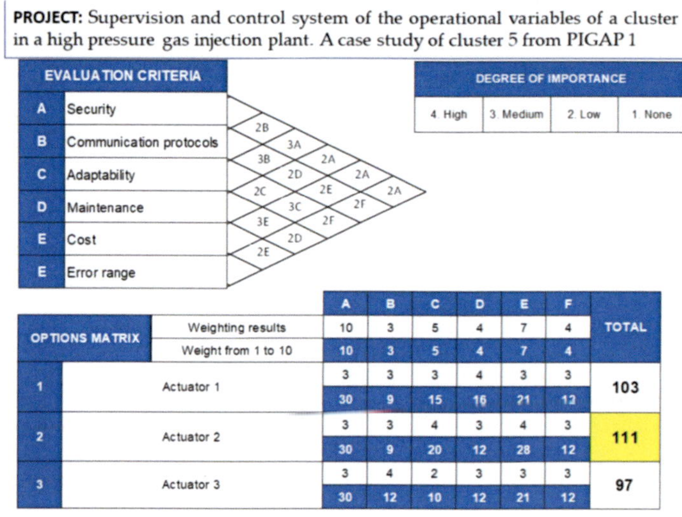


**Figure 4.** Evaluation matrix for the actuators. Yellow Highlight: This is to highlight the highest score achieved in the technical comparison.

Selection of devices and equipment

According to the results obtained in the evaluation of the different devices, it is proposed for the following devices and equipment to be part of the system architecture:

- Controller 2.
- Multivariate transmitter 1.
- Actuator 2.

It is considered for use relay 1 to be used since it is one of the most used by the company, PDVSA, and due to its characteristics described above, it supplies the minimum needs for the system architecture, allowing the change in the required logical states, which are local and remote.

### 3.2.3. Elaboration of the Proposal for the System Architecture

Taking into account the current situation and the needs of the process to be addressed to carry out the supervision of the operational variables and to regulate the flow of gas to be injected into the injector wells of cluster 5, a structure was defined to guarantee the integration of the elements of the system, ensuring the functionality and adaptability to the process. This architecture contemplates the instrumentation to the control system that allows the supervision of the variables. In this order of ideas, the elements of the system architecture are shown in Figure 5:

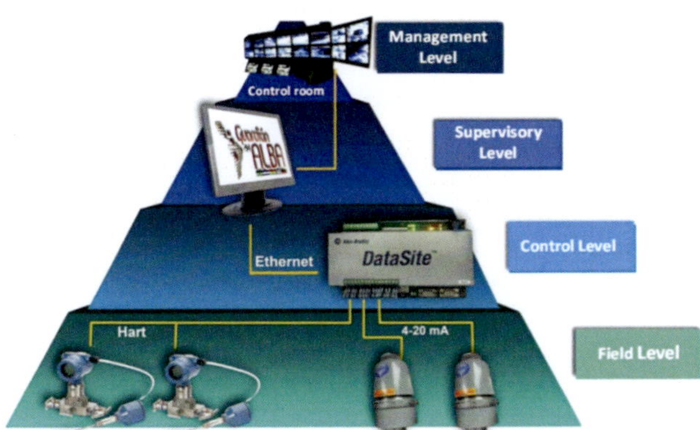


**Figure 5.** System architecture.

As shown in Figure 5, the proposed architecture contemplates a supervision system, specifically the GALBA SCADA, developed in Venezuela to carry out PDVSA's supervision processes, through equipment with free software located in the control room, and in the particular case of this study, the software will be managed from AIT's offices to carry out the corresponding operation and evaluation tests.

In turn, the control level consists of a remote terminal unit (RTU), specifically a data site [32], which facilitates control actions in harsh and extreme industrial environments, such as oil, gas, pipeline, and electrical applications [33,34]. The data site controller can be programmed using Isagraf Workbench software, which allows for the development of control logic according to the programming standards of IE34,35C 61131-3. This software allows for the development of the control logic that will be subsequently transferred or downloaded to the RTU, with it performing the necessary compilations to verify its correct operation.

Finally, for the field level in which the actions or commands are executed, for the control and display of variables ordered by the RTU, it is necessary to establish the instrumentation equipment or field devices, which interact directly with the variable to be measured or regulated. For this reason, the devices to be used in the proposed architecture are listed below:

- A multivariate transmitter for MUC 67.
- A multivariate transmitter for MUC 68.
- An actuator for the flow control valve for MUC 67.
- An actuator for the flow control valve for MUC 68.

It is necessary to mention that the installation of both the transmitters and actuators for the flow control valves is carried out with their corresponding instrumentation wiring, and in this process, the devices are tested in the laboratory.

### 3.3. Phase III: Define

The objective of this phase was the development of detailed engineering to carry out the execution of a project; these details are, among others, the flow diagram of the system, which graphically represents the stages of the process; the connection diagram, which allows for understanding the connection of the different elements and the signal flow of the process; the list of system signals, which identifies and defines how the data transmission is carried out and is taken as a starting point for the elaboration of the GALBA SCADA® database; and finally, the elaboration of the displays that will be used as part of the SCADA system. The following is a description of each of the activities involved in this phase:

### 3.3.1. Developing the System Flow Diagram

This diagram shows the relationship of the different elements of the system and, in addition, helps to understand the sequence of steps from the start or access to the GALBA SCADA® system, user identification that carries out the supervision action of the variables perceived in the field (pressure, temperature, and gas flow), and action that consecutively grants the administrator or operator the option to control the values displayed through the total or partial opening/closing of the control valves, with the purpose of guaranteeing the stability, safety, and productivity of the process. Figure 6 shows a flow chart of the proposed system.

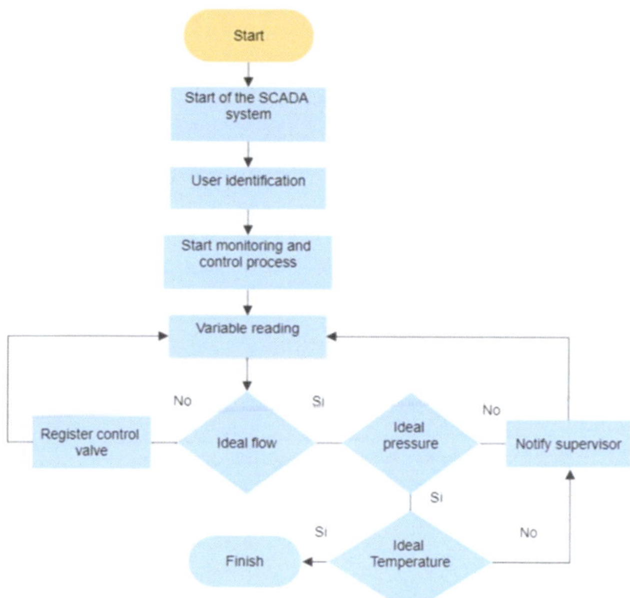


**Figure 6.** System flow chart.

### 3.3.2. Making the Connection Diagram

In order to understand the connection of the different elements and the signal flow of the process (analog, digital, and HART), the connection diagram of the system was developed, which shows the wired connections of the data site remote terminal unit with the different actuators and multivariables implemented in the proposed architecture, for which it was necessary to identify the specific location of each of the communication elements of controller 2.

### 3.3.3. Develop a List of System Signals

Once the proposed system architecture was defined, which represents a generalized vision of the process, it was necessary to identify and define how data transmission is carried out in the system. For this purpose, a list of signals must be elaborated, which represents the basis on which the integration of the proposed structure is achieved.

In this sense, the starting point for the development of the database is the list of signals representative of the process of supervision and control of the operational variables of cluster 5. It is necessary that the signals sent and received by the remote terminal unit used (the data site) are defined in the GALBA SCADA® database, in order to be integrated into the displays, considering the TAGS or tags in addition to the registers or addresses of the devices.

### 3.3.4. Development of the Displays for Interfacing with the SCADA System

For the development of this activity, the 3D design software Blender® was used, since it is the one used by PDVSA for the elaboration of displays and graphic representations of devices and equipment, to achieve a visual integration that allows for contemplation of a general scheme of the process. Graphic representations of each of the devices and equipment that will be part of the representative scheme of cluster 5 were made.

General display for graphical visualization of the devices and equipment.

This section presents an overview and graphic representation of cluster 5, which has two gas injector wells and a gas injection manifold that is fed through HPGIP I.

Figure 7 shows the operational display of cluster 5 where the upper part of the collector is focused, the section where the control action is executed. The operational screen allows for detailed observation of the distribution of the instruments in the field; finally, this screen will be converted into a PNG format image that represents the visual part of the SCADA® system.

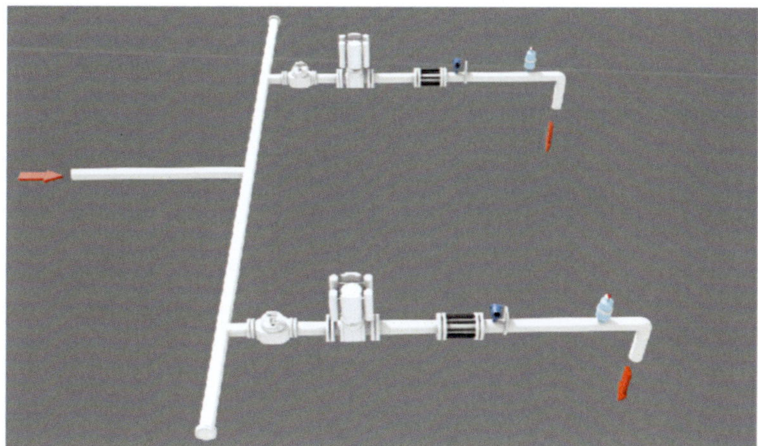

**Figure 7.** Cluster rendering five.

Subsequently, animations were assigned to the control valves, since they are the devices in the field that will change their status according to actions carried out by the operator; for graphic purposes, when the actuator is red, this indicates the total closure of the valve; on the other hand, when the actuator is green, the valve is partially or totally open.

### 3.4. Phase IV: Deploy

### 3.4.1. Configure Equipment and Devices

In this phase the corresponding configuration of each of the multivariable transmitters was performed; then, the configurations for the RTU data site were established. The following is a list of the configurations performed.

- Configuration of the multivariable transmitters: This activity starts with the configuration of the transmitters for the assignment of a HART address to the device, which is carried out with a tool called hand help. Subsequently, a TAG previously created in the signal list is associated with it, which is nothing more than a tag that associates a specific variable read by the transmitter.
- Configuration of the HART address: The HART address identifies the device, in order to bring a sequence of the arrival of signals to the control device and an order of membership through which the HART address is associated with the device.

### 3.4.2. Connection and Testing of the Devices to the Data Site

Once the equipment was configured, it proceeded to connect them, taking as a starting point and reference the connection diagram previously made. First, the multivariable transmitter is identified and connected to the RTU data site, with it locating the corresponding pins, then the actuators are connected to the corresponding outputs to finally start testing the devices through MODSCA® software, which has an interface that allows reading and/or writing in the data site registers. First, the communication between the control device and the MODSCAN® must be established through the Modbus TCP protocol.

### 3.4.3. Construction of the Control Logic

Once the connection of the equipment has been made, the communication between the RTU and the computer has been established, and the signals are identified in the signal list, we proceed with the construction of the control logic of the RTU data site; the development of the same is completed with DATASITE WORKBENCH® software in version 5.22 distributed by Rockwell Automation, with this being the default software for the development of control logic.

Actuator logic: For the elaboration of the actuator control logic, we proceeded the construction of a scaling of the write and read values to be interpreted, remembering that the write and read from the MODSCAN environment are from 10,000 to 50,000, where 10,000 represents 0% (valve fully closed) and 50,000 represents 100% (valve fully open); the logic was developed in the functional block diagram language, as shown in Figure 8.

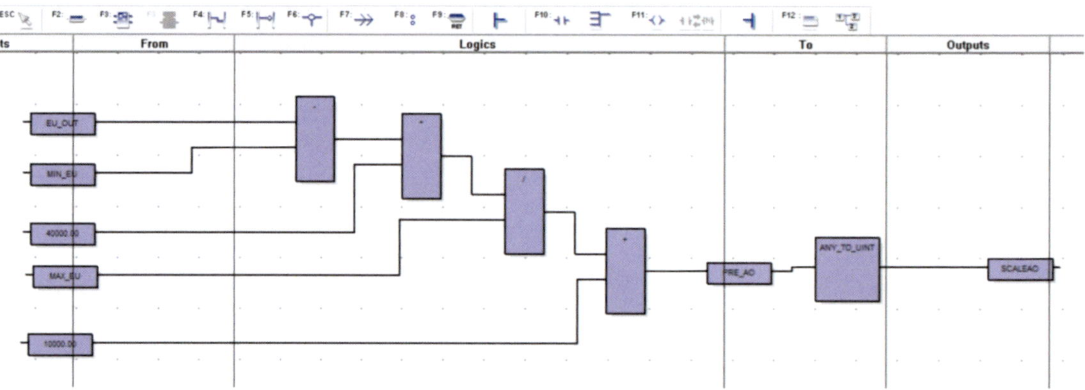

**Figure 8.** Logic of analog output signals.

Once the scaling was created, both the writing and reading of data were restricted, denoting that the maximum value that the operator can write to regulate the control valve actuator is 100% and the minimum value is 0% in the same way for reading data. If the operator enters a value greater than the established value, the logic will interpret it as 100%; otherwise, if they enter a negative value, the control logic will interpret it as 0%. Finally, all of the logic presented above is a set of subroutines that are mentioned in the main program. See Figures 9 and 10. Later in Figure 11, we can see the logic for remote and local status control.

The logic presented above allows the reading of the analog input signal, which the operator manipulates in the local state, to become the valve position command when it returns to the remote state.

*Reading of operational variables:* To obtain the operational variables from the register to which they are associated, the DS settings software allows for assigning of an address in a range in which the variables can be observed; for this case, we want to monitor the pressure, temperature, and flow. Through the logic shown in the Figure 12, you can obtain the reading of the variables and associate them with an output address.

Simulation of the control logic: Once the control logic is completed, we proceed to verify if it works correctly; first, a compilation of the logic was performed. Usually the programming software indicates if any error occurs in the programming when compiling the same; if it was verified that no error had occurred, we proceed to the simulation of the control logic as shown in Figure 13.

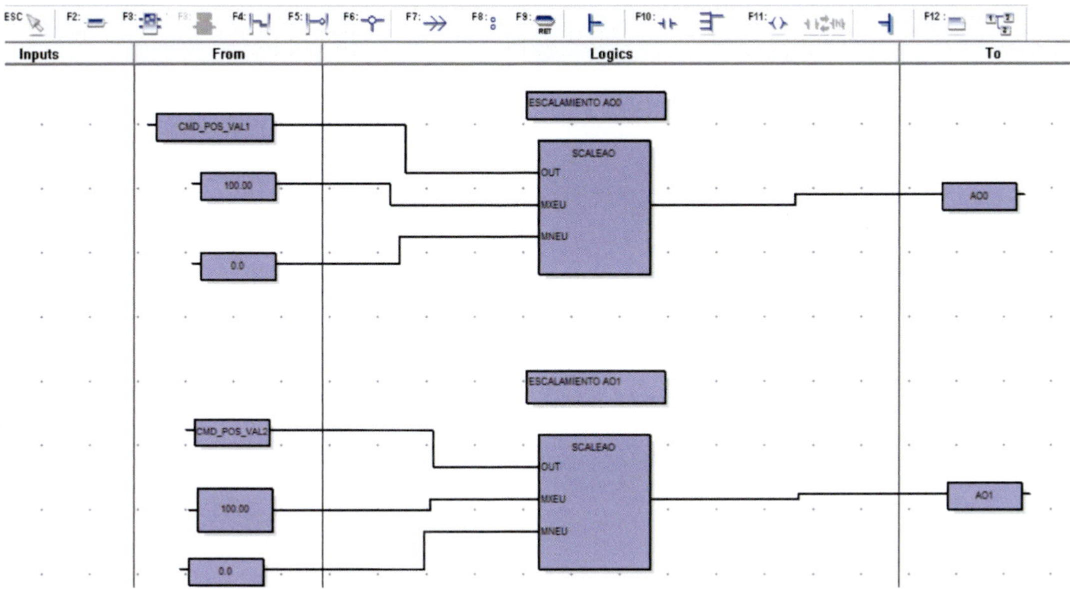

**Figure 9.** Main program, analog output signals.

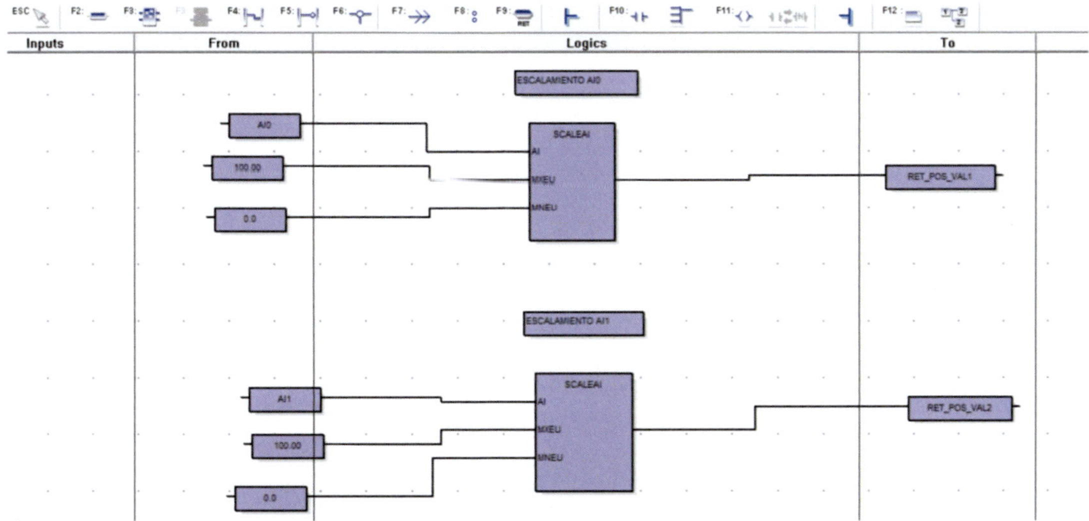

**Figure 10.** Main program and analog input signals.

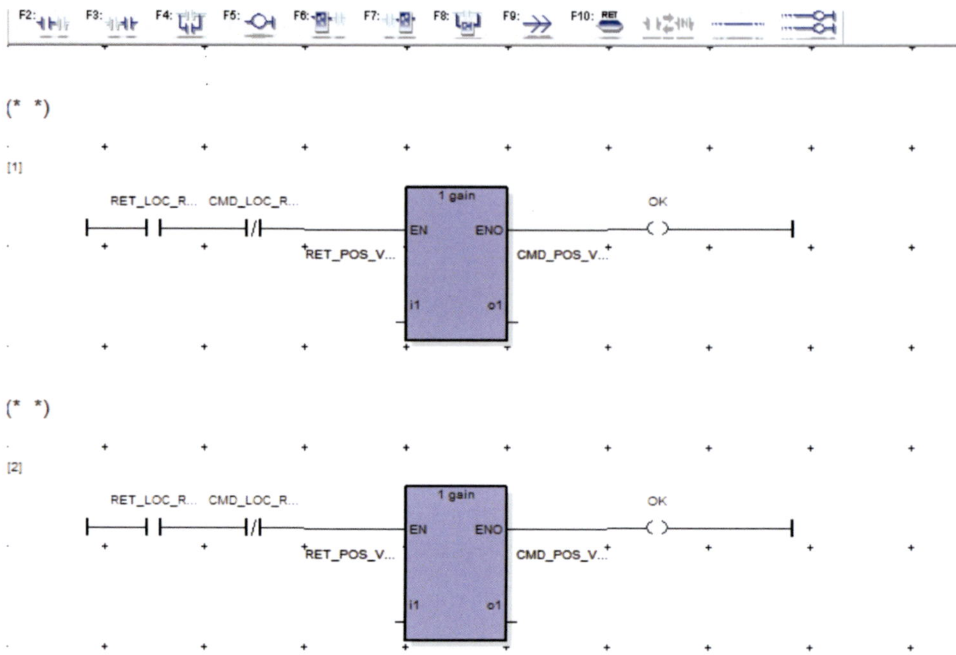

Figure 11. Logic for remote and local status control.

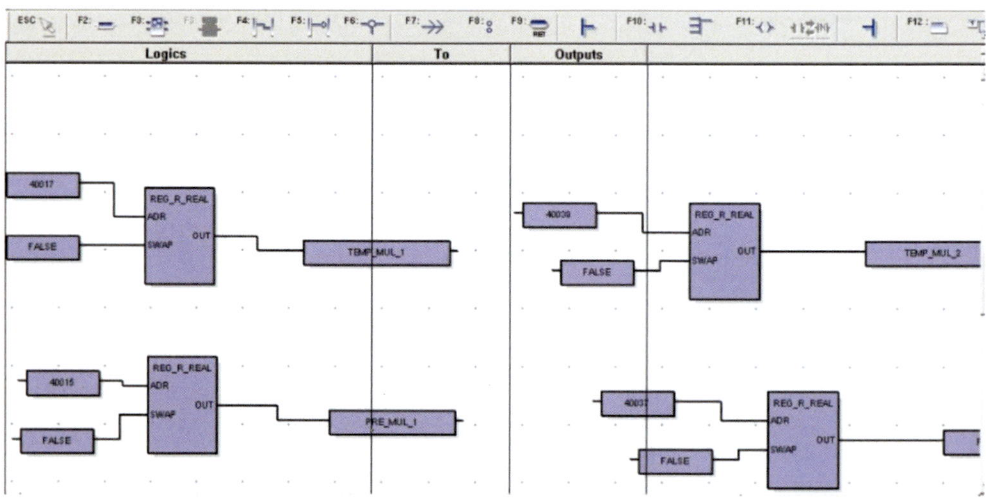

Figure 12. Logic for reading operational variables.

Next, in Figure 14, the programming in the ladder diagram carried out in the PLC of the main execution routine is shown; in it, the call of the different subroutines is integrated for its subsequent execution. Later in Figure 15, the programming is shown to obtain the scaling of the signals. Figure 16 details the programming for assigning the analog and digital input to the operation variables and Figure 17 shows the corresponding programming for the selection of the operation that needs to be performed.

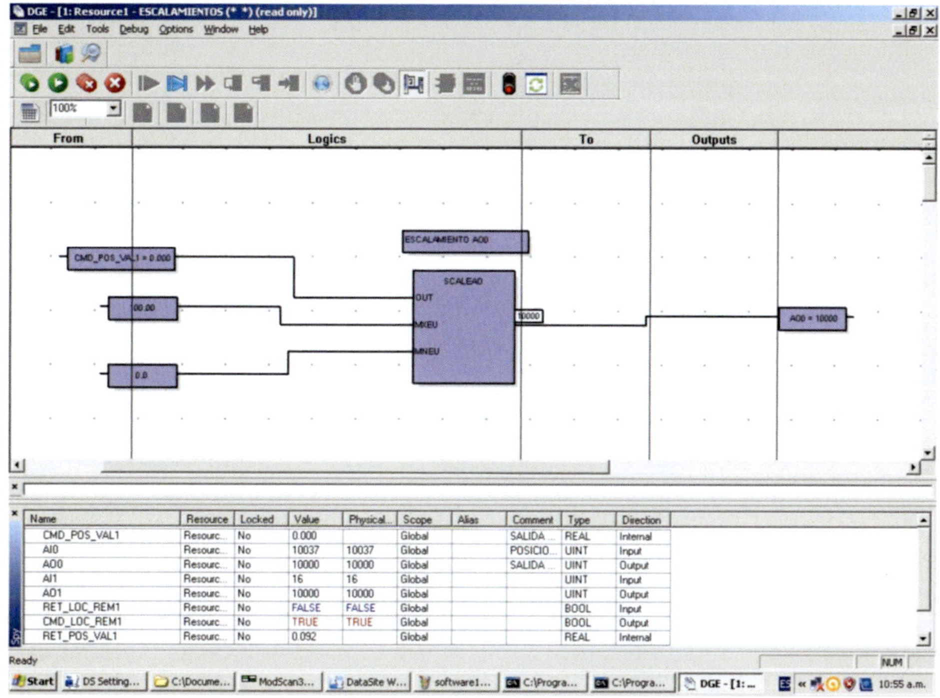

**Figure 13.** Control logic simulation.

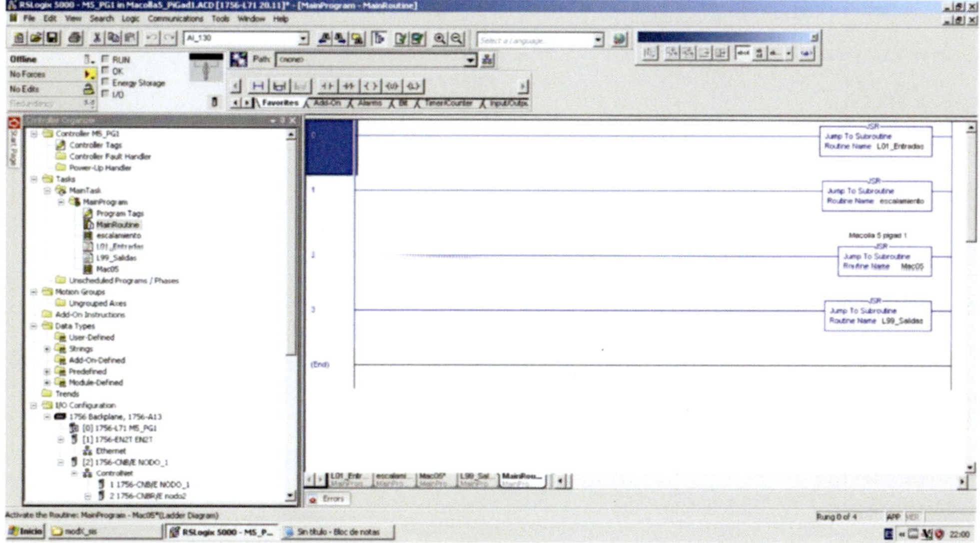

**Figure 14.** Main execution routine.

### 3.4.4. GALBA SCADA Deployment Development

For the development of the GALBA SCADA®, first, a new project must be created through JCONFIG, which is assigned the name CLUSTER_5; then, a user is created, which will be used to enter the JDESKTOP later on. Then, we proceeded to the creation of the controller to which the signals to be integrated into the database will be associated once the device is configured as shown in the following figure, where a name is assigned to the device DATASITE_ETHERNET and the IP address with which it is connected to the network is assigned.

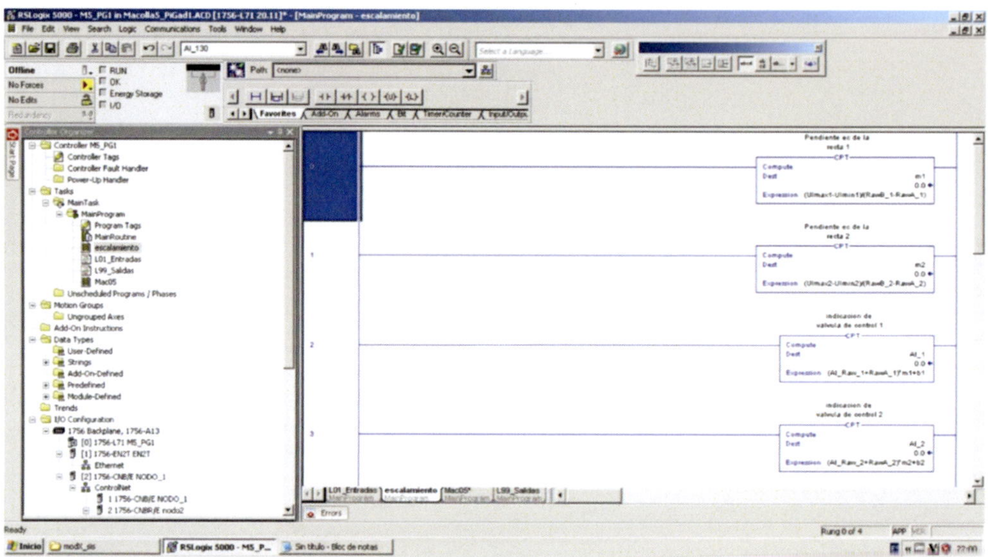

**Figure 15.** Signal scaling.

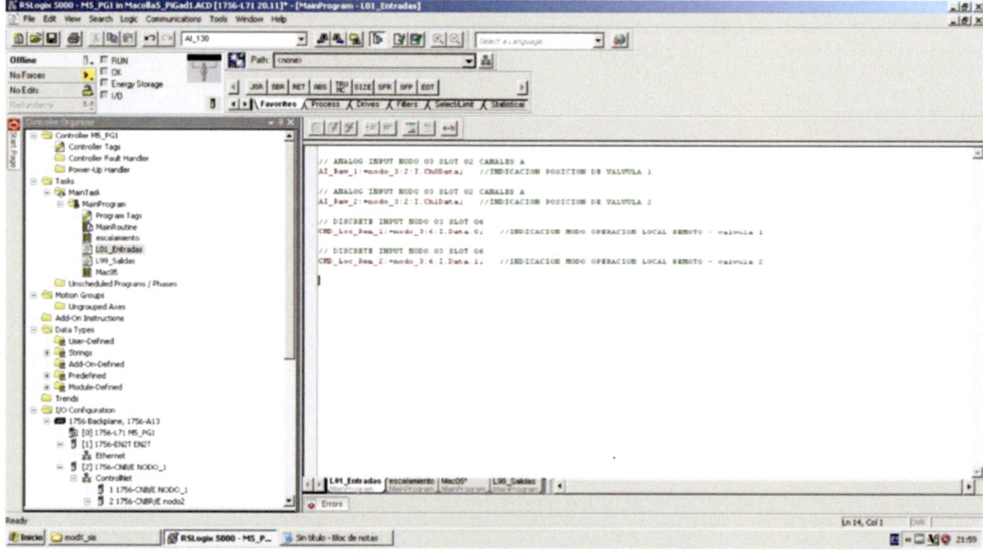

**Figure 16.** Analog and digital input assignment to variables.

Subsequently, the analog and digital signals were integrated into the JCONFIG database; this process can be carried out in two ways: the first one is through the creation of a document with the extension. csv where all the necessary data for the database can be stored, one document for the analog signals and another one for the digital signals to be imported to the database. Then, in the JEDITION environment, we proceeded to configure and design all of the elements for the GALBA SCADA® in order to obtain a screen for supervision and control of the processes, which is nothing more than a man-machine interface of cluster 5.

The Figure 18 shows the reading of the operational variables and the corresponding MeterEx to the perform control actions on the actuator.

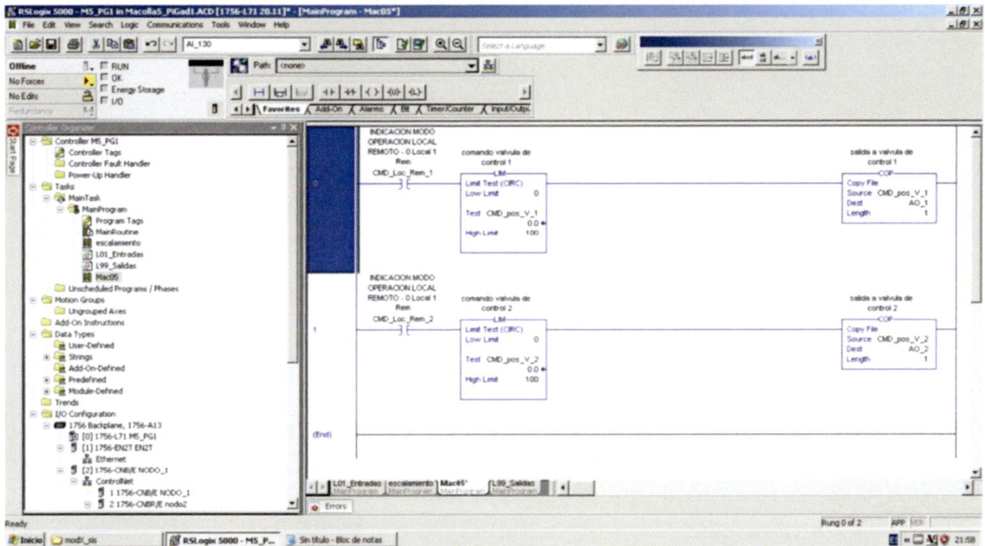

**Figure 17.** Operation selection.

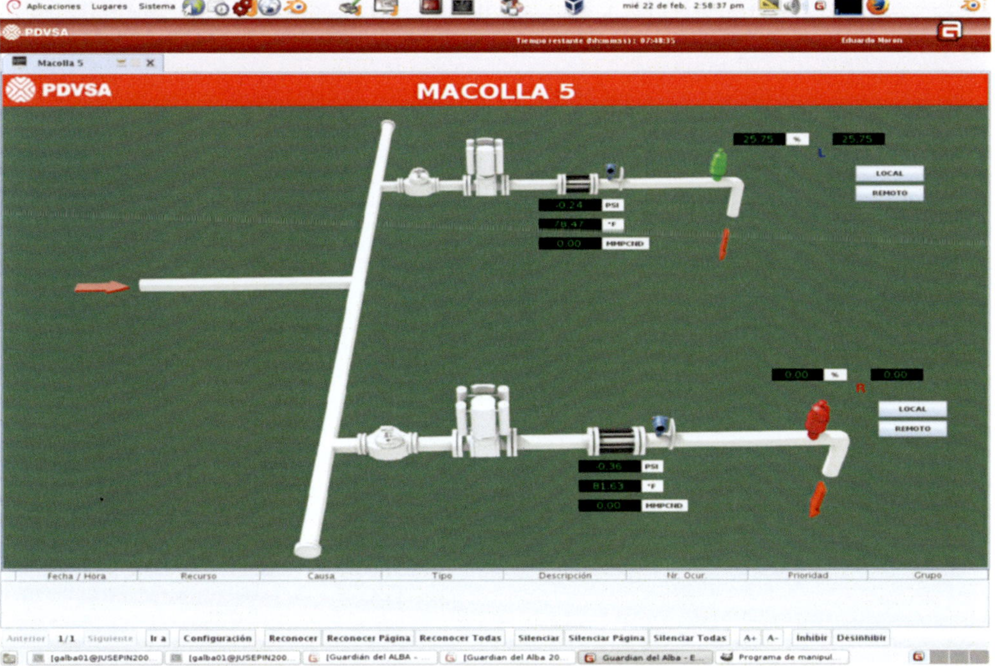


**Figure 18.** Display in JEDITION and the palette tool.

Through the interface offered by SCADA GALBA®, it will be possible to facilitate the work of operators in the field and it will be possible to regulate the amount of gas injected into each of the gas injection wells by opening or closing the actuator, a process that is carried out in the PDVSA industry.

## 4. Conclusions

The evaluation of the current situation, through the operating philosophy of PDVSA, allowed for the establishment of comparisons of the process and, at the same time, highlighted the existing problems in terms of reading the parameters associated with gas injection, in order to understand and face situations by diagnosing needs, which allowed for development of a proposal to optimize the process and consequently reduce operating costs.

The identification of the technology used in the production process, in addition to the necessary requirements, suggested a technological architecture proposal for the supervision of variables such as flow, temperature, and pressure, while it is possible to control the pressure of the gas injected into the wells. The instruments chosen for the measurement and/or control of the operating variables have the necessary characteristics adapted to the PDVSA process.

Regarding the architecture design and its technical specifications, two multivariate transmitters were used that encapsulate a HART communication protocol, which allows for its multipoint connection, which is ideal for the characteristics of the remote terminal unit of the site of data; in the case of the actuators, they handle a protocol of 4–20 mA. The use of relays in the system allows the change of local–remote status through digital signals.

The connection and adaptation of the different equipment, such as multivariable transmitters and actuators to the remote terminal unit, made it possible to apply functional tests with satisfactory results. Among these results, the measurement of variables and the regulation or opening/closing of the actuator stand out for the design and development of the control logic for the two process states, including the local one, so that the corresponding field operator of cluster 5 performs corrective actions manually, and the remote one, in which supervision and control of the parameters are carried out performed from the control room.

The proposed system allows us to control the opening percentage of the gas flow control valve, the opening percentage depends on the well conditions (flow, temperature, and pressure). These conditions change and this generates the need for a greater or lesser opening percentage of the control valve. With manual valves, an operator adjusts the flow measurement and the opening necessary for its control, and they need to carry out on-site monitoring at each key point in the system. The opening and closing of the valve is carried out manually until a proper balance is achieved with variations of two to five percent precision, with the variation affecting the diameter of the pipe which directly affects the injection of gas, causing the affectation of the crude fluid and ther affectation of the reservoir. Currently, the MUC-67 and MUC-68 wells that make up cluster 5 require a control valve opening of 20% and 5%, respectively, and this percentage is directly affected by the average valve opening error when performed in the manual way. Cluster 5 is located 10 km from the HPGIP 1 plant, which makes it difficult to open the valve on time. Automatic opening allows timely decisions to improve well production without affecting the useful life of the reservoir. The present investigation serves as a reference for the implementation of a technological architecture that allows for the supervision of variables such as the flow, temperature, and pressure of the wells, while it is possible to control the pressure of the gas injected into the wells, through the automatic opening of the control valves proposed for gas injection wells present in the oil industry, specifically in high-pressure gas injection plants.

In PDVSA, the study can be replicated in other clusters that contain gas injection wells as a secondary recovery method, since the Venezuelan oil industry maintains an approved homologated platform where the equipment considered takes into account the standards established by PDVSA. Oil companies that use secondary recovery processes through gas injection may replicate the study considering the nature and particularity of their platform, which could lead to the use of the proposed control logic and possible change in the equipment and/or devices used in the present investigation.

**Author Contributions:** Conceptualization, C.R. and J.V.; Methodology, C.R, P.L., J.F. and E.U.; Software, J.M., J.V. and J.F.; Validation, J.M. and J.V.; Formal analysis, C.R., J.M., J.V., P.L. and J.F.; Investigation, C.R., J.M. and E.U.; Resources, P.L. and J.F.; Data curation, J.M., J.V. and J.F.; Writing—original draft, C.R.; Writing—review & editing, P.L. and E.U.; Visualization, C.R.; Supervision, E.U.; Project administration, C.R. All authors have read and agreed to the published version of the manuscript.

**Funding:** This research received no external funding.

**Institutional Review Board Statement:** Not applicable.

**Informed Consent Statement:** Not applicable.

**Data Availability Statement:** Data available on request from the authors.

**Conflicts of Interest:** The authors declare no conflict of interest.

# References


1. Acedo, B.E.I.; Wright, C.; Ghys, T.; Cools, P. Imaginaries of Robotization: Automation from the employee's perspective. *Sociol. Stud.* **2020**, *38*, 567–600. [CrossRef]
2. Mendoza, L. Automatic Verification of Critical Industrial Processes with Automata. *Tech. J. Fac. Eng. Univ. Zulia* **2016**, *39*, 121–129. Available online: http://ve.scielo.org/scielo.php?script=sci_arttext&pid=S0254-07702016000300004&lng=es&tlng=es (accessed on 13 January 2022).
3. Muqeet, M.A. PLC and SCADA Based Control of Continuous Stirred Tank Reactor (CSTR). *IJIREEICE* **2015**, *3*, 185–189. [CrossRef]
4. Acharya, V.; Sharma, S.K.; Gupta, S.K. Analyzing the factors in industrial automation using analytic hierarchy process. *Comput. Electr. Eng.* **2018**, *71*, 877–886. [CrossRef]
5. Pino, J. *Automation and Control Proposal for the Salt Water Injection Plant of the Bared- 8 Discharge Station, Belonging to the Macura District*; Eastern University: Barcelona, Venezuela, 2013.
6. Serrano, G. *Automation of the Exit Manifold of the Crude Oil Storage and Transportation Center (ATC) Naughty Tank Yard. PDVSA-Punta de Mata District*; Eastern University: Barcelona, Venezuela, 2014.
7. El-Sayed, A.; Zawawi, A. Integration of DCS and ESD through an OPC application for upstream oil and Gas. In Proceedings of the IEEE Power and Energy Society General Meeting, San Diego, CA, USA, 22–26 July 2012; pp. 1–5.
8. Fletcher, S.R.; Johnson, T.; Adlon, T.; Larreina, J.; Casla, P.; Parigot, L.; Alfaro, P.J.; Otero, M.D.M. Adaptive Automation Assembly: Identifying System Requirements for Technical Efficiency and Worker Satisfaction. *Comput. Ind. Eng.* **2020**, *139*, 105772. [CrossRef]
9. García, M.V.; Irisarri, E.; Pérez, F.; Estévez, E.; Marcos, M. Automation Architecture Based on Cyber-Physical Systems for Flexible Manufacturing in the Oil and Gas Industry. *Rev. Iberoam. de Automática e Inf. Ind.* **2018**, *15*, 156. [CrossRef]
10. Bortolini, M.; Galizia, F.; Mora, C. Reconfigurable manufacturing systems: Review of the literature and research trend. *J. Manuf. Syst.* **2018**, *49*, 93–106. [CrossRef]
11. Maganha, I.; Silva, C.; Ferreira, L.M.D. Understanding reconfigurability of manufacturing systems: An empirical analysis. *J. Manuf. Syst.* **2018**, *48*, 120–130. [CrossRef]
12. Suarez, R. *Design of a Punta de Mata gas Pipeline Network Infrastructure That Guarantees an Optimal Suction Pressure to the Injection Plant HPGIP II, North District, PDVSA, Edo Monagas*; Eastern University: Barcelona, Venezuela, 2013.
13. Velásquez, C. *Automation of Oil Transfer Pump Room, of the Flow Station of the Amana Operational Complex (AOC), PDVSA-District Punta de Mata, Monagas State*; University of the East Nucleus of Monagas: Maturín, Venezuela, 2015.
14. Estrada, L.; Chacin, J.; Pérez, K. Gas Injection Control and Monitoring System for Oil Wells. Private University Dr. Rafael Belloso Chacín. 2017. Available online: http://renati.sunedu.gob.pe/handle/sunedu/2028714 (accessed on 7 June 2022).
15. Lashin, M. Different Applications of Programmable Logic Controller (PLC). *Int. J. Comput. Sci. Eng. Inf. Technol.* **2014**, *4*, 27–32. [CrossRef]
16. Cotrina, M. Instrumentation Design for the Automation and Control System of the Bio-Fuel Supply and Dispatch Plant in the Puerto Maldonado Region. Technological University of Peru. 2018. Available online: https://hdl.handle.net/20.500.12867/2375 (accessed on 25 May 2022).
17. Díaz, J. Design of a Control System Based on Scada in a Compressed Natural Gas Decompression Station for the Reduction of Production Costs in Brick Manufacturing Plants. National University of Engineering. 2019. Available online: http://cybertesis.uni.edu.pe/handle/uni/21572 (accessed on 10 February 2022).
18. Rodríguez, A.M.B.; Urrego, O.C. Review of the use of biosurfactants for implementation in enhanced oil recovery processes. *Inventum* **2021**, *16*, 4–14. [CrossRef]

19. PDVSA. Annual Management Report 2016. Available online: http://www.pdvsa.com/images/pdf/Inversionistas/INFORMEDEGESTION2016.pdf (accessed on 10 December 2021).

20. Tavara, J. *System Sizing for High Pressure Gas Injection to Remove oil Remnants*; Universidad de Piura: Piura, Peru, 2016; Available online: https://pirhua.udep.edu.pe/bitstream/handle/11042/2649/MAS_IME_SEM_011.pdf?sequence=1&isAllowed=y (accessed on 4 February 2022).

21. Garcia, A. Design of the Architecture of Wells for Macollas of the Carabobo Division Petroindependencia area, Orinoco Oil Belt. Central University of Venezuela. 2012. Available online: http://hdl.handle.net/10872/2865 (accessed on 2 March 2022).

22. Morales, K. Optimization of the Productivity of the Horizontal Wells of Clusters 2, 3 and 4 Drilled in the Morichal Field of the Cerro Negro Field. Central University of Venezuela. 2012. Available online: http://hdl.handle.net/10872/3425 (accessed on 20 March 2022).

23. Figueroa, A. Cost Control of Basic and Detailed Engineering Projects. Tillers N2 and N3. Yucal Placer Field, Guárico State, Venezuela. Central University of Venezuela. 2013. Available online: http://hdl.handle.net/10872/4591 (accessed on 15 March 2022).

24. Santiago, A.; Arana, J.; Matías, V. Valuation of an enhanced oil recovery project in Mexico through the binomial method. *Econ. Análisis* **2020**, *35*, 229–253. Available online: http://www.scielo.org.mx/scielo.php?script=sci_arttext&pid=S2448-665520200003 00229&lng=es&tlng=es (accessed on 5 March 2022).

25. Libertador Experimental Pedagogical University. Manual of Specialization and Master's Degree Projects and Doctoral Theses. In *Vice President of Research and Postgraduate Studies of Libertador Experimental Pedagogical University*, 5th ed.; Libertador Experimental Pedagogical University: Caracas, Venezuela, 2016.

26. MGCIP. *Management Guide for Capital Investment Projects (GGPIC)*; PDVSA Operations Committee: Caracas, Venezuela, 1999.

27. Quispe, R. Design of the Scada System for the Reception, Storage and Distribution Area of Liquid Hydrocarbons in the Petroperú Plant—CUSCO. National University of San Antonio Abad of Cusco. 2019. Available online: http://hdl.handle.net/20.500.12918/4177 (accessed on 3 March 2022).

28. Grilo, A.M.; Chen, J.; Diaz, M.; Garrido, D.; Casaca, A. An Integrated WSAN and SCADA System for Monitoring a Critical Infrastructure. *IEEE Trans. Ind. Inform.* **2014**, *10*, 1755–1764. [CrossRef]

29. Chan, D. *Development of the ALBA Guardian Comprehensive System for the Supervision and Control of the Jusepín200 Compressor Plant Located in the Hato Nuevo Limón of PDVSA—Furrial Division*; Eastern University: Monagas, Venezuela, 2015.

30. Heidari, P.; Alizadeh, N.; Kharrat, R.; Ghazanfari, M.; Laki, A.S. Experimental Analysis of Secondary Gas Injection Strategies. *Pet. Sci. Technol.* **2013**, *31*, 797–802. [CrossRef]

31. Chai, X.; Tian, L.; Wang, G.; Zhang, K.; Wang, H.; Peng, L.; Wang, J. Integrated Hierarchy–Correlation Model for Evaluating Water-Driven Oil Reservoirs. *ACS Omega* **2021**, *6*, 34460–34469. [CrossRef] [PubMed]

32. Cunha, A.D.L.; Neto, S.R.D.F.; de Lima, A.G.B.; Barbosa, E.S. Secondary Oil Recovery by Water Injection: A numerical study. *Defect Diffus. Forum* **2013**, *334–335*, 83–88. [CrossRef]

33. Chaparro, J.; Barrera, N.; León, F. Remote Terminal Module, for data acquisition, monitoring and control of Agroindustrial processes—Agricul-TIC. *Ingeniare Chil. Eng. Mag.* **2021**, *29*, 245–264. [CrossRef]

34. Villalba, E. Development and Analysis of a DCS System and Industrial Protocols. National University of San Agustin of Arequipa. 2019. Available online: http://repositorio.unsa.edu.pe/handle/UNSA/10779 (accessed on 12 February 2022).

*Review*

# Review on Energy Conservation of Construction Machinery for Pumping Concrete

Huiyong Liu [1,*] and Qing Zhao [2]

[1]   School of Mechanical Engineering, Guizhou University, Guiyang 550025, China
[2]   College of Civil Engineering, Guizhou University, Guiyang 550025, China
*   Correspondence: hyliu1@gzu.edu.cn

**Abstract:** The excessive consumption of fossil fuel, energy shortage and global warming along with environmental deterioration have increasingly become a global issue. In order to deal with the energy crisis, energy conservation has been developed and applied in vehicles and construction machineries, i.e., excavators, loaders and forklifts. Due to the shortcoming of low efficiency, high-energy consumption and bad exhaust, the energy conservation of construction machinery for pumping concrete is necessary and urgent. This paper aims to carry out a review on energy conservation of construction machinery for pumping concrete. The research methodology comprises a quantitative analysis method and literature investigation method. First, the structure and working principle of construction machinery for pumping concrete are expounded, and energy consumption ways of construction machinery for pumping concrete are analyzed. Then, research developments in the energy conservation of construction machinery for pumping concrete are summarized. Finally, challenges with the energy conservation of construction machinery for pumping concrete are presented.

**Keywords:** energy conservation; concrete pump; hybrid power; control strategy; parameter matching; energy management

**Citation:** Liu, H.; Zhao, Q. Review on Energy Conservation of Construction Machinery for Pumping Concrete. *Processes* **2023**, *11*, 842. https://doi.org/10.3390/pr11030842

Academic Editor: Sergey Y. Yurish

Received: 24 January 2023
Revised: 3 March 2023
Accepted: 9 March 2023
Published: 11 March 2023

## 1. Introduction

With the rapid development of the global economy, governments have been focusing on the construction of infrastructures such as high-rise buildings, water power stations and long-span bridges where fresh concrete are required to be transported quickly. Reducing time, as the mark of differentiating modern and conventional construction modes, is a feasible way to realize rapid construction [1]. In order to satisfy the increasing requirement of rapid construction, pumping concrete known as a common technique for transporting fresh concrete has been extensively used in various construction sites worldwide. Pumping concrete can make the construction of many infrastructures feasible, i.e., high-rise buildings and long-span bridges, within a very short time period [2–8]. It not only can significantly improve the work efficiency and shorten the construction process duration, but also can notably lighten labor intensity and ensure good construction quality [9,10].

In recent years, the excessive consumption of fossil fuel, energy shortage and global warming as well as environmental deterioration have increasingly become a global issue. In order to deal with the energy crisis, energy conservation has been developed and applied in vehicles and construction machineries, i.e., excavators [11–14], forklifts [15–17], cranes [18–22] and loaders [23–26]. Concrete pumps, a kind of construction machinery for pumping concrete, have been widely used in many fields such as high-rise buildings, water power stations and long-span bridges. Traditional concrete pump has some disadvantages including low efficiency, high-energy consumption and bad exhaust. The application of energy conservation to concrete pumps will improve engine efficiency, reduce energy consumption and promote engine exhaust; thus, carrying out research on energy conservation for concrete pumps is necessary and urgent.

With the application of automatic control technology, visualization technology, fault diagnosis and monitoring technology, etc., the concrete pump has become the high knowledge-intensive and high value-added product. Many concrete pump manufactures and research institutions have been carrying out a number of studies on concrete pumps. Now, they are aware of the importance and urgency of energy conservation for concrete pumps and have put much effort into this. To date, in the literature, many contributions toward energy conservation for concrete pumps can be found.

This paper aims to carry out a review on the energy conservation of construction machinery for pumping concrete. The research methodology of this paper comprises a quantitative analysis method and literature investigation method. The remainder of this paper is organized as follows: In Section 2, the methodology of this study is described. In Section 3, the structure and working principle of concrete pumps are expounded, and energy consumption ways for concrete pumps are analyzed. In Section 4, recent research developments in the energy conservation of concrete pumps are summarized. The challenges with energy conservation for concrete pumps are presented in Section 5. Finally, conclusions are drawn in Section 6.

## 2. Methodology

The methodology for this study includes a quantitative analysis method and literature investigation method. Figure 1 shows the overview of the basic research process for this study, which includes 6 steps as follows.

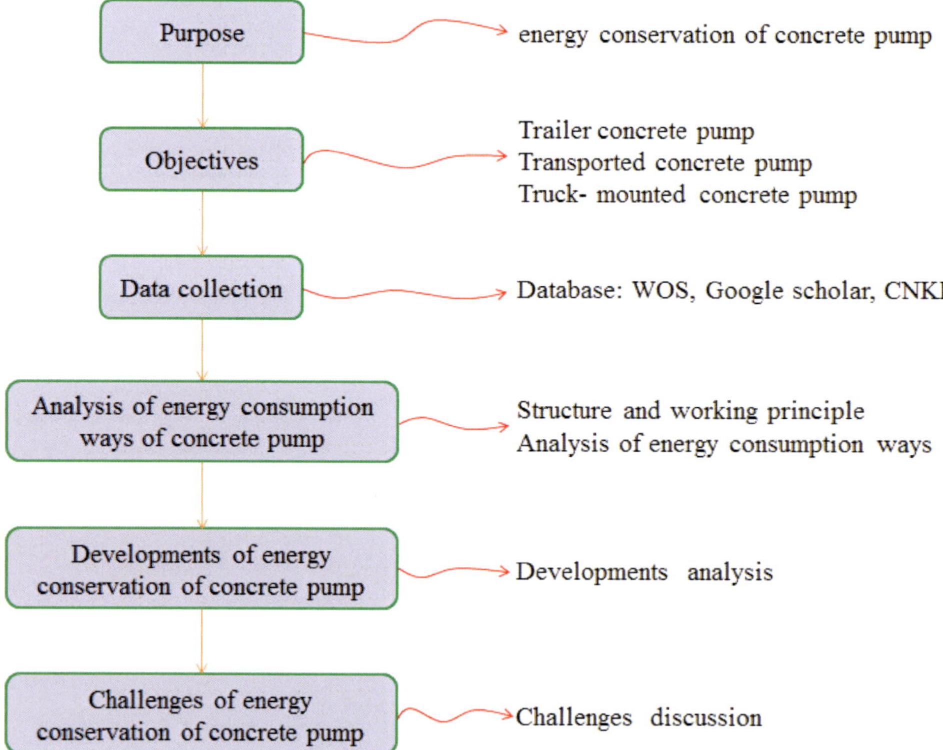


**Figure 1.** The basic research process.

Step 1: Purpose

First, it is clear that the purpose of this study is to comprehensively and deeply understand the issue of energy conservation for concrete pumps.

Step 2: Objectives

At present, there are three types of concrete pumps widely used in the world, namely trailer concrete pumps, transported concrete pumps and truck-mounted concrete pumps. Therefore, this study selected three types of concrete pumps as the research objectives.

Step 3: Data collection

Currently, there are numerous documents related to concrete pumps. Effective collection of documents required for this study ensured the integrity and credibility of this study. The research literature was retrieved from academic databases including WOS, Google Scholar and CNKI using the search strategy of ("concrete pump" AND "energy conservation" OR "energy saving").

Step 4: Analysis of energy consumption ways for concrete pumps

The three types of concrete pumps are composed of multiple components, and there is a conversion between different types of energy. There is an energy loss in the process of energy conversion and transmission. This paper describes the basic structure and working principle of three types of concrete pumps, and analyzes the energy consumption ways for three types of concrete pumps.

Step 5: Developments in energy conservation of concrete pumps

To date, researchers and manufacturers of concrete pumps have developed a variety of energy conservation methods and technologies for concrete pumps, each of which has its own characteristics. This paper analyzes the research developments in the energy conservation of concrete pumps.

Step 6: Challenges with energy conservation of concrete pumps

Due to the limitations of relevant technologies, there are still some challenges with the energy saving of concrete pumps. Moreover, this paper discusses the challenges with the energy conservation of concrete pumps.

## 3. Analysis of Energy Consumption Ways for Concrete Pumps

### 3.1. Structure and Working Principle

There are three types of concrete pumps, namely trailer concrete pumps, transported concrete pumps and truck-mounted concrete pumps, as shown in Figure 2.

(a)        (b)        (c)

**Figure 2.** Three types of concrete pumps. (**a**) Trailer concrete pumps [27]. (**b**) Transported concrete pumps [28]. (**c**) Truck-mounted concrete pumps [29].

As shown in Figure 3, trailer concrete pumps mainly consist of an engine, a hydraulic pump, control valves, a main system and auxiliary system. The main system including a pumping system, oscillating system and agitating system is used to realize pumping concrete. The auxiliary system including outrigger system, cleaning system, lubrication system and cooling system is used for the maintenance of the concrete pump.

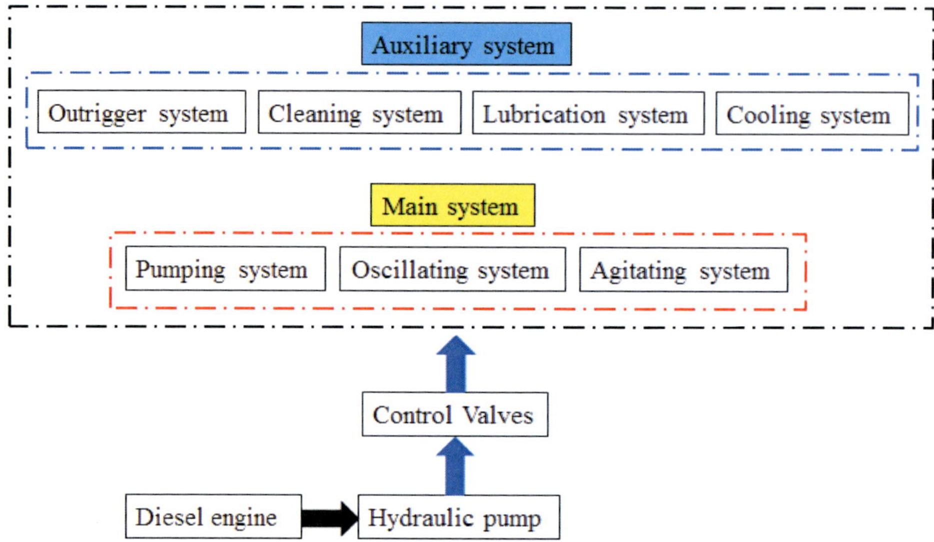

**Figure 3.** Structure diagram of trailer concrete pump.

Mechanical power offered by the engine is inputted into the hydraulic pump, and then the hydraulic pump rotates and sucks oil from the tank, consequently discharging oil into the oil pipe. By this mechanism, mechanical power provided by the engine is transferred to hydraulic power. The pressure oil from hydraulic pump flows through control valves into cylinders or hydraulic motors. Cylinders or hydraulic motors are driven by the pressure oil. By this mechanism, the hydraulic power in the cylinders or hydraulic motors is converted to mechanical power; cylinders or hydraulic motors carry out corresponding pumping work, oscillating work, agitating work or cooling work, etc.

As shown in Figure 4, besides the structure of the trailer concrete pump, the gearbox, transfer box and vehicle chassis are included in the structure of the transported concrete pump. The transfer box is used to distribute mechanical power from the gearbox between vehicle chassis and hydraulic pump.

The transported concrete pump has two work conditions: driving condition and pumping condition. When working in the driving condition, by operating the transfer box, the channel from the gearbox to vehicle chassis is linked but the channel from the gearbox to the hydraulic pump is meanwhile cut off. When working in the pumping condition, by operating the transfer box, the channel from the gearbox to the hydraulic pump is linked but the channel from the gearbox to the vehicle chassis is meanwhile cut off.

As shown in Figure 5, besides the structure of the transported concrete pump, the rotating system and boom system are included in the structure of the truck-mounted concrete pump. Figure 6 shows that boom system consists of booms and cylinders together with conveying pipes for delivering fresh concrete. The boom system is mounted on a rotating platform, and it can rotate 365° with the rotating platform. Each boom can rotate around its own axis. By combining the boom system and rotating system, fresh concrete can be poured to any expected position.

Similar to the transported concrete pump, the truck-mounted concrete pump also has two work conditions: driving condition and pumping condition. In addition, when working in the pumping condition, the hydraulic motor will drive the rotating platform to rotate and the cylinders linking booms will drive the corresponding boom to extend or retract so that it can ensure that fresh concrete can be transported to an expected position.

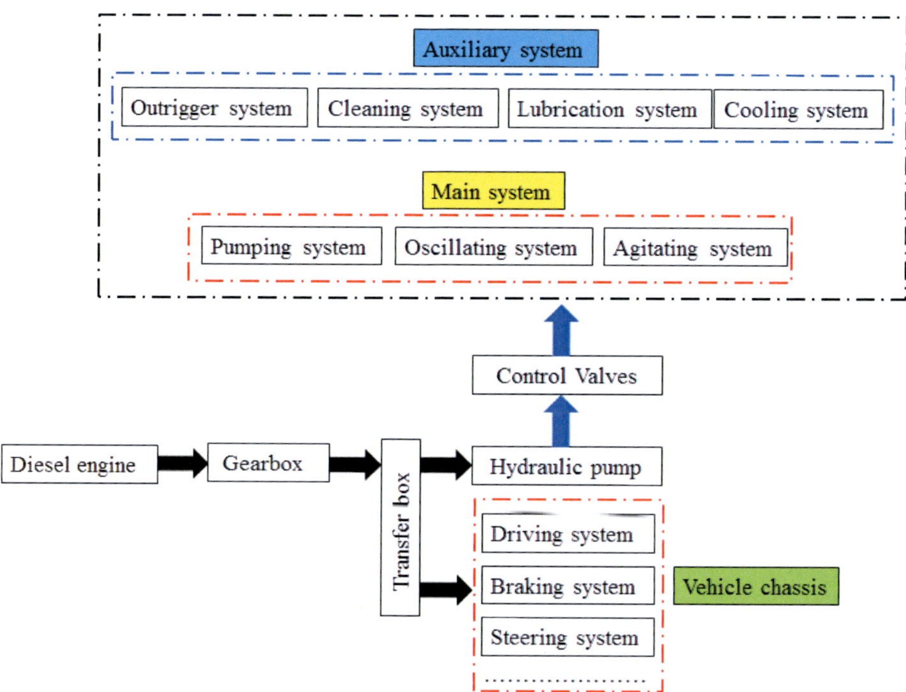

**Figure 4.** Structure diagram of transported concrete pump.

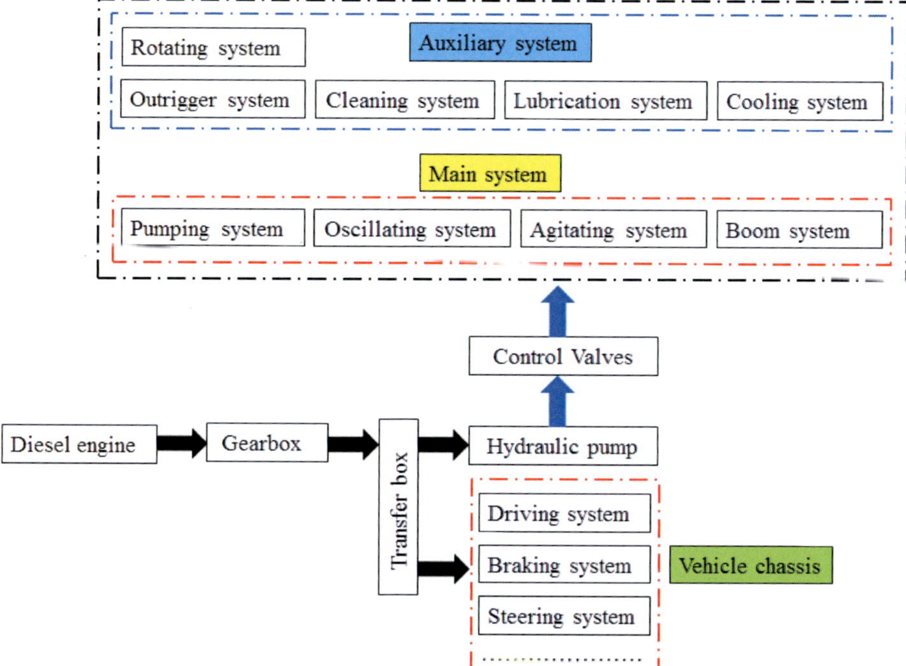

**Figure 5.** Structure diagram of truck-mounted concrete pump.

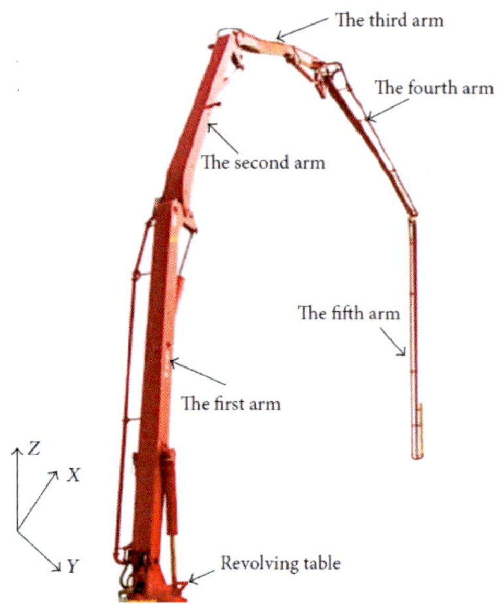



**Figure 6.** Boom system of truck-mounted concrete pump [30].

*3.2. Analysis of Energy Consumption Ways*

As mentioned above, the trailer concrete pump just has the pumping condition. Both transported concrete pump and truck-mounted concrete pump have two work conditions: driving condition and pumping condition. Since when working in driving conditions, there have been a number of studies on energy conservation about vehicles, this paper only focuses on analyzing energy consumption ways when working in pumping conditions.

The energy flow route of the trailer concrete pump, transported concrete pump and truck-mounted concrete pump power flow route can be seen in Figures 7 and 8. It can be seen that the energy provided by diesel engine flows through the hydraulic pump into the concrete pumping system, and finally drives the external load. With regard to the trailer concrete pump, the diesel engine and hydraulic pump are directly linked. As for transported concrete pump and truck-mounted concrete pump, the diesel engine is linked with the gearbox, then with the transfer box and then with the hydraulic pump. When mechanical power is inputted into the hydraulic pump, it thus rotates so that mechanical power is converted to hydraulic power. Oil, as the medium of the hydraulic system realizing energy converting, flows along oil pipes and through control valves and into cylinders or hydraulic motors to drive the external load. The energy consumption includes mechanical friction loss, impact loss, path pressure loss, local pressure loss, leakage loss, throttling loss and overflow loss.

### 3.2.1. Mechanical Friction Loss

When diesel engine works normally, there is friction that happens between the cylinder liner and piston ring. Commonly, an axial plunger pump is used in concrete pumps, so friction occurs on three key friction pairs: plunger and cylinder bore, cylinder block and port plate, slipper and swash plat. For the cylinder, there are two friction pairs, namely cylinder barrel and piston, piston rod and cylinder head, where friction happens. With regard to hydraulic motors used in the agitating system and rotating system, friction occurs between the motion part and static part. The boom system cylinders are mounted in rotary joints to connect two adjacent booms; therefore, friction happens in these rotary joints. Since the diesel engine is linked to the hydraulic pump directly or through the gearbox and transfer box by coupling, friction definitely occurs. In a word, mechanical friction loss mainly exists in the internal of the diesel engine and hydraulic pump, cylinders and hydraulic motors and rotary joints. The energy flowing path is from diesel engine to hydraulic pump (Figure 7) or from diesel engine to gearbox to transfer box to hydraulic pump (Figure 8).

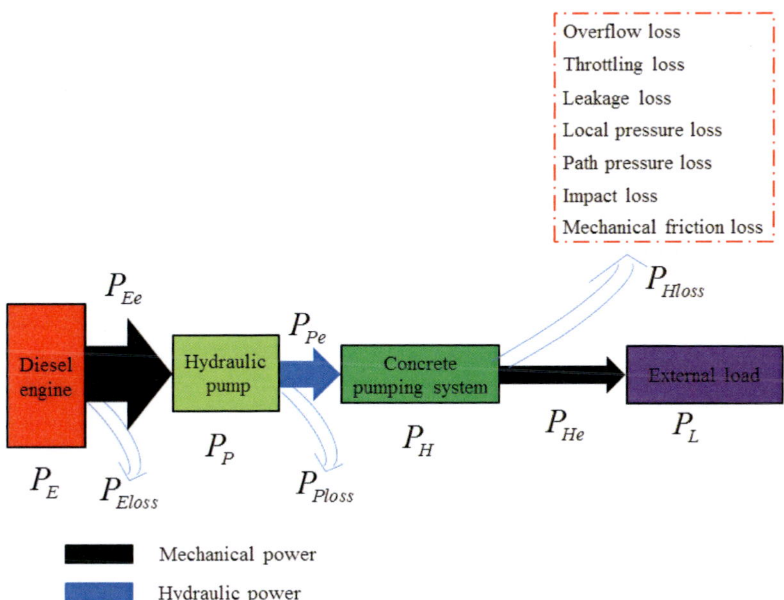


**Figure 7.** Energy flow route of trailer concrete pump.

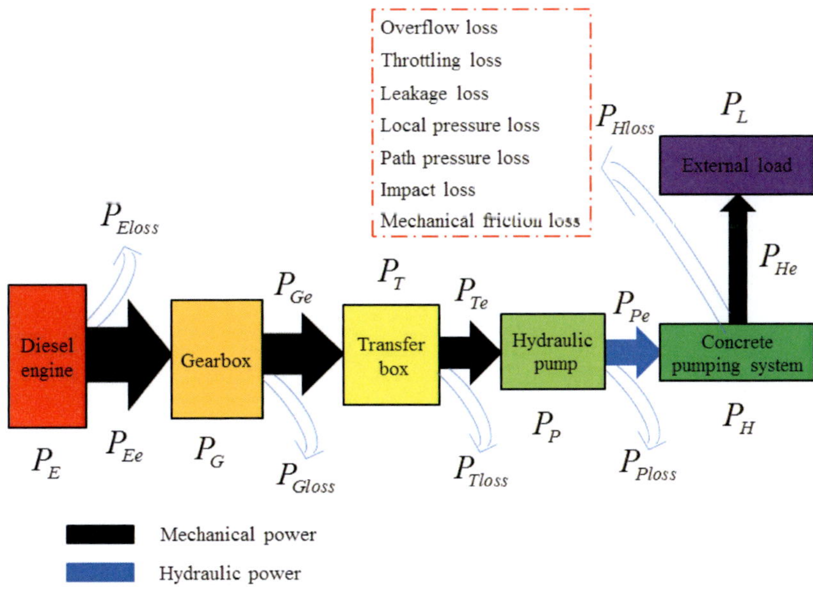


**Figure 8.** Energy flow route of transported/truck-mounted concrete pump.

### 3.2.2. Impact Loss

At present, the pumping system of concrete pumps is mainly a double-cylinder reciprocating piston pump driven by hydraulic pressure. The successful pumping of fresh concrete is accomplished by alternating two cylinders. When changing the control signal of the control valve to change its working position quickly, one cylinder pumping concrete would shift to sucking fresh concrete, and the other cylinder sucking fresh concrete would shift to pumping concrete. At the time when the control valve changes the working position, since the sudden changing of the external load, and oil in oil pipes cannot change their direction rapidly, hydraulic pressure in the corresponding oil pipes would abruptly shift from high value to low value or from low value to high value. The instantaneous high pressure caused by the rapid changing of oil flow is much higher than normal working pressure, even several times higher than the normal working pressure, which forms a hydraulic impact. After the oil absorbs part of the hydraulic impact, its temperature would increase quickly, thus causing system leakage and energy loss.

The oscillation system is used to control the swing of the S-tube to connect the concrete cylinder of the pumping system by operating two oscillation cylinders under hydraulic pressure. The direction changing time of the oscillation system is very short. When changing the working position of the control valve, the inertial force caused by fresh concrete in the S-tube during the swing process would lead to pressure in the oscillation system rising abruptly, thus resulting in a hydraulic impact.

Moreover, when the rotating system works in the braking mode and the agitating system changes its agitating direction and outrigger system, cleaning system, lubrication system, etc., and they shift their direction, an impact would inevitably occur, consequently consuming some energy and causing impact loss.

### 3.2.3. Path Pressure Loss

When oil flows through the same diameter pipes, i.e., pipes linking hydraulic pump and control valves and cylinders or hydraulic motors, due to friction between oil and pipes together with the interaction of oil particles under the viscosity of oil, a path pressure loss would inevitably occur in these pipes.

### 3.2.4. Local Pressure Loss

When the cross section area of the hydraulic components of concrete pumps change abruptly, such as elbow of pipes, joint of pipes, cylinder inlet or outlet, once hydraulic oil flows through these sections, since its direction and velocity would change abruptly and thus forms some whirlpool and cavitation, oil particles would also crash with each other, so that results in local pressure loss.

### 3.2.5. Leakage Loss

Leakage exists in all hydraulic systems of concrete pumps mainly including a pumping system, oscillation system, boom system, outrigger system and agitating system. Generally, the leakage amount increases with the increase in temperature and pressure. Especially, under the high external load thus the high pressure of system, leakage loss would share a certain proportion of energy consumption.

### 3.2.6. Throttling Loss

When the rotating system works and the rotating platform rotates to drive the boom system to an expected position, at this time, the rotating system needs to be braked. Since boom system has a large inertial force, during the braking process, the inertial of pressure oil and motion mechanism also compel the actuator hydraulic motor to keep rotating. Meanwhile it acts upon the oil in the return chamber and thus increases pressure in the return chamber. In fact, the pressure in the return chamber can be several times the normal working pressure. Under high pressure, oil would be squeezed from the opening gap of the control valve, thus a throttling loss occurs and converts the inertial force of the motion mechanism to heat consumption. In addition, for adjusting the extending or retracting velocity of cylinder linking two adjacent booms, throttle valves are employed so that throttling loss exists.

### 3.2.7. Overflow Loss

Relief valves are indispensably employed in the pumping system, oscillation system, agitating system, etc., to assure safety of the subsystem mentioned above. Once the system pressure exceeds a set value, relief valves are opened and overflow begins. Furthermore, balance valves are widely used in a rotating system and boom system to lock the working position of the rotating platform and booms. If pressure caused by the load of the rotating system and boom system is higher than the set value of corresponding balance valves, it thus results in overflow loss.

### 3.2.8. Power Matching Loss

Currently, no matter what work condition concrete pump is in, diesel engine always works in the same fuel consumption mode, but cannot be regulated reasonably according to different load, which results in diesel engine works in the area of high fuel consumption ratio long term. The original reason is that the output power of diesel engine cannot match with the absorption power of hydraulic pump, thus causes power of diesel engine cannot be utilized fully and results in energy loss.

## 4. Developments of Energy Conservation of Concrete Pump

According to the results of the literature search, the methods for energy conservation of concrete pumps can be classified into three types: the power matching approach, dual-fuel approach and dual power approach, as shown in Table 1.

**Table 1.** Classification of methods for energy conservation of concrete pumps.

| Type | Fuel-Saving Rate | Title |
|---|---|---|
| Power matching approach | 10% | Research on Power Matching and Energy Saving Control of the Truck Mounted Concrete Pump [31] |
| | 16% | Energy Optimization by Parameter Matching for a Truck Mounted Concrete Pump [32] |
| | 16% | Global power matching on truck- mounted concrete pump based on genetic algorithm [33] |
| | 5% | Double-parameter control on concrete pumps for pressure-difference-sensed inversion [34] |
| | 20% | Analysis and Research of Energy Saving Technology for Concrete Pump Truck [35] |
| | 15% | Research and conduct of power saving strategy on closed hydraulic system for mounted concrete pump truck [36] |
| | 11% | Study on Automatic Gear Control of Gearbox of Concrete Pump Truck [37] |
| | 10% | Research on the Energy Saving Technology of the Truck Mounted Concrete Pump [38] |
| | 20% | Study on power-matching control of the concrete pump truck's power system [39] |
| | 20% | Research on power-matching of energy-saving for power system of the concrete pump truck [40] |
| | 11.67% | Modeling of universal characteristics and optimization of operating conditions of concrete pump truck based on neural network [41] |
| | 10% | An Energy Saving Control Strategy for Operation of Concrete Pump Trucks [42] |

**Table 1.** *Cont.*

| Type | Fuel-Saving Rate | Title |
|---|---|---|
| | 31.79% | Research of Energy-saving Control Methods for Concrete Pump Trucks [43] |
| | 26.5% | Study on Energy Saving Control Strategy of Concrete Pump Truck [44] |
| | 10% | Study on the energy-saving parameter matching of dynamic system of the truck-mounted concrete pump [45] |
| Dual-fuel approach | 63% | Research and Application on the Truck Mounted Concrete Pump with Dual Fuel Mixed Combustion and Energy-saving [46] |
| Dual-power approach | — | A New Type Dual-power Hydraulic System and Its Application to Concrete Pump Trucks [47] |

### 4.1. Power Matching Approach

The core idea of the power matching approach is to match the diesel engine, the hydraulic pump and the load by adjusting the speed of the diesel engine and the displacement of the hydraulic pump, so that the diesel engine can work near the economic curve, and the hydraulic pump can fully absorb the power of the diesel engine, so as to improve the working efficiency of the whole machine and reduce fuel consumption. The schematic of the power matching approach of energy conservation for concrete pumps is shown in Figure 9.

Yang et al. [31] proposed the power matching control strategy of hydraulic pumps and diesel engines based on the analysis of the complex operating conditions of the truck-mounted concrete pump and the working characteristics of the diesel engine. When the load changes, the speed induction fuzzy control is adopted. By adjusting the speed of the diesel engine and the displacement of the hydraulic pump, the diesel engine can work near the economic curve, and the hydraulic pump can fully absorb the power of the diesel engine. By properly controlling the hydraulic pump and diesel engine, the power utilization rate is improved, thus improving the efficiency of the whole machine and the reliability of the diesel engine. According to the working condition analysis and fuel consumption test results of the truck-mounted concrete pump in the assessment, the proposed power matching control strategy is expected to reduce the fuel consumption of the whole machine by more than 10%.

Ye et al. [32,33] proposed a global power matching strategy based on engine fuel consumption rate and hydraulic pump efficiency to solve the problems of low efficiency and high-energy consumption of truck–mounted concrete pumps. The law of load change was summarized by mathematical statistics. The fuel consumption curve of diesel engine was drawn through bench test. The efficient working area of the hydraulic pump under different pressure, rotating speed and displacement was defined through the efficiency test. Based on a genetic algorithm, the combination optimization of engine fuel consumption rate and hydraulic pump efficiency was realized, and the engine speed and hydraulic pump displacement were jointly adjusted, so that the output control index changed adaptively in real time with the change in load conditions, so as to ensure that all components worked in the high-efficiency zone at the same time. The comparison test results of the fuel consumption for the new and old power matching strategies showed that in the area where the pumping capacity is 10~80 m$^3$/h, the truck-mounted concrete pump has remarkable energy-saving effect, the fuel-saving rate could reach 22~46%, and the comprehensive fuel-saving rate was 16%.

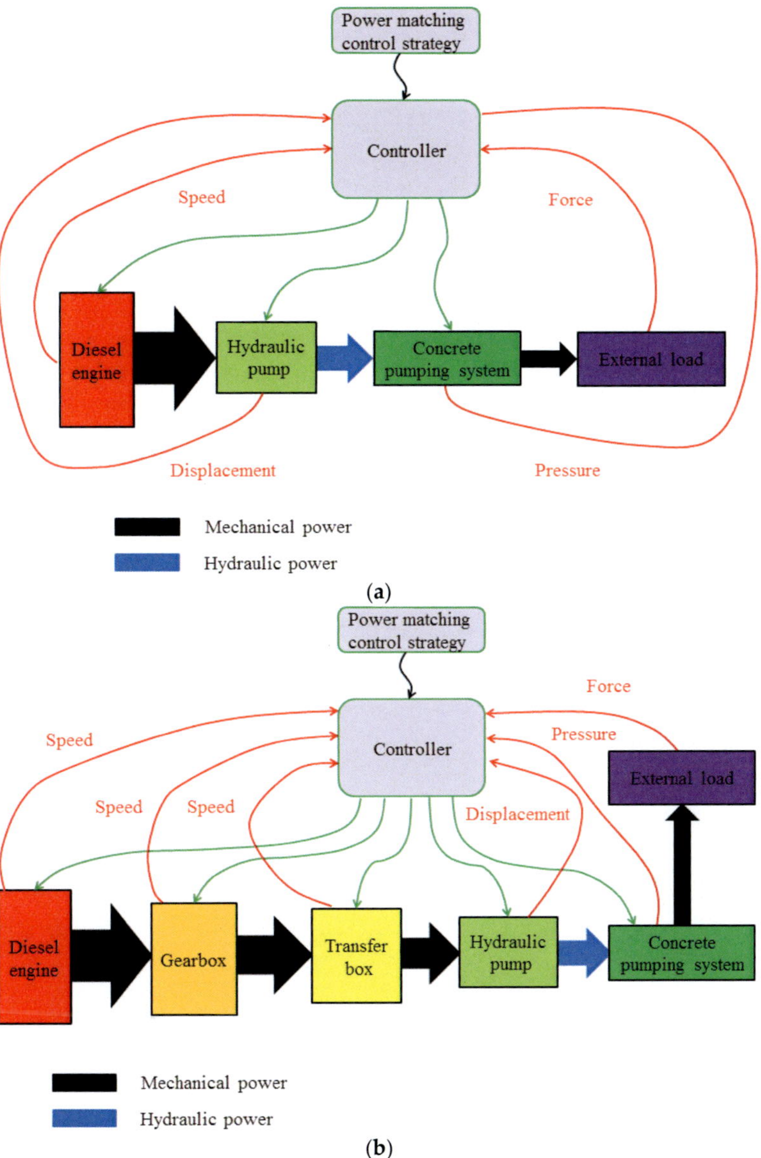



**Figure 9.** Schematic of power matching approach of energy conservation for concrete pumps. (a) Power matching approach of trailer concrete pump. (b) Power matching approach of transported/truck-mounted concrete pump.

In order to overcome the time varying and uncertain factors of the parameters of the concrete pumping system and the sudden change in the load, Ye et al. [34] proposed a two-parameter control strategy of engine speed and hydraulic pump displacement, which improved the rapidity and stability of the response of the control system. Through the field test, it was concluded that the load of the pumping system of the truck-mounted concrete pump had the characteristics of periodicity and mutation. Through the fuel consumption test, it was found that the fluctuation of the engine speed had a serious impact on the fuel consumption. Based on the dynamic equation of the system, a pressure differential induction reversing two-parameter control system was designed, which realized the double closed-loop control of the engine speed and the hydraulic pump displacement of the truck-mounted concrete pump. The test showed that the average fuel consumption was reduced by 5% by using the dual-parameter control strategy.

Xia et al. [35] designed the energy-saving control system of truck-mounted concrete pump based on the power matching strategy of engine and hydraulic pump. The system took energy saving as the control goal to realize the joint regulation of the engine and hydraulic pump. The controller of the concrete pump truck could automatically adjust the output power of the engine according to the change in the actual working conditions, and maintained a good match with the load, so that the engine always ran at the best working point or the best working area, so that the output power of the engine could be fully utilized. The test data showed that when the working displacement of the concrete pump truck was in the range of 10~70%, the energy-saving effect was significant, and the fuel-saving rate could reach 22~46%. When the working displacement of the concrete pump truck was within 70~80%, the energy saving was average, and the fuel-saving rate was about 13%. When the working displacement of the concrete pump truck was 80~100%, the energy saving was not obvious, at about 1%. The average fuel-saving rate of the whole region was more than 20%.

According to the characteristics of the closed hydraulic system of the concrete pump truck, Shi et al. [36] proposed the energy-saving method of adopting adaptive PID control to control the engine speed and the displacement of the closed hydraulic pump on the basis of ensuring the system load flow and system power demand. The results showed that, according to the power and fuel consumption curve of the engine, the composite control of the engine speed and the displacement of the variable displacement pump under different working conditions could achieve good energy-saving effect on the premise of meeting the requirements of working conditions, especially in the middle and small power range of the load, the energy-saving effect was obvious, and the average energy saving was more than 15%.

In view of the shortage that the concrete pump truck uses a fixed gear for pumping, Shi et al. [37] proposed a method to select the optimal transmission ratio by automatically switching the gear of the gearbox according to the load during the pumping process. The comparison test of fuel consumption between the new and the old pumping control scheme was carried out. The test showed that when the pumping volume of the new scheme was 50~100%, the energy-saving effect was obvious, and the energy-saving was up to 11%.

Based on the research on the working characteristics of the truck-mounted concrete pump under various working conditions and the universal characteristics of the engine, Wang et al. [38] developed the power matching control technology of the whole machine based on the adaptive variable torque model, and incorporated the service subsystems of the whole machine (pumping, distribution, mixing, boom, etc.) into the power control system to achieve the perfect match between the power system of the whole machine and the multi-load external working conditions, which could ensure that the engine had a stable load rate under any working conditions, making the engine always work at the economical fuel consumption point. The test results showed that the proposed energy-saving technology could reduce the fuel consumption of truck-mounted concrete pump by more than 10%.

Cheng [39] proposed a power matching strategy by controlling diesel engine and hydraulic pump simultaneously. When the load changes, regulating the flowrate of the variable displacement pump reasonably and timely so that when the revolving speed of the diesel engine varies, the torque of the diesel engine can be adjusted and the diesel engine can work nearby the economical point. The test results showed that the average fuel-saving rate was about 20% by employing the proposed energy-saving technology.

Li et al. [40] proposed to use the interval setting matching method and speed induction control to carry out a power energy-saving matching control on the engine and hydraulic system to reduce the useless power loss of the engine, so that the matching control of the engine and hydraulic pump could be carried out automatically with the size of the load, thus reducing fuel consumption. The actual vehicle test showed that this power matching control method had a significant energy-saving effect, which made the truck-mounted concrete pump deliver the same amount of concrete under the same working conditions with a fuel-saving rate of 20%.

He et al. [41] used the BP neural network to establish the universal characteristic model for the engine of a certain type of truck-mounted concrete pump, and optimized the working conditions of this type of truck-mounted concrete pump and obtained the engine speed that best matched the hydraulic system under four working conditions. The test results showed that the optimized engine working conditions based on the model had an obvious fuel-saving effect, with a maximum fuel saving of 11.67%.

Jie et al. [42] proposed a global power matching energy-saving control strategy of concrete pump trucks based on the load operation mode. The engine speed and hydraulic pump displacement are jointly adjusted by a dedicated controller to achieve the matching between the engine, hydraulic pump and load. In the real vehicle application, the proposed strategy achieved more than 10% fuel-saving rate under 80% load pumping condition.

Zeng et al. [43] proposed to realize the energy-saving control strategy of matching an engine hydraulic pump load by adjusting the relationship between engine speed, hydraulic pump swashplate opening, hydraulic pump pressure and the number of commutations per minute, so as to improve the working efficiency of the whole machine and reduce fuel consumption. The test results show that the maximum fuel-saving rate can reach 31.79%.

Wu et al. [44] proposed an energy-saving control strategy to match the power of the engine and the hydraulic pump of the truck-mounted concrete pump. When the load changed, the engine could work near the economic curve by adjusting the engine speed and the displacement of the hydraulic pump, and the hydraulic pump could fully absorb the power of the engine. The results showed that the average oil-saving rate reached 26.5%.

Pu [45] put forward an energy-saving parameter matching strategy of a load adaptive power system, which dynamically adjusted the speed of engine and displacement of the hydraulic pump according to the change in load conditions, so that the engine always worked near the economic working curve. The test results showed that the proposed parameter matching strategy could effectively reduce the fuel consumption of the power system, and the average fuel-saving rate was more than 10%.

*4.2. Dual-Fuel Approach*

The dual-fuel approach means that two fuels are simultaneously fed into the internal combustion engine and mixed and burned in the internal combustion engine. The proportion of the two fuels is adjusted by the controller in real time according to the external load conditions, so as to reduce the consumption of fossil energy and reduce the emissions generated by burning fossil energy. The schematic of the dual-fuel approach of energy conservation for concrete pumps is shown in Figure 10.

Li et al. [46] studied the key technology in the application of oil and gas blended combustion technology on the truck-mounted concrete pump, in which, on the basis of the original diesel engine, a set of fuel supply and electric control systems was added, and the automatic matching energy-saving control strategy of the engine load rate was adopted, so that the truck-mounted concrete pump could realize the diesel and natural gas-blended combustion switching operation mode under both driving and pumping conditions. According to the statistics of all test data, the blended combustion engine consumed 770.5 L of diesel oil and 1360.53 cubic meters of gas at all speeds, and 2078.3 L of diesel oil in the pure diesel mode under the same conditions.

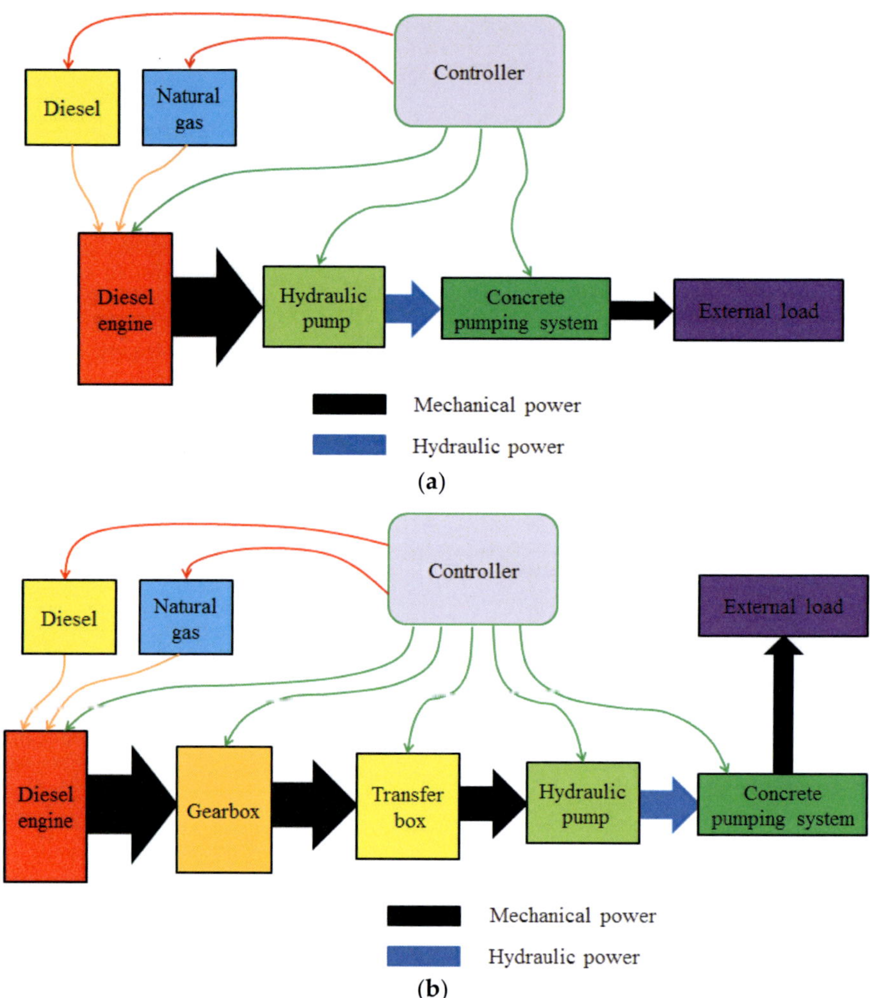


**Figure 10.** Schematic of dual-fuel approach of energy conservation for concrete pumps. (**a**) Dual-fuel approach of trailer concrete pump. (**b**) Dual-fuel approach of transported/truck-mounted concrete pump.

### 4.3. Dual-Power Approach

He et al. [47] stated that the dual-power approach referred to the method of using two power sources (internal combustion engine and electric motor). When the vehicle was operating in the field or it was inconvenient to use electricity, the engine provided power to drive the working device; When the vehicle worked in a narrow, closed space or flammable and explosive conditions, it could be switched to the electric motor working mode by connecting AC power and performing the corresponding operation, and the electric motor provided power to drive the working device, which could greatly improve the economy and exhaust emissions on the premise of meeting the safety and reliability of the equipment. The schematic of the dual-power approach of energy conservation for concrete pumps is shown in Figure 11.

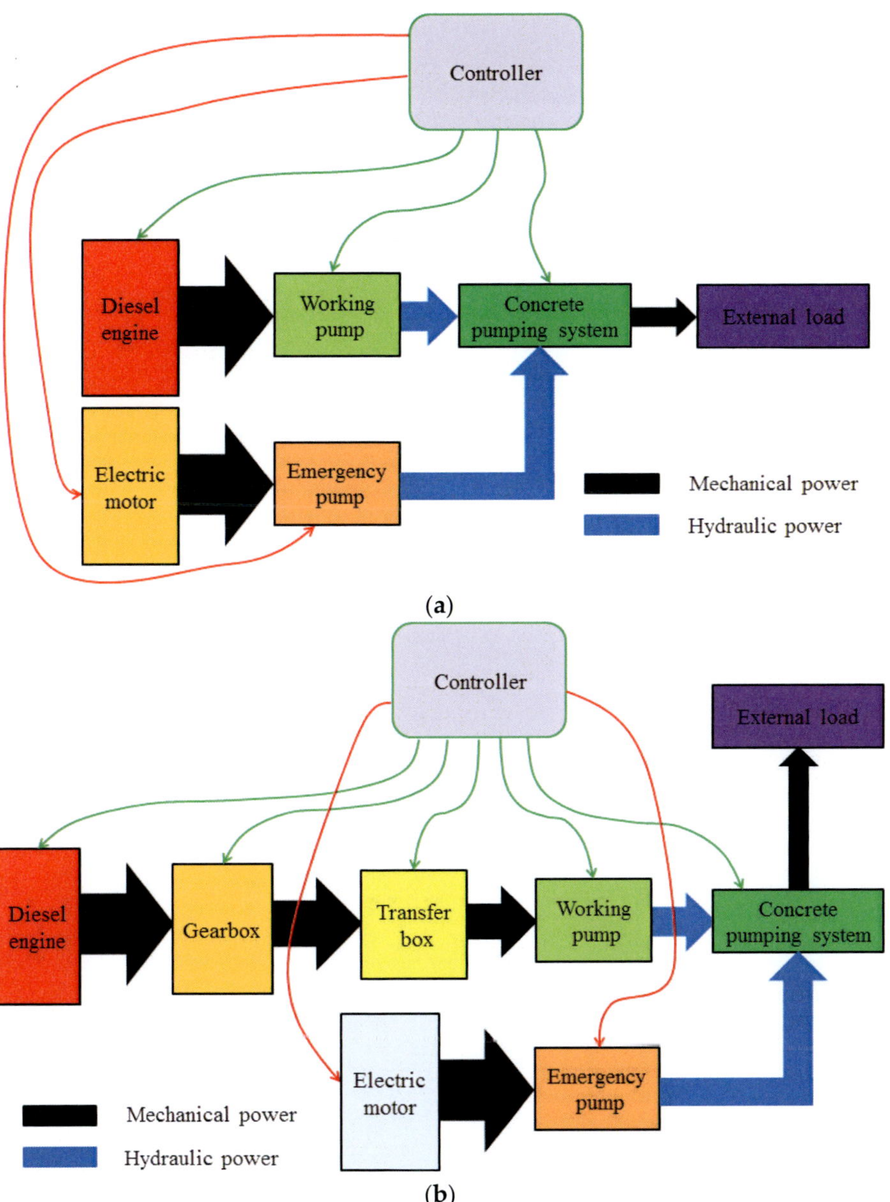



**Figure 11.** Schematic of dual-power approach of energy conservation for concrete pumps. (**a**) Dual-power approach of trailer concrete pump. (**b**) Dual-power approach of transported/truck-mounted concrete pump.

In view of the problems existing in the application of emergency power, He et al. [31,47] proposed a dual-power hydraulic system composed of a diesel engine and electric motor. When the conventional hydraulic system driven by electric motor cannot work, the diesel engine provides emergency power to the emergency hydraulic system by the manual operation of the remote controller. Through the control of the hydraulic valve group, the coordinated action of each actuator is realized, and the boom retraction, concrete pumping and cleaning are completed. The prototype commissioning and construction site application show that the parameters and indicators of the proposed dual-power hydraulic system meet the expected design requirements, meet the emergency power requirements in construction, have high cost performance and convenient operation.

According to the above statistical results, it can obviously be seen that the power matching approach is the most widely used in energy conservation for concrete pumps, which indicates that this is a research hot spot. In addition, although there is less research on the dual-power approach and dual-fuel approach, it also provides a new research idea for this area of study. With the breakthrough of fuel technology and power technology, it is believed that the dual-power approach and dual-fuel approach will gradually be successfully applied.

## 5. Challenges with Energy Conservation of Concrete Pumps

A pumping system is a common part of trailer concrete pumps, transported concrete pumps and truck-mounted concrete pumps, and it works in a continuous periodical pumping process. From different working conditions of concrete pumps, it can obviously be concluded that load power varies periodically with different loads in a large range, and accordingly, the working condition of diesel engine also changes periodically, and therefore, diesel engine cannot always work in a high-efficiency region. This is the main reason that concrete pumps have low-fuel economy.

A hybrid power system including Petrol–Hydraulic hybrid and Petrol–Electric hybrid systems has the potential to improve fuel economy by operating diesel engine in a high-efficiency region. Petrol–Hydraulic hybrid consists of a diesel engine converting chemical energy to mechanical energy and a hydraulic motor converting hydraulic energy to mechanical energy, and Petrol–Electric hybrid consists of a diesel engine converting chemical energy to mechanical energy and an electric motor converting electrical energy to mechanical energy. Because hybrid power system can significantly improve fuel economy it has been applied successfully in buses [48–52], vehicles [53–59], hydraulic excavators [60–66], hydraulic cranes [67–69], hydraulic forklifts [70–73], etc. Equipping concrete pumps with a hybrid power system offers a feasible way to achieve better energy conservation efficiency. In this paper, we forecast some trends on energy conservation for concrete pumps.

### 5.1. Reasonable Selecting of Structure Type

The working conditions of concrete pumps are complex, and different concrete pumps have different load characteristics; thus, it may suit for a different type of hybrid power structure. The serial or parallel structure type of concrete pumps should be comprehensively compared according to different working conditions and load characteristics so that concrete pumps can achieve better energy conservation efficiency.

### 5.2. Reasonable Optimizing of Control Strategy

The control strategy plays an important role in energy conservation and emission reduction in concrete pumps. There are so many types of concrete pumps, and the same type of concrete pump has different working characteristics under different working conditions. Therefore, the control strategy of hybrid concrete pumps should be studied on the basis of analyzing the load characteristics of concrete pumps so that concrete pumps can obtain the anticipated efficiency of energy conservation.

### 5.3. Reasonable Matching of Component Parameters

A Petrol–Hydraulic hybrid or Petrol–Electric hybrid power concrete pump is a mechanic–electric–hydraulic integration product, and it exists for the definite coupling of mechanic, electric and hydraulic power. The working performance of the whole machine is to be improved greatly if the reasonable matching of each component parameters can be realized. Thus, it should analyze the coupling characteristic among components and further study the parameter matching method on the basis of analyzing the characteristics of each component which is used to compose a hybrid power concrete pump.

### 5.4. Reasonable Effectiveness of Energy Storage Component

The energy storage component which includes a storage battery, super capacitor or accumulator is the important part for a hybrid concrete pump, and its performance has great influence on the working performance of a hybrid concrete pump. Due to the working characteristics of large-load changing, frequent working, a bad working environment, etc., the requirements such as power density, energy density, storage capacity, working lifetime and reasonable use are raised as to the effectiveness of the energy storage component.

### 5.5. Reasonable Design of Energy Management System

The energy required by a hybrid concrete pump varies because of different working conditions and load characteristics during the working process. If a reliable energy management system distributing energy in different working stages can reasonably be designed, the target of energy conservation is to be eventually realized.

## 6. Conclusions and Future Work

This paper expounds the structure and working principle of concrete pumps, analyzes the energy consumption ways for concrete pumps, summarizes the developments in energy conservation for concrete pumps, and presents the challenges with energy conservation of concrete pumps.

The research results of this paper indicate that the power matching approach is the most widely used approach in energy conservation of concrete pumps. In addition, both the dual-power approach and dual-fuel approach provide a new research idea for energy conservation of concrete pumps even if the research on the two approaches is less. It is expected that the dual-power approach and dual-fuel approach will gradually be successfully applied with the breakthrough of fuel technology and power technology.

By selecting the structure type, optimizing the control strategy, matching component parameters, enhancing the effectiveness of the energy storage component and designing a reasonable energy management system, it can achieve the goal of energy conservation for concrete pumps.

In our future research work, we will deeply study the technology with regard to applying a hybrid power system for concrete pumps. Specifically, we will carefully study the structure type of a hybrid system suitable for concrete pumps, further study and optimize the control strategy, study the parameter matching method and study the energy storage system of a hybrid system for concrete pumps, as well as study its energy management system.

**Author Contributions:** Conceptualization, H.L.; methodology, H.L.; validation, Q.Z.; formal analysis, Q.Z.; investigation, Q.Z.; writing—original draft preparation H.L.; writing—review and editing, Q.Z.; supervision, Q.Z.; funding acquisition, H.L. All authors have read and agreed to the published version of the manuscript.

**Funding:** This research received no external funding.

**Data Availability Statement:** The data presented in this study are available on request from the corresponding author.

**Acknowledgments:** This work is supported by the National Natural Science Foundation of China (grant number 51365008); the Joint Foundation of Science and Technology Department of Guizhou Province (grant number Qiankehe LH Zi (2015)7658). The authors are grateful to them for their support.

**Conflicts of Interest:** The authors declare no conflict of interest.

## Nomenclature

| | |
|---|---|
| $P_E$ | Power of diesel engine |
| $P_{Eloss}$ | Power loss of diesel engine |
| $P_{Ee}$ | Effective power of diesel engine |
| $P_P$ | Power of hydraulic pump |
| $P_{Ploss}$ | Power loss of hydraulic pump |
| $P_{Pe}$ | Effective power of hydraulic pump |
| $P_H$ | Power of concrete pumping system |
| $P_{Hloss}$ | Power loss of concrete pumping system |
| $P_{He}$ | Effective power of concrete pumping system |
| $P_L$ | Power of external load |
| $P_G$ | Power of gearbox |
| $P_{Gloss}$ | Power loss of gearbox |
| $P_{Ge}$ | Effective power of gearbox |
| $P_T$ | Power of transfer box |
| $P_{Tloss}$ | Power loss of transfer box |
| $P_{Te}$ | Effective power of transfer box |

## References


1. Mechtcherine, V.; Nerella, V.N.; Kasten, K. Testing pumpability of concrete using Sliding Pipe Rheometer. *Constr. Build. Mater.* **2014**, *53*, 312–323. [CrossRef]
2. Kim, J.S.; Kwon, S.H.; Jang, K.P.; Choi, M.S. Concrete pumping prediction considering different measurement of the rheological properties. *Constr. Build. Mater.* **2018**, *171*, 493–503. [CrossRef]
3. Sun, X.; Ye, H.; Fei, S. A closed-loop detection and open-loop control strategy for booms of truck-mounted concrete pump. *Autom. Constr.* **2013**, *31*, 265–273. [CrossRef]
4. Choi, M.S.; Kim, Y.J.; Jang, K.P.; Kwon, S.H. Effect of the coarse aggregate size on pipe flow of pumped concrete. *Constr. Build. Mater.* **2014**, *66*, 723–730. [CrossRef]
5. Choi, M.S.; Kim, Y.S.; Kim, J.H.; Kim, J.S.; Kwon, S.H. Effects of an externally imposed electromagnetic field on the formation of a lubrication layer in concrete pumping. *Constr. Build. Mater.* **2014**, *61*, 18–23. [CrossRef]
6. Wu, Y.; Li, W.; Liu, Y. Fatigue life prediction for boom structure of concrete pump truck. *Eng. Fail. Anal.* **2016**, *60*, 176–187. [CrossRef]
7. Choi, M.; Roussel, N.; Kim, Y.; Kim, J. Lubrication layer properties during concrete pumping. *Cem. Concr. Res.* **2013**, *45*, 69–78. [CrossRef]
8. Choi, M.; Ferraris, C.F.; Martys, N.S.; Lootens, D.; Bui, V.K.; Hamilton, H.R.T. Metrology Needs for Predicting Concrete Pumpability. *Adv. Mater. Sci. Eng.* **2015**, *2015*, 10. [CrossRef]
9. Nerella, V.N.; Mechtcherine, V. Virtual Sliding Pipe Rheometer for estimating pumpability of concrete. *Constr. Build. Mater.* **2018**, *170*, 366–377. [CrossRef]
10. Tang, H.B.; Ren, W. Research on rigid-flexible coupling dynamic characteristics of boom system in concrete pump truck. *Adv. Mech. Eng.* **2015**, *7*, 7. [CrossRef]
11. Jung, T.; Raduenz, H.; Krus, P.; De Negri, V.J.; Lee, J. Boom energy recuperation system and control strategy for hydraulic hybrid excavators. *Autom. Constr.* **2022**, *135*, 104046. [CrossRef]
12. Yu, Y.X.; Do, T.C.; Park, Y.; Ahn, K.K. Energy saving of hybrid hydraulic excavator with innovative powertrain. *Energy Convers. Manag.* **2021**, *244*, 114447. [CrossRef]
13. Li, J.; Zhao, J. Energy recovery for hybrid hydraulic excavators: Flywheel-based solutions. *Autom. Constr.* **2021**, *125*, 103648. [CrossRef]
14. Do, T.C.; Dang, T.D.; Dinh, T.Q.; Ahn, K.K. Developments in energy regeneration technologies for hydraulic excavators: A review. *Renew. Sustain. Energy Rev.* **2021**, *145*, 111076. [CrossRef]
15. Atashi Khoei, A.; Süral, H.; Tural, M.K. Energy minimizing order picker forklift routing problem. *Eur. J. Oper. Res.* **2023**, *307*, 604–626. [CrossRef]
16. Zajac, P.; Rozic, T. Energy consumption of forklift versus standards, effects of their use and expectations. *Energy* **2022**, *239*, 122187. [CrossRef]
17. Cheng, L.L.; Zhao, D.X.; Li, T.Y.; Wang, Y. Modeling and simulation analysis of electric forklift energy prediction management. *Energy Rep.* **2022**, *8*, 353–365. [CrossRef]
18. Zhao, N.; Fu, Z.R.; Sun, Y.; Pu, X.N.; Luo, L. Digital-twin driven energy-efficient multi-crane scheduling and crane number selection in workshops. *J. Clean. Prod.* **2022**, *336*, 130175. [CrossRef]
19. Vlahopoulos, D.; Bouhouras, A.S. Solution for RTG crane power supply with the use of a hybrid energy storage system based on literature review. *Sustain. Energy Technol. Assess.* **2022**, *52*, 102351. [CrossRef]

20. Tan, C.; Yan, W.; Yue, J. Quay crane scheduling in automated container terminal for the trade-off between operation efficiency and energy consumption. *Adv. Eng. Inform.* **2021**, *48*, 101285. [CrossRef]
21. Kusakana, K. Optimal energy management of a retrofitted Rubber Tyred Gantry Crane with energy recovery capabilities. *J. Energy Storage* **2021**, *42*, 103050. [CrossRef]
22. Kusakaka, K.; Phiri, S.F.; Numbi, B.P. Optimal energy management of a hybrid diesel generator and battery supplying a RTG crane with energy recovery capability. *Energy Rep.* **2021**, *7*, 4769–4778. [CrossRef]
23. Shafikhani, I.; Åslund, J. Energy management of hybrid electric vehicles with battery aging considerations: Wheel loader case study. *Control. Eng. Pract.* **2021**, *110*, 104759. [CrossRef]
24. Comino, F.; Ruiz de Adana, M.; Peci, F. Energy saving potential of a hybrid HVAC system with a desiccant wheel activated at low temperatures and an indirect evaporative cooler in handling air in buildings with high latent loads. *Appl. Therm. Eng.* **2018**, *131*, 412–427. [CrossRef]
25. Oh, K.; Yun, S.; Ko, K.; Ha, S.; Kim, P.; Seo, J.; Yi, K. Gear ratio and shift schedule optimization of wheel loader transmission for performance and energy efficiency. *Autom. Constr.* **2016**, *69*, 89–101. [CrossRef]
26. Oh, K.; Kim, H.; Ko, K.; Kim, P.; Yi, K. Integrated wheel loader simulation model for improving performance and energy flow. *Autom. Constr.* **2015**, *58*, 129–143. [CrossRef]
27. SANY. 80 series trailer-mounted concrete pump Hbt8018c-5s. In *SANY Trailer-Mounted Concrete Pump*; S.H.I.C. Ltd., Ed.; SANY: Changsha, China, 2016.
28. SANY. C8 series line pump SY5128THB-10020C-8S. In *C8 Series Line Pump*; S.H.I.C. Ltd., Ed.; SANY: Changsha, China, 2016.
29. SANY. C8 series truck-mounted concrete pump SYM5330THBDZ 470C-8S. In *SANY C8 Series Truck-Mounted Concrete Pump*; S.H.I.C. Ltd., Ed.; SANY: Changsha, China, 2016.
30. Jin, M.; Wu, D. Collision-Free and Energy-Saving Trajectory Planning for Large-Scale Redundant Manipulator Using Improved PSO. *Math. Probl. Eng.* **2013**, *2013*, 8. [CrossRef]
31. Yang, Y.; Gao, R. Research on Power Matching and Energy Saving Control of the Truck Mounted Concrete Pump. *Constr. Mach. Technol. Manag.* **2017**, *1*, 111–114.
32. Ye, M.; Yi, X.; Jiao, S. Energy Optimization by Parameter Matching for a Truck-mounted Concrete Pump. *Energy Procedia* **2016**, *88*, 574–580.
33. Ye, M.; Yi, X.G.; Pu, D.L.; Jiao, S.J. Global power matching on truck-mounted concrete pump based on genetic algorithm. *J. Jilin Univ. (Eng. Technol. Ed.)* **2015**, *45*, 820–828.
34. Ye, M.; Yi, X.G.; Pu, D.L.; Jiao, S.J. Double-parameter control on concrete pumps for pressure-difference-sensed inversion. *Chin. J. Constr. Mach.* **2015**, *13*, 217–223.
35. Xia, Y.M.; Shi, P.F.; Zhao, Y.F.; Xie, X.Z.; Yang, X. Analysis and Research of Energy Saving Technology for Concrete Pump Truck. *Mech. Res. Appl.* **2014**, *3*, 20–22.
36. Shi, F.; Shen, Q.; An, D. Research and conduct of power saving strategy on closed hydraulic system for mounted concrete pump truck. *Constr. Mach.* **2013**, *1*, 59–61, 66.
37. Shi, P.; Zhu, J.; Deng, X. Study on Automatic Gear Control of Gearbox of Concrete Pump Truck. *Road Mach. Constr. Mech.* **2013**, *4*, 84–86.
38. Wang, J.; Wan, L.; Chen, Q. Research on the Energy Saving Technology of the Truck Mounted Concrete Pump. *Constr. Mach. Technol. Manag.* **2012**, *6*, 98–101.
39. Cheng, Z. Study on Power-Matching Control of the Concrete Pump Truck's Power System. Master Thesis, Jiangsu University, Zhenjiang, China, 2011.
40. Li, Z.X.; Wang, S.X.; Jiang, H.; Cheng, Z.S. Research on power-matching of energy-saving for power system of the concrete pump truck. *Mach. Des. Manuf.* **2011**, *6*, 76–78.
41. He, S.; Yang, Y. Modeling of universal characteristics and optimization of operating conditions of concrete pump truck based on neural network. *J. Cent. South Univ. (Sci. Technol.)* **2010**, *41*, 1398–1404.
42. Jie, L.; Zen, Y. An Energy Saving Control Strategy for Operation of Concrete Pump Trucks. *Constr. Mach. Equip.* **2010**, *41*, 20–23.
43. Zeng, Y.P.; Zeng, F.L.; Wang, C.; Liu, L.; Cai, Y. Research of Energy-saving Control Methods for Concrete Pump Trucks. *Constr. Mach. Equip.* **2010**, *41*, 20–24.
44. Wu, S.; Zeng, F. Study on Energy Saving Control Strategy of Concrete Pump Truck. *Heavy Truck.* **2008**, *4*, 9–11.
45. Pu, D. Study on the Energy-Saving Parameter Matching of Dynamic System of the Truck-Mounted Concrete Pump. Master Thesis, Chang'An University, Chang'An, China, 2009.
46. Li, X.; Shu, Z.; Li, H. Research and Application on the Truck Mounted Concrete Pump with Dual Fuel Mixed Combustion and Energy-saving. *Constr. Mach. Technol. Manag.* **2014**, *1*, 94–98.
47. He, J.; Zhang, W. A New Type Dual-power Hydraulic System and Its Application to Concrete Pump Trucks. *Constr. Mach.* **2018**, *49*, 52–57.
48. Agrawal, A.; Gupta, R. Optimized sensor charge controller for bus voltage stabilization in hybrid battery-supercapacitor fed islanded microgrid system. *J. Energy Storage* **2023**, *59*, 106482. [CrossRef]
49. Wang, Z.G.; Wei, H.Q.; Xiao, G.W.; Zhang, Y.T. Real-time energy management strategy for a plug-in hybrid electric bus considering the battery degradation. *Energy Convers. Manag.* **2022**, *268*, 116053. [CrossRef]

50. Roth, M.; Franke, G.; Rinderknecht, S. Decentralised multi-grid coupling for energy supply of a hybrid bus depot using mixed-integer linear programming. *Smart Energy* **2022**, *8*, 100090. [CrossRef]
51. Liu, Z.; Zhao, J.; Qin, Y.J.; Wang, G.W.; Shi, Q.; Wu, J.Y.; Yang, H. Real time power management strategy for fuel cell hybrid electric bus based on Lyapunov stability theorem. *Int. J. Hydrogen Energy* **2022**, *47*, 36216–36231. [CrossRef]
52. Li, J.W.; Yang, L.M.; Yang, Q.Q.; Wei, Z.B.; He, Y.T.; Lan, H. Degradation adaptive energy management with a recognition-prediction method and lifetime competition-cooperation control for fuel cell hybrid bus. *Energy Convers. Manag.* **2022**, *271*, 116306. [CrossRef]
53. Ye, Y.M.; Zhang, J.F.; Pilla, S.; Rao, A.M.; Xu, B. Application of a new type of lithium-sulfur battery and reinforcement learning in plug-in hybrid electric vehicle energy management. *J. Energy Storage* **2023**, *59*, 106546. [CrossRef]
54. Xu, B.; Wang, H. A comparative analysis of adaptive energy management for a hybrid electric vehicle via five driving condition recognition methods. *Energy* **2023**, *269*, 126732. [CrossRef]
55. Xiao, C.W.; Wang, B.; Zhao, D.; Wang, C.H. Comprehensive Investigation on Lithium Batteries for Electric and Hybrid-Electric Unmanned Aerial Vehicle Applications. *Therm. Sci. Eng. Prog.* **2023**, *38*, 101677. [CrossRef]
56. Sun, X.L.; Fu, J.Q.; Yang, H.Y.; Xie, M.K.; Liu, J.P. An energy management strategy for plug-in hybrid electric vehicles based on deep learning and improved model predictive control. *Energy* **2023**, *269*, 126772. [CrossRef]
57. Nassar, M.Y.; Shaltout, M.L.; Hegazi, H.A. Multi-objective optimum energy management strategies for parallel hybrid electric vehicles: A comparative study. *Energy Convers. Manag.* **2023**, *277*, 116683. [CrossRef]
58. Huang, Y.; Hu, H.Q.; Tan, J.Q.; Lu, C.L.; Xuan, D.J. Deep reinforcement learning based energy management strategy for range extend fuel cell hybrid electric vehicle. *Energy Convers. Manag.* **2023**, *277*, 116678. [CrossRef]
59. Huang, R.; He, H. Naturalistic data-driven and emission reduction-conscious energy management for hybrid electric vehicle based on improved soft actor-critic algorithm. *J. Power Sources* **2023**, *559*, 232648. [CrossRef]
60. Ranjan, P.; Wrat, G.; Bhola, M.; Mishra, S.K.; Das, J. A novel approach for the energy recovery and position control of a hybrid hydraulic excavator. *ISA Trans.* **2020**, *99*, 387–402. [CrossRef] [PubMed]
61. Yu, Y.; Ahn, K.K. Optimization of energy regeneration of hybrid hydraulic excavator boom system. *Energy Convers. Manag.* **2019**, *183*, 26–34. [CrossRef]
62. Chen, Q.H.; Lin, T.L.; Ren, H.L.; Fu, S.J. Novel potential energy regeneration systems for hybrid hydraulic excavators. *Math. Comput. Simul.* **2019**, *163*, 130–145. [CrossRef]
63. Chen, Q.; Lin, T.; Ren, H. Parameters optimization and control strategy of power train systems in hybrid hydraulic excavators. *Mechatronica* **2018**, *56*, 16–25. [CrossRef]
64. Cao, T.F.; Russell, R.L.; Durbin, T.D.; Cocker, D.R.; Burnette, A.; Calavita, J.; Maldonado, H.; Johnson, K.C. Characterization of the emissions impacts of hybrid excavators with a portable emissions measurement system (PEMS)-based methodology. *Sci. Total Environ.* **2018**, *635*, 112–119. [CrossRef]
65. Lin, T.L.; Huang, W.P.; Ren, H.L.; Fu, S.J.; Liu, Q. New compound energy regeneration system and control strategy for hybrid hydraulic excavators. *Autom. Constr.* **2016**, *68*, 11–20. [CrossRef]
66. Kim, H.; Yoo, S.; Cho, S.; Yi, K. Hybrid control algorithm for fuel consumption of a compound hybrid excavator. *Autom. Constr.* **2016**, *68*, 1–10. [CrossRef]
67. Corral-Vega, P.J.; García-Triviño, P.; Fernández-Ramírez, L.M. Design, modelling, control and techno-economic evaluation of a fuel cell/supercapacitors powered container crane. *Energy* **2019**, *186*, 115863. [CrossRef]
68. Corral-Vega, P.J.; Fernández-Ramírez, L.M.; García-Triviño, P. Hybrid powertrain, energy management system and techno-economic assessment of rubber tyre gantry crane powered by diesel-electric generator and supercapacitor energy storage system. *J. Power Sources* **2019**, *412*, 311–320. [CrossRef]
69. Ovrum, E.; Bergh, T.F. Modelling lithium-ion battery hybrid ship crane operation. *Appl. Energy* **2015**, *152*, 162–172. [CrossRef]
70. Hsieh, C.-Y.; Pei, P.C.; Bai, Q.; Su, A.; Weng, F.-B.; Lee, C.-Y. Results of a 200 hours lifetime test of a 7 kW Hybrid–Power fuel cell system on electric forklifts. *Energy* **2021**, *214*, 118941. [CrossRef]
71. Girbés, V.; Armesto, L.; Tornero, J. Path following hybrid control for vehicle stability applied to industrial forklifts. *Robot. Auton. Syst.* **2014**, *62*, 910–922. [CrossRef]
72. Hosseinzadeh, E.; Rokni, M.; Advani, S.G.; Prasad, A.K. Performance simulation and analysis of a fuel cell/battery hybrid forklift truck. *Int. J. Hydrogen Energy* **2013**, *38*, 4241–4249. [CrossRef]
73. Keränen, T.M.; Karimaki, H.; Viitakangas, J.; Vallet, J.; Ihonen, J.; Hyotyla, P.; Uusalo, H.; Tingelof, T. Development of integrated fuel cell hybrid power source for electric forklift. *J. Power Sources* **2011**, *196*, 9058–9068. [CrossRef]

MDPI AG
Grosspeteranlage 5
4052 Basel
Switzerland
Tel.: +41 61 683 77 34
www.mdpi.com

*Processes* Editorial Office
E-mail: processes@mdpi.com
www.mdpi.com/journal/processes

Printed by Libri Plureos GmbH in Hamburg,
Germany